Fundamentals of Temperature, Pressure, and Flow Measurements

Handbook of Generalized Gas Dynamics
Plenum Press, 1966 (with W. G. Steltz)

Handbook of Specific Losses in Flow Systems
Plenum Press, 1966 (with N. A. Carlucci)

*Manual on the Use of Thermocouples in
Temperature Measurement*
STP470 & STP470A, American Society for
Testing and Materials, 1970 & 1974 (Editor)

Journey Away From God
Fleming H. Revell Co., 1972

Fundamentals of Temperature, Pressure, and Flow Measurements

SECOND EDITION

Robert P. Benedict

Fellow Mechanical Engineer
Westinghouse Electric Corporation
Steam Turbine Division

Adjunct Professor
of Mechanical Engineering

Drexel University
Evening College
Philadelphia, Pennsylvania

A Wiley-Interscience Publication

JOHN WILEY & SONS

New York . London . Sydney . Toronto

Library of Congress Cataloging in Publication Data:

Benedict, Robert P.
  Fundamentals of temperature, pressure, and flow measurements.

  Bibliography: p.
  Includes indexes.
  1. Thermometers and thermometry. 2. Pressure—
Measurement. 3. Fluid dynamic measurements. I. Title.

QC271.B47   1976   536′.5   76-54341
ISBN 0-471-06561-7

Printed in the United States of America

10 9 8 7 6 5 4 3 2

To My Wife
Ruth

"... I often say that when you can measure *what you are speaking about, and express it in numbers, you know something about it; but when you cannot express it in numbers, your knowledge is of a meagre and unsatisfactory kind; it may be the beginning of knowledge, but you have scarcely, in your thoughts, advanced to the stage of Science, whatever the matter may be* ...."

Sir William Thomson
(Lord Kelvin)

PREFACE
to the First Edition

Someone has said that science and the history of science are inseparable, and I believe it. How can one appreciate Michelson's and Morley's feat in determining the speed of light if the many attempts and frustrations of others to do the same are overlooked? How can one grasp the profundity of Max Planck's quantum theory if the paths of Maxwell, Kirchoff, and Wien are not followed first? So it is with thermal and flow measurements; unless we struggle through with Galileo, da Vinci, Newton, Bernoulli, Pitot, Joule, Kelvin, Weisbach, Herschel, and others in their efforts to sense temperature, pressure, and flow, we cannot appreciate our present-day array of fine instruments and techniques.

Behind every successful man, they say, there is an understanding woman. So, behind every successful project, sale, or test involving machinery there is an engineer who understands measurements. Spend what you will on design and manufacture, in the end it is the field test that justifies your efforts. Were the heat rates met? Did the engine perform up to specs? Was the cooling arrangement adequate? The list could be long and varied.

This book is intended as a working reference for practicing engineers and associated workers involved in the measurement of temperature, pressure, and flow rate and as a text for engineering students. It should be especially useful to mechanical engineers in the fields of fluid mechanics, thermodynamics, and heat transfer. It provides a thorough review of techniques for measuring temperature, pressure, and flow rate and also includes practical data. Each topic is treated in depth, and applications are illustrated by many practical worked-out numerical examples. The book is self-contained in the sense that all necessary equations, corrections, schematic diagrams, curves, and tables of data are included.

vii

The subject matter with which this book is concerned has been gathered through the years in the course of my work at the Westinghouse Steam and Aviation Gas Turbine Divisons. It has been strongly influenced by my participation in various ASME, ASTM, and ASA Committees on measurements. The material contained here is essentially that presented in a one-term course in instrumentation which I have taught for a number of years at the Evening College of the Drexel Institute of Technology. There has always been much interest in this subject, as evidenced by the students' response to the material.

Parts of this work have appeared in a number of other publications. Three articles I have written in particular form the basis of this book. "Temperature and Its Measurement," July 1963; "Fluid Flow Measurement," January 1966; and "Pressure and Its Measurement," October 1967; all of which appeared in the Science and Engineering Series of *Electro-Technology*.

As we part company personally and get down to the business at hand, namely the study of thermal and flow instrumentation, I note with a twinge of conscience that the contributions of others have not been recognized adequately in this preface. I should like to acknowledge here that any merit or order that is found in this book must be credited largely to the work of others. Wherever possible references are made to the original workers at appropriate places in the text. In addition, I especially want to thank my wife for her patience and skill in typing the manuscript.

I hope that you, kind reader will enjoy the perusal of this material as I have enjoyed gathering it.

R. P. BENEDICT

Holly Hill, Pennsylvania
October 1968

PREFACE

In preparing this edition, I have endeavored to include the major changes that have occurred in the areas of temperature, pressure, and flow measurements over the past decade. Specifically, the International Practical Temperature Scale has changed. This new scale (IPTS-68) is now the law of the land, and Chapter 4 has been completely rewritten to reflect this. Chapter 6, on Resistance Thermometry, also reflects changes in the IPTS. New Thermocouple Reference Tables were issued by the National Bureau of Standards in 1974. Accordingly, Chapter 7 has been revised to include these latest tables of temperature versus electromotive force for the thermocouple types most commonly used in industry. Also, along these same lines, the National Bureau of Standards has issued new methods for generating the new Reference Table values for computer applications. These power series relationships, giving emf as a function of temperature, are now included in Chapter 7. In Part II, much new material has been added in Chapter 17 to account for pressure tap errors, probe blockage effects, probe performance in pressure gradients, and so on. In Part III, on Flow Measurement, Chapters 21, 22, and 23 have been rewritten to include new material on ASME discharge coefficients, expansion factors, and the like. A new section, Chapter 10, has been added on the important subject of Statistics and Uncertainties as applied to engineering measurements. I have further attempted to correct any gross errors in the First Edition, and to provide a more complete and updated bibliography. Finally, many additional worked-out examples have been included in each chapter, and problems have been added at the end of each chapter to aid in the teaching of this subject material.

Once again, I wish to thank my wife for her continued patience and skill in typing this manuscript again. And finally, my best wishes go along to the students, teachers, and researchers who are the ultimate users of this work.

Holly Hill, Pennsylvania   ROBERT P. BENEDICT
August 1976

CONTENTS

Fundamentals of Temperature, Pressure, and Flow Measurements

Part I
Temperature and Its Measurement

". . . to fix on a unit or degree for the numerical measurement of temperature, we may . . . call some definite temperature, such as that of melting ice, unity, or any other number we please . . ."

William Thomson (1854)

INTRODUCTION TO PART I

Part 1 concerns temperature: what it was thought to be; early attempts to measure it; what it is now thought to be; how it may be realized in the laboratory; and various conventional methods for measuring it.

Temperature is today recognized as one of the basic variables in science. Thermodynamics, fluid mechanics, heat transfer, aerodynamics, space aeronautics, chemistry, and physics would be lost without the concept of temperature. Yet, as we examine the basis of temperature, we find a great deal of confusion as to its importance and meaning; for example, although early scientists did not at all question the fixity of the human body temperature, they most certainly did question and consider variable the freezing point of water. Or again, consider the terms "latent heat," "phlogiston," and "caloric"; all of these terms were introduced to describe the subtle "spirits" thought to be involved in the study of heat and temperature.

George Sarton, the great American historian of science has said, "No one should be recognized a master in any subject who does not know at least the outline of its history." In accord with this, it is our firm belief that to be adept at temperature measurements we must be as aware of the historical development of thermometry as we are familiar with the various instruments and techniques that are presently available in thermometry.

Thus our purpose in Part I is to sketch the growth of a satisfying temperature concept, passing quickly over the main stepping stones of thermometry until we arrive at the International Practical Temperature Scale; to indicate in some detail various practical means for realizing temperatures on the IPTS such as liquid-in-glass thermometers, resistance thermometers, thermoelectric thermometers, and optical pyrometers; to discuss the ideas of thermal response and thermal recovery as they modify the steady-state, nonflow temperature concept; and to give special consideration to the installation problems encountered when we attempt to measure temperature in moving fluids or in solids.

Chapter 1

EARLY ATTEMPTS
TO MEASURE
DEGREES OF HEAT

"... when I read that water boils at a definite degree of heat, I immediately felt a great desire to make myself a thermometer...."

Daniel Gabriel Fahrenheit (1706)

1.1 Universal Awareness of Hot and Cold

As far back in time as mind can imagine, man has been aware of degrees of heat: of blazing sun, of frigid waters, of burning desert, of cool forests, of scalding oil, and chilling ice; indeed, of a seemingly infinite variation of hot and cold. Yet, acute as man's senses are, they cannot distinguish between extremes in heat and cold outside of a very limited range because of pain. Even in the range where pain is not controlling, man's reaction is only to relative temperature, and thus, from a scientific viewpoint, is an inadequate temperature-sensing instrument.

1.2 Historical Resume

According to several old sources, credit for the discovery of the first instrument for measuring degrees of heat and cold appears to belong to Galileo Galilei [1].* One of his pupils, Vincenzo Viviani, in his *Life of*

---

* Much of the early history of thermometry can be found in the valuable little book by H. C. Bolton cited in reference [1].

Galileo (1718) says, "... that about the time Galileo took the Chair of Mathematics in Padua at the end of 1592, he invented the thermometer, a glass containing air and water ...." In addition, Francesco Sagredo of Venice wrote to Galileo on May 9, 1613 stating, "... The instrument for measuring heat, which you invented, I have made in several convenient styles, so that differences in temperature between one place and another can be determined...." Instruments of the type used by Galilei were influenced by barometric pressure and are now called barothermoscopes.

The word, thermometer, first appeared in the literature in 1624 in a book by J. Leurechon entitled *La Récréation Mathématique*. The author describes a thermometer as, "... an instrument of glass which has a little bulb above and a long neck below, or better a very slender tube, and it ends in a vase full of water ... Those who wish to determine changes by numbers and degrees draw a line all along the tube and divide it into 8 degrees, according to the philosophers...."

At about 1654, Ferdinand II, Grand Duke of Tuscany, made thermoscopes of the usual form, filled them with alcohol, and then hermetically sealed the thermometric fluid within the bulb and stem. These were the first temperature-sensing instruments to be independent of pressure.

Robert Hooke, in 1664, placed the zero of his thermometer, "... at the point which the liquid stood when the bulb was placed in freezing distilled water ...."

The Dutch scientist and mathematician, Christian Huygens recognized the dilemma that early pyrometrists faced. In a letter dated January 2, 1665, he wrote, "... It would be well to have a universal and determinate *standard* for heat and cold, securing a definite proportion between the capacity of the bulb and the tube, and then taking for the commencement the degree of cold at which water begins to freeze or better the temperature of boiling water ...."

Again in 1665, Robert Boyle stated, "... we are greatly at a loss for a *standard* ... not only the several differences of this quantity (temperature) have no names assigned them, but our sense of feeling cannot therein be depended upon; and thermometers are such very variable things that it seems morally impossible for them to settle such a measure of coldness as we have of *time, distance, weight*...."

Illustrating the complete arbitrariness of the times is the description of the construction of a thermometer scale by L. Magalotti [2] of the Acadimia del Cimento in 1667: "... The next thing is to divide the Neck of the Instrument or Tube into Ten equal parts with Compasses, marking each of them with a knob of white Enamel, and you may mark the intermediate Divisions with green Glass, or black Enamel: These lesser Divisions are best made by the Eye, which Practice will render easie ...."

Not until 1694 did Carlo Renaldini, who held the same Chair of Mathematics at Padua as had Galilei, suggest taking the melting point of ice and the boiling point of water for two *fixed points* on a thermometer scale. He divided the space between them into 12 equal parts. Good news does not always travel quickly; Renaldini's contribution to thermometry was unappreciated and forgotten.

Newton, in 1701, independently defined a temperature scale based on two fixed points (reproducible) so that at these "landmarks" no ambiguity in the scales of various thermometers was apparent. For one fixed point, he chose the melting point of ice, and labeled this as zero on his temperature scale. As the second fixed point he chose the armpit temperature of a healthy Englishman, and labeled this 12. Based on what Newton termed "equal parts of heat," boiling water stood at 34 on this scale.

In 1706, Daniel Gabriel Fahrenheit, an instrument maker in Amsterdam, began making thermometers. He wrote [3], "...when I read that water boils at a definite degree of heat, I immediately felt a great desire to make myself a thermometer so that I might examine with my own eyes this beautiful phenomenon of nature and be convinced of the truth of the experiment...." Fahrenheit first used the Florentine scale (90, 0, 90); next he arbitrarily contrived the scale 0, 12, 24; finally, in favor of finer divisions, he settled on the scale 0, 48, 96, where, as he stated, "...48 on my thermometer is midway between the most intense cold artificially produced by a mixture of water, ice, sal-ammoniac or even common salt, and that which is found in the blood of a healthy man...." All of Fahrenheit's instruments had this in common, "...The degrees of their scales *agree* with one another, and their variations are within fixed limits...." Fahrenheit found that, on his scale, the melting and boiling points of water at normal atmospheric pressure were approximately 32 and 212, respectively. These were soon adopted as the most reliable landmarks on the Fahrenheit scale, and were defined to be exact.

In 1742 Anders Celsius, Professor of Astronomy at the University of Uppsala, proposed a scale with zero at the boiling point of water and 100 at the melting point of ice. The following year, Christin of Lyons independently suggested the familiar centigrade scale (now called the Celsius scale). See Figure 1.1 for some relationships between the Fahrenheit and Celsius scales.

It was gradually realized that neither one, nor two, nor any finite number of fixed points, no matter how judiciously chosen, could define an acceptable temperature scale. A standard interpolation instrument and a standard interpolation procedure were essential if ambiguity at other than the fixed points was to be avoided. By the end of the eighteenth century,

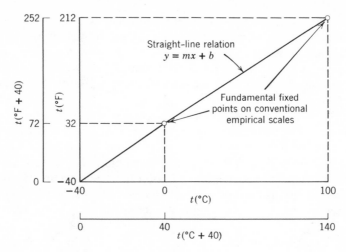

Figure 1.1 Relationships between common empirical temperature scales. For the F and C scales $y = mx + b$, where $m = \Delta F/\Delta C = 180/100 = \frac{9}{5}$ and $b = 32$. It follows that $F = \frac{9}{5}(C) + 32$ and $C = \frac{5}{9}(F - 32)$ (familiar relations). For the $(F + 40)$ and $(C + 40)$ scales, $y = mx + b$, where $m = \Delta F/\Delta C = 180/100 = \frac{9}{5}$ and $b = 0$. It follows that $F = \frac{9}{5}(C + 40) - 40$, and $C = \frac{5}{9}(F + 40) - 40$ (symmetrical relations).

there were as many instruments for interpolating between the fixed points, making use of as many thermometric substances, as there were glass-blowing scientists; and, in general, no two empirically labeled thermometers in thermal equilibrium with each other gave the same indication of temperature [4].

1.3   References

[1] H. C. Bolton, *Evolution of the Thermometer*, Chem. Pub. Co., Easton, Pa., 1900.

[2] H. Guerlac, *Selected Readings in the History of Science*, Vol. 2, Pt. 1, Cornell University, Ithaca, N.Y., 1953, p. 128.

[3] *Ibid.*, Pt. 2, p. 382.

[4] R. P. Benedict, "Review of Practical Thermometry," *ASME Paper 57-A-203*, August 1957.

Problems

1.   If Renaldini's scale had been adopted, what would be the relationship between this and the Fahrenheit scale?

Ans.   R = 1/15(F − 32).

2. At what point would the Renaldini and the Fahrenheit scales coincide?

   Ans. R = −32/14.

3. What is the relationship between the original Celsius scale and the centigrade scale?

   Ans. Cel = −Cen + 100.

4. Derive the symmetrical relation between the original Celsius scale and the centigrade scale.

   Ans. Cel = −(Cen − 50) + 50, Cen = −(Cel − 50) + 50.

5. If a Fahrenheit thermometer indicates 59°F, determine the Celsius temperature by means of the conventional and the symmetrical relations.

   Ans. 15°C.

Chapter 2

THE AIR THERMOMETER

"... I shall not venture to determine whether or no the intimated theory will hold universally and precisely ... all that I shall now urge being the hypothesis that supposes the pressures and expansions to be in reciprocal proportion ..."

Robert Boyle (1662)

In the first half of the nineteenth century, a thermometer was developed, based on the work of Boyle, Gay-Lussac, Clapeyron, and Regnault, and founded on the expansion of air. The so-called air-thermometer was soon recognized as the instrument least liable to uncertain variations, and was generally accepted as the standard for comparing all types of thermometers.

2.1 The Boyle-Mariotte Law

In 1662 Robert Boyle had cautiously observed that, over a limited pressure range, the product of pressure and volume of a fixed mass of a gas was essentially a constant independent of pressure level under isothermal conditions [1]. On the continent of Europe, Edme Mariotte announced this same observation in 1676. The Boyle-Mariotte law can be stated as:

$$(pv)_t = K_t,$$ (2.1)

where the subscript t signifies that changes of state are allowed only under conditions of constant temperature. The symbol K represents a constant, but it bears the subscript t to indicate that although the isothermal pv product does always remain a constant, its value changes as temperature changes. The temperature we are speaking of here can, of course, be sensed by any empirical thermometer, for example, the liquid-in-glass thermometer.

We know today that Boyle's caution was quite in order, for no real gas satisfies (2.1) exactly.

2.2 The Charles-Gay-Lussac Law

Jacques-Alexandre César Charles, in 1787, and Joseph Louis Gay-Lussac, in 1802, had found that equal volumes of real gases (such as oxygen, nitrogen, hydrogen, carbon dioxide, and air) would expand the same amount for a given temperature increase under isobaric conditions. The Charles–Gay-Lussac law can be stated as:

$$\frac{1}{v_0}\left(\frac{v-v_0}{t-t_0}\right)_p = \alpha_{0p}, \qquad (2.2)$$

where the subscript p signifies that changes of state are allowed under conditions of constant pressure only, and the subscript 0 signifies that the variable is taken with respect to a definite reference state (usually the ice point). The constant represented by the symbol α bears the subscript $0p$ to indicate that, although the isobaric cubical coefficient of expansion of any gas remains a constant, its value changes as the reference state or the pressure level changes. The symbol t represents temperature as measured on any empirical scale.

Similarly, when the volume of a given amount of any gas is held constant, the change in pressure is proportional to the change in temperature. Thus the Charles–Gay-Lussac law can also be given as:

$$\frac{1}{p_0}\left(\frac{p-p_0}{t-t_0}\right)_v = \alpha_{0v}, \qquad (2.3)$$

where the subscript v signifies that changes of state are allowed under conditions of constant volume only, and the subscript $0v$ to the symbol α indicates that, although the isochoric pressure coefficient of any gas remains a constant, its value changes as the reference state or the volume changes.

We know today that no real gas follows the Charles–Gay-Lussac law exactly.

2.3 Clapeyron's Equation of State

Emile Clapeyron, in 1834, first combined the Boyle–Mariotte and Charles–Gay-Lussac laws to give the equation of state of a gas [2].

Alternate forms of Clapeyron's equation can be given, depending on whether pressure or volume is considered the dependent variable.

In one development we consider for a single-phase, single-component substance:

$$v = f(p, t)$$

or

$$dv = \left(\frac{\partial v}{\partial p}\right)_t dp + \left(\frac{\partial v}{\partial t}\right)_p dt. \tag{2.4}$$

But from (2.1):

$$\left(\frac{\partial v}{\partial p}\right)_t = -\frac{v}{p}. \tag{2.5}$$

And from (2.2):

$$\left(\frac{\partial v}{\partial t}\right)_p = \alpha_{0p} v_0. \tag{2.6}$$

Combining (2.4), (2.5), and (2.6) yields:

$$\frac{dp}{p} + \frac{dv}{v} = \frac{dt}{v/(\alpha_{0p} v_0)}. \tag{2.7}$$

But by (2.2):

$$\frac{v}{\alpha_{0p} v_0} = \left(t - t_0 + \frac{1}{\alpha_{0p}}\right). \tag{2.8}$$

Thus (2.7) can be integrated to yield:

$$\ln p + \ln v = \ln \left(t - t_0 + \frac{1}{\alpha_{0p}}\right) + \ln R_p, \tag{2.9}$$

where R is a constant of integration whose subscript p signifies that the isobaric form of the Charles–Gay-Lussac law has been used. Taking antilogarithms yields

$$pv = R_p \left(t - t_0 + \frac{1}{\alpha_{0p}}\right), \tag{2.10}$$

where the constant of integration can be evaluated at the reference state as

$$R_p = p_0 v_0 \alpha_{0p}, \tag{2.11}$$

and where the bracketed expression in (2.10) is known as the temperature from the zero of the air thermometer (in terms of the constant pressure cubical coefficient of expansion of a gas). This result follows because the quantity $[t - t_0 + (1/\alpha_{0p})]$ is equivalent to expressing temperature on a new scale, whose zero is lower than that of the empirical scale of t by the quantity $[(1/\alpha_{0p}) - t_0]$, but whose unit temperature interval (i.e., degree) is the same as that of the empirical scale of t.

The development can also be given in terms of pressure by beginning with

$$p = f(v, t)$$

or

$$dp = \left(\frac{\partial p}{\partial v}\right)_t dv + \left(\frac{\partial p}{\partial t}\right)_v dt. \tag{2.12}$$

By similar reasoning to that above, we obtain:

$$pv = R_v \left(t - t_0 + \frac{1}{\alpha_{0v}}\right), \tag{2.13}$$

where R is again a constant of integration; the subscript v signifies that now the isochoric form of the Charles–Gay-Lussac law has been used; and R can be evaluated at the reference state as

$$R_v = p_0 v_0 \alpha_{0v}. \tag{2.14}$$

The bracketed expression in (2.13) is again known as temperature from the zero of the air thermometer (now in terms of the constant volume pressure coefficient of the gas). Of course, these air thermometer temperatures are in no way unique, since they vary with the gas used to determine α, the method of determining α, and the empirical thermometer used to determine t and t_0.

Equations 2.10 and 2.13 are forms of Clapeyron's equation of state of a gas. We know today that there is no *real* gas whose thermodynamic state is described exactly by the Clapeyron equation.

2.4   Regnault's Ideal Gas

Victor Regnault, in 1845, found that the mean isobaric cubical coefficient of expansion of any real gas, when heated at atmospheric pressure from the ice point to the steam point, was approximately 1/273 per Celsius degree, in contrast with Gay-Lussac's evaluation of 1/267 per Celsius degree. Similarly, Regnault found that the mean isochoric pressure coefficient of any real gas, when heated at constant volume from the ice point to the steam point, was likewise approximately 1/273 per Celsius degree [3], [4], [5].

Regnault realized, however, that the isobaric cubical coefficients of expansion of all permanent gases were only approximately equal, and that these volume coefficients were only approximately equal to the respective pressure coefficients. For simplicity, he proposed an imaginary substance that perfectly fulfilled all conditions of the Boyle–Mariotte and Charles–Gay-Lussac laws; that is, he defined an *ideal gas*, whose thermodynamic state satisfied either form of Clapeyron's equation of state. This, of course, required that $\alpha_{0p} = \alpha_{0v} = \alpha^0$, and as a consequence, $R_p = R_v = R$. The equation of state of Regnault's ideal gas is:

$$pv = R\left(t - t_0 + \frac{1}{\alpha^0}\right), \tag{2.15}$$

where the bracketed expression represents temperature on the ideal gas absolute temperature scale; this is similar to the air thermometer scale in that its zero is lower than that of the empirical scale of t by the constant $(1/\alpha^0 - t_0)$, and its degree is of the same size as that of the empirical scale of t. Still, unique temperatures cannot be realized experimentally on the ideal gas scale, since the use of real empirical thermometers is required.

The integration constant for the ideal gas $(R = p_0 v_0 \alpha^0)$ can also be viewed in a different light. If both sides of (2.15) are multiplied by the weight of one mole of the particular gas (i.e., the gas molecular weight, M.W.), then the M.W. $\times v$ product represents the volume of a mole of the gas. By Avogadro's hypothesis, this mole volume and hence the M.W. $\times R$ product on the other side of the equality sign *are the same* for any gas at the same pressure and temperature. Furthermore, since both the molecular weight and the integration constant R are individually independent of pressure and temperature, their product will be *a constant* for all gases at all pressures and temperatures. Thus the M.W. $\times R$ product represents a universal gas constant that we denote by **R**. Some typical values of M.W., **R**, and R are given in Table 2.1.

Today we know that no real gas satisfies the requirements of Regnault's ideal gas exactly. Regnault, however, found only insignificant variations between the indications of different gas thermometers when used according to rigorously prescribed experimental procedures. Thus the gas thermometer made available a series of reference temperatures that constituted a practical "standard" of thermometry. But, since a reference scale could be defined only in terms of a definite thermometric substance, under a definite pressure for the expansion-type gas thermometer, or at a definite specific volume for the pressure-type gas thermometer, while following definite test procedures, a truly universal temperature scale was still wanting [6].

Table 2.1   *Gas Constants for Several Common Gases*

Gas	Molecular Weight	Specific Gas Constant, R $\dfrac{\text{ft lbf}}{°R \text{ lbm}}$	Specific Gas Constant, R $\dfrac{J}{\text{kg } K}$
Air —	28.96	53.36	287.1
Carbon Dioxide, CO_2	44.01	35.11	188.9
Nitrogen, N_2	28.01	55.17	296.8
Oxygen, O_2	32.00	48.29	259.8
Water, H_2O	18.02	85.77	461.4

$$\text{where} \quad R = \frac{\mathbf{R}}{\text{MW}}$$

$$\text{and} \quad \mathbf{R} = 1545 \, \frac{\text{ft lbf}}{\text{lbm mol} \, °R}$$

$$= 8315 \, \frac{J}{\text{kg mol } K} \, .$$

2.5   References

[1] H. Guerlac, *Selected Readings in the History of Science*, Vol. 2, Pt. 1, Cornell University, Ithaca, N.Y., 1953, p. 138.

[2] R. Roseman and S. Katzoff, "The Equation of State of a Perfect Gas," *J. Chem. Ed.*, **2**, June 1934, p. 350.

[3] H. Guerlac, *Selected Readings in the History of Science*, Vol. 2, Pt. 2, Cornell University, Ithaca, N.Y., 1953, p. 382.

[4] E. Clapeyron, "Memoir on the Motive Power of Heat," in *Reflections on the Motive Power of Fire* (E. Mendoza, translator), Dover, New York, 1960, p. 73.

[5] E. Edser, *Heat for Advanced Students*, Macmillan, London, 1927, p. 99.

[6] R. P. Benedict, "Essentials of Thermodynamics," *Electro-Technol.*, July 1962, p. 107.

Nomenclature

Roman

d exact differential
f function of
K constant
p absolute pressure
R specific gas constant
$\mathbf{R}$ universal gas constant

t empirical temperature
v specific volume

Greek

α either isobaric volume coefficient or isochoric pressure
        coefficient

Subscripts

0   at reference state (usually the ice point)
p at constant pressure
t at constant temperature
v at constant volume

Superscript

0   signifies ideal gas property

Problems

1. Find the specific gas constant (R) for water vapor, air, and carbon
    dioxide, in U.S. Customary units, given that the universal gas con-
    stant is 1545 ft lbf/(lbm mol °R).

    Ans. $R_{H_2O} = 85.76$, $R_{air} = 53.35$, $R_{CO_2} = 35.11$ ft lbf/°R lbm.

2. Using Regnault's cubical coefficient of expansion, determine T_{ice} and
    T_{steam} on the air thermometer scale.

    Ans. $T_{ice} = 273°A$, $T_{steam} = 373°A$.

3. Using Gay-Lussac's cubical coefficient of expansion, determine T_{ice}
    and T_{steam} on the air thermometer scale.

    Ans. $T_{ice} = 267°A$, $T_{steam} = 367°A$.

4. Find the final pressure according to Boyle's law if a gas initially at 1
    atmosphere expands to twice its volume.

    Ans. $p_2 = 0.5$ atmosphere.

5. If Clapeyron's equation of state is accepted, along with Regnault's
    evaluation of the cubical coefficient of expansion, find the volume
    occupied by 1 kg of air at a temperature of 127°C and at a pressure
    of 1 N/m².

    Ans. $V = 114, 840$ m³.

Chapter 3

THERMODYNAMIC
VIEWPOINTS OF
TEMPERATURE

"... The definition of a temperature on the absolute scale does not depend on any assumption about the pressure or volume of a gas becoming zero at absolute zero, nor does it depend on the existence of a hypothetical ideal gas, nor does it involve any statement about the absence of all molecular motion at absolute zero, nor does it imply that a temperature of absolute zero or less is unattainable ..."

Francis Weston Sears (1950)

3.1 Kelvin's Thermodynamic Temperature Scale

William Thomson (later Lord Kelvin) recognized in 1848 that Sadi Carnot's analysis of 1824 of a reversible heat engine operating between two isotherms and two adiabats [1] provided a basis for defining an absolute thermometric scale, since the efficiency of the Carnot engine was a function only of the two empirical temperatures, as it was independent of the working substance (see Figure 3.1). Thomson's proposed absolute thermodynamic temperature function Θ can be given in terms of the reversible Carnot heats as [2]

$$\frac{\delta Q}{Q} = \frac{d\Theta}{\Theta} = f(t)\,dt, \tag{3.1}$$

where Θ is any arbitrary function of the empirical temperature t. Acting on a suggestion made by Joule, Kelvin patterned the new temperature function after the temperature from the zero of the air thermometer (i.e., he took $\Theta \approx [t - t_0 + (1/\alpha_{0p})] \approx [t - t_0 + (1/\alpha_{0v})]$ [3]. He used 273 as the

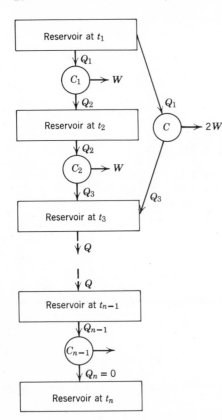

Figure 3.1 Kelvin's definition of the absolute thermodynamic temperature scale is based on the Carnot cycle efficiency, defined as follows:

$$\eta_c = 1 - \frac{Q_2}{Q_1}$$

where Q_1 = heat absorbed from reservoir at empirical temperature t_1; Q_2 = heat rejected to reservoir at t_2; and W = work performed by Carnot engine C. Then,

$$Q_1/Q_2 = f(t_1, t_2)$$
$$Q_2/Q_3 = f(t_2, t_3)$$
$$Q_1/Q_3 = f(t_1, t_3)$$

OR

$$\frac{Q_1}{Q_2} = \frac{\phi(t_1)}{\phi(t_2)} = \frac{T_1}{T_2}.$$

The absolute thermodynamic temperature ratio (T_1/T_2) of two reservoirs equals the ratio of the heats absorbed and rejected (Q_1/Q_2) by a Carnot engine operating between these two reservoirs. Absolute zero on this thermodynamic scale is defined as the temperature of that reservoir to which a Carnot engine rejects no heat.

idealized constant for $1/\alpha_0$, based on Regnault's observations on the coefficients of expansion of various gases. Thus Kelvin's absolute temperature scale was related to the ordinary Celsius scale by the simple relation

$$\Theta_c = t(°C) + 273, \tag{3.2}$$

since the reference temperature (t_0) was taken as 0°C. By further relating the fundamental temperature interval of his Θ scale to that of the empirical scale (i.e., by taking $\Theta_{\text{steam}} - \Theta_{\text{ice}} = t_{\text{steam}} - t_{\text{ice}}$), Kelvin succeeded in defining completely the absolute thermodynamic temperature scale. In summary, he said [4]: "To fix on a unit or degree for the numerical measurement of temperature, we may either call some definite temperature, such as that of melting ice, unity, or any other number we please; or we may choose two definite temperatures, such as that of melting ice and that of saturated vapor of water under the pressure 29.9218 inches Hg in latitude 45, and call the difference of these temperatures any number we please, 100 for instance. The latter assumption is the only one that can be made conveniently in the present state of science, on account of the necessity of retaining a connection with practical thermometry as hitherto practiced; but the former is far preferable in the abstract, and must be adopted ultimately."

Kelvin knew that his thermodynamic scale would be of little importance in practical thermometry unless some means were found to realize these absolute temperatures experimentally. He turned to a gas thermometer, that is, he took

$$\frac{\Theta_2}{\Theta_1} \approx \left(\frac{p_2}{p_1}\right)_v \approx \left(\frac{v_2}{v_1}\right)_p. \tag{3.3}$$

He understood that the differences in the α_{0p}s and the α_{0v}s of all real gases meant that no two gas thermometers would give exactly the same label (absolute temperature) to a common isotherm. Thus, although in theory Kelvin defined a thermodynamic temperature scale independent of the thermometric substance, in effect, he simply redefined an ideal gas whose thermodynamic state was now given by

$$pv = R\Theta. \tag{3.4}$$

Today we know that no real gas follows (3.4) exactly. We also know that a gas thermometer, used as Kelvin suggested, fails to provide a unique experimental determination of the thermodynamic temperature. Nevertheless, in 1887, it was agreed internationally that the constant-volume hydrogen gas thermometer should define the absolute temperature scale, although such a scale (uncorrected) did not agree exactly with Kelvin's thermodynamic scale, or even with similar scales defined by other gas thermometers, and was therefore in no sense universal. The use of this scale and this thermometer continued until 1927. Today, both empirical and absolute temperatures are defined and distinguished by thermodynamic axioms [5]–[11].

3.2 Thermodynamic Axioms and Identities

The zeroth axiom states, in effect, that if two bodies are in thermal equilibrium with a third body, they are in equilibrium with each other. This axiom asserts that each state of thermal equilibrium between any given substance and a reference substance can be labeled. These entirely arbitrary numerical labels are known as empirical temperatures (t). This basic concept was used by the earliest workers in the field of thermometry and is tacitly assumed to this day in every direct measurement of temperature. Empirical temperatures can be calculated from any convenient base point. The scale will depend on the choice of the thermometric substance, the choice of the thermometric property of the substance, and the choice of the function relating the empirical temperature to the change in property of the substance.

The first axiom (i.e., first law of thermodynamics) states, in effect, that the work done in any adiabatic change of state of a substance is fixed by the end states of the substance. This axiom introduces the concepts of a thermally isolated (adiabatic) container and an internal energy function (U), and then provides a precise definition of the heat transferred (Q) in a diabatic change of state. But the first axiom does not itself add anything new to the concept of temperature.

The second axiom (i.e., second law of thermodynamics) states, in effect, that, in any adiabatic system, all processes requiring a decrease in entropy are impossible. This axiom asserts as a mathematical consequence that an absolute temperature function (T) exists along with its conjugate function, the entropy (S). Absolute temperatures serve as extremely useful numerical labels for the various isotherms of a substance. They are calculated from a fixed zero base point. The scale will be independent of the choice of thermometric substance. The scale is arbitrary only to the extent that a free choice of a multiplicative constant must be made. In practice, a definite number must be assigned to only one arbitrary thermodynamic fixed point. Once this is provided, the scale is fixed.

From these axioms many thermodynamic identities can be derived. One in particular provides the basis for realizing temperatures on the absolute scale experimentally, that is, the Maxwell relation

$$\left(\frac{\partial h}{\partial p}\right)_t = v - T\left(\frac{\partial v}{\partial T}\right)_p, \tag{3.5}$$

where T is the universal absolute temperature function of thermodynamics (which coincides, as will be seen, with Kelvin's Θ temperature function). The pressure and volume in (3.5) are approximated by the polynomial equation of state of a real gas

$$pv = A + Bp + Cp^2 + \cdots, \tag{3.6}$$

where A, B, and C are functions of temperature. Similarly, the enthalpy ($h = u + pv$) of (3.5) may be approximated by:

$$h = a + bp + cp^2 + \cdots, \tag{3.7}$$

where a, b, and c are functions of temperature. It is presumed (and has never been contradicted by experiment) that, in the limit, as the pressure of any real gas approaches zero along any isotherm, the pv product and its derivative, and the enthalpy and its derivative, remain finite [12].

When (3.6) and (3.7) are operated on according to (3.5), the result is

$$0 + b \cdots = \frac{1}{p}\left[A - T\left(\frac{dA}{dT}\right)\right] + \left[B - T\left(\frac{dB}{dT}\right)\right] + \cdots . \qquad (3.8)$$

For (3.8) to hold for all pressures and temperatures, the terms made up of like coefficients must be equal. Of specific interest are the first terms; thus, with variables separated,

$$\frac{dA}{A} = \frac{dT}{T}, \qquad (3.9)$$

from which we obtain by integration

$$\ln A = \ln T + \ln R, \qquad (3.10)$$

where R is the familiar gas constant of integration. From (3.6), $A = (pv)^0$ and represents the finite pv product at the zero pressure intercept. On taking antilogarithms, there results the important relation

$$(pv)^0 = RT. \qquad (3.11)$$

We conclude that absolute temperature is proportional not simply to the pv product, as in Kelvin's ideal gas, but to the limiting value of the pv product as p approaches zero along the respective isotherm. This presumption is supported by all existing experimental data. Hence

$$\frac{T_2}{T_1} = \lim_{p \to 0} \frac{(pv)_2}{(pv)_1} = \frac{(pv)_2{}^0}{(pv)_1{}^0} = \frac{\Theta_2}{\Theta_1}. \qquad (3.12)$$

Once a definite value is assigned to a particular (reference) isotherm (T_{ref}), or once a definite temperature difference is assigned between two reproducible isotherms ($dT = dt$), all other temperatures on the absolute temperature scale can be determined in principle by (3.12). Thus a basis is provided for realizing Kelvin's absolute thermodynamic temperature function Θ experimentally.

3.3 Practical Realization of Absolute Temperature

There are many experimental methods for determining the absolute temperature ratios associated with the various reproducible isotherms. All make use of the important relation (3.12) and rely on the truth of the axioms and presumptions listed.

Berthelot's Method

Using constant volume or constant pressure gas thermometers, pressure-volume observations are made at successively lower pressures, first along

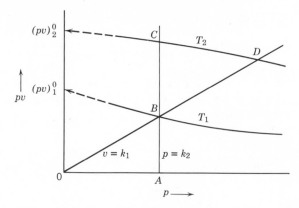

Figure 3.2 Berthelot's method for establishing absolute fixed-point temperature ratios.

one isotherm, then along another. Linear extrapolation of these isotherms to zero pressure (usually by algebraic means) yields the $(pv)^0$ intercepts, and thus the ratio of the two absolute temperatures directly. Let it be emphasized that these zero pressure intercepts are functions of their corresponding absolute temperatures only, will be the same regardless of the thermometric gas employed, and will be independent of the initial charge of gas in the thermometer [13]. According to Figure 3.2, for a constant volume thermometer

$$\frac{T_2}{T_1} = \lim_{p \to 0} \left(\frac{OD}{OB}\right)_v = \frac{(pv)_2^{\ 0}}{(pv)_1^{\ 0}}, \tag{3.13}$$

whereas, for a constant pressure thermometer

$$\frac{T_2}{T_1} = \lim_{p \to 0} \left(\frac{AC}{AB}\right)_p = \frac{(pv)_2^{\ 0}}{(pv)_1^{\ 0}}. \tag{3.14}$$

Goff's Method

A method that avoids the practical difficulties encountered in progressively decreasing the charge of gas in the gas thermometer has been described by J. A. Goff [14]. A Burnett apparatus [15], in which volume measurements are avoided, is used in conjunction with a constant volume gas thermometer whose charge need not be varied, to make available the ratio of the zero pressure pv products for any two experimentally realizable isotherms. Thus the ratio of the absolute temperatures is ultimately obtained, just as in the Berthelot method (see Figure 3.3); for example, by experiment, the ratio

$$\frac{T_{\text{steam}}}{T_{\text{ice}}} = 1.36609 \tag{3.15}$$

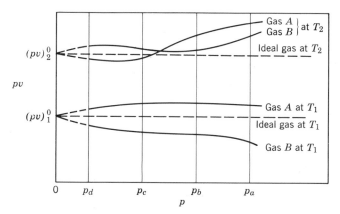

Figure 3.3 One method for determining absolute-temperature ratios. Pressures are progressively reduced in a constant-volume container immersed in a constant-temperature bath. The pressure p_d represents the lowest pressure attainable. Values of pv extrapolated to zero pressure are the same for any gas at the same temperature. Example:

$$\frac{(pv)^0_{\text{steam point}}}{(pv)^0_{\text{ice point}}} = 1.36609 = \frac{T_{\text{steam}}}{T_{\text{ice}}}.$$

can be and has been confirmed in many laboratories throughout the world. Once a definite number is assigned to one arbitrary state (as 273.15 K to the ice point) or once a definite temperature difference is assigned between two reproducible reference states (as 100 to $T_{\text{steam}} - T_{\text{ice}}$), all other temperatures on the absolute scale can be determined in principle. In practice, however, no continuous absolute temperature scale is forthcoming, since only a finite number of reliable fixed-point environments exist where such temperature ratios can be defined [16].

3.4 References

[1] S. Carnot, *Reflections on the Motive Power of Fire* (R. H. Thurston, translator), Dover, New York, 1960.

[2] W. Thomson, "On an Absolute Thermometric Scale Founded on Carnot's Theory of the Motive Power of Heat, and Calculated from Regnault's Observations," *Cambridge Phil. Soc. Proc.*, June 5, 1848.

[3] W. Thomson, "On the Dynamical Theory of Heat, with Numerical Results Deduced from Mr. Joule's Equivalent of a Thermal Unit, and M. Regnault's Observations in Steam," Pt. 2, *Trans. Roy. Soc. Edinburgh*, March 1851.

[4] J. P. Joule and W. Thomson, "On the Thermal Effects of Fluids in Motion," Pt. 2, *Phil. Trans. Roy. Soc. London*, **144,** June 15, 1854, p. 350.

[5] F. W. Sears, *An Introduction to Thermodynamics, The Kinetic Theory of Gases, and Statistical Mechanics*, Addison-Wesley, Reading, Mass., 1950, p. 9.

[6] R. P. Benedict, "Temperature and Its Measurement," *Electro-Technol.*, July 1963, p. 71.

[7] P. T. Landsberg, "Foundations of Thermodynamics," *Rev. Mod. Phys.*, October 1956, p. 363.

[8] L. Balamuth, H. C. Wolfe, and M. W. Zemansky, "The Temperature Concept from the Macroscopic Point of View," *Am. J. Phys.*, August 1941, p. 199.

[9] J. H. Kennan and A. H. Shapiro, "History and Exposition of the Laws of Thermodynamics," *Mech. Eng.*, November 1947, p. 915.

[10] A. R. Miller, "The Concept of Temperature," *Am. J. Phys.*, November 1952, p. 488.

[11] H. C. Wolfe, "The Temperature Concept," in *Temperature, Its Measurement and Control in Science and Industry*, Vol. 2, Reinhold, New York, 1955, p. 3.

[12] J. A. Beattie, "The Thermodynamic Temperature of the Ice Point," in *Temperature, Its Measurement and Control in Science and Industry*, Vol. 1, Reinhold, New York, 1941, p. 76.

[13] H. T. Wensel, "Temperature," *Ibid.*, p. 15.

[14] J. A. Goff, "Thermodynamic Notes," University of Pennsylvania, 4th ed., 1947, p. 77.

[15] E. S. Burnett, "Compressibility Determinations Without Volume Measurements," *J. App. Mech.*, **3**, 1936, p. A-136.

[16] R. P. Benedict, "Essentials of Thermodynamics," *Electro-Technol.*, July 1962, p. 107.

Nomenclature

Roman

a, b, c constants
A, B, C constants
d exact differential
f function of
h specific enthalpy
p absolute pressure
Q quantity of heat
R specific gas constant
S entropy
t empirical temperature
T absolute temperature
u specific internal energy
U internal energy
v specific volume

Greek

α either isobaric volume coefficient or isochoric pressure coefficient
δ inexact differential
Θ absolute temperature function

Subscripts

0   at reference state (usually the ice point)

p   at constant pressure

v   at constant volume

Superscript

0   signifies ideal gas property

Problems

1.   Using a constant volume gas thermometer, it is found experimentally that $(pv)^\circ_X/(pv)^\circ_S = 1.5000$, whereas $(pv)^\circ_S/(pv)^\circ_Y = 1.5000$. If a new absolute temperature scale is defined such that $T_S - T_Y = 150^\circ A$, find the absolute temperature of X, Y, and S on this scale.

    Ans. $T_X = 675^\circ A$, $T_Y = 300^\circ A$, $T_S = 450^\circ A$.

2.   What would be the predicted freezing point of platinum, in degrees Celsius, if the ratio $T_{platinum}/T_{steam} = 5.4700$ were found experimentally via a constant-volume gas thermometer?

    Ans. $T_{pt} = 1767.98^\circ C$.

3.   If the freezing point of mercury is $-38.841^\circ C$, what ratio of zero pressure (pv)s would be expected when a constant-pressure gas thermometer was exposed to constant-temperature baths of ice and mercury?

    Ans. $(pv)^\circ_{ice}/(pv)^\circ_{mercury} = 1.1658$.

Chapter 4

THE INTERNATIONAL PRACTICAL TEMPERATURE SCALE

"... The IPTS-68 is a practical, standard, empirical scale chosen in such a way that temperatures measured on it closely approximate thermodynamic temperatures..."

George W. Burns (1974)

There are natural limitations in rigorously defining a complete thermodynamic temperature scale. For example, we have already seen that only a finite number of reliable fixed-point environments exist from which temperature ratios can be formed. In addition, there is the very practical fact that one cannot afford to determine absolute temperature ratios each time a temperature measurement is required.

There is a definite need for standard interpolation instruments and standard interpolation procedures so that temperatures intermediate between the thermodynamic fixed points can also be determined.

These limitations and needs led to the establishment and widespread acceptance in 1927 of an international empirical temperature scale [1]. This scale went through several minor revisions in 1948 [2], 1954 [3], 1960 [4], [5], and most recently in 1968.

4.1 The Fixed Points

The International Practical Temperature Scale of 1968, in the range from about −200°C to 1000°C, is based on six fixed and reproducible equilibrium states to which are assigned numerical values corresponding to the respective Celsius temperatures. These primary fixed points, as defined in 1968 by the Advisory Committee on Thermometry of The General Conference on Weights and Measures, are the following:

Oxygen Point (equilibrium temperature between liquid oxygen and its vapor) -182.962

Triple Point of Water (Fundamental)* (equilibrium temperature between ice, liquid water, and water vapor) $+0.01$

Steam Point (equilibrium temperature between liquid water and its vapor)   100

Zinc Point (equilibrium temperature between solid and liquid zinc)
419.58

Silver Point (equilibrium temperature between solid and liquid silver)
961.93

Gold Point (equilibrium temperature between solid and liquid gold)
1064.43

All of these primary fixed points are given in degrees Celsius, and all are at a pressure of one standard atmosphere except the triple point of water.

In addition to the six primary fixed points, many reliable secondary fixed points have been defined. Some of these are given later.

4.2 The Interpolating Instruments and Equations

Between the primary fixed points, temperature is defined by certain standard interpolation instruments used with certain interpolation equations. These have changed over the years, but one of the basic instruments and one of the basic equations were first proposed in 1887 by Hugh Longbourne Callendar (following an idea of C. W. Siemens).

Callendar proposed that the resistance variation of a standard platinum wire could serve as a dependable interpolating instrument [6]. He suggested the following parabolic equation to define temperature in the range 0 to 500°C:

$$t = \left(\frac{R_t - R_0}{R_{100} - R_0}\right) 100 + \delta \left(\frac{t}{100} - 1\right)\left(\frac{t}{100}\right), \qquad (4.1)$$

*The triple point of water is susceptible of being a more precise thermometric reference point than the "melting point of ice." Hence the triple point is now made one of the defining fixed points on the International Scale. The ice point, however, still stands at about 0°C by any practical measurement.

where R_t = platinum resistance at temperature t,

R_0 = platinum resistance measured with the thermometer immersed in an air-saturated ice–water mixture when the ice-point temperature is unaffected (±0.001°F) by barometric pressure variations from 28.50 in. to 31.00 in. Hg, and is independent of immersion depths up to 6 in.,

R_{100} = platinum resistance measured with the thermometer immersed in saturated steam at atmospheric pressure (by means of a hypsometer); however, the steam-point temperature is greatly affected by barometric pressure variations (typical necessary corrections are given in Table 4.1),

δ = a characteristic constant of the particular thermometer, defined at the zinc point.

Equation 4.1 represents the only second-power curve that may be passed through the ice, steam, and zinc points. (Until 1968, the Callendar

Table 4.1   Variation of Steam Point Temperature with Barometric Pressure

(Inches of mercury versus degree Fahrenheit)

Pressure (″Hg)	0	2	4	6	8
			t, (°F)		
29.5	211.287	211.321	211.355	211.389	211.423
.6	.457	.491	.525	.559	.593
.7	.627	.661	.695	.729	.763
.8	.796	.830	.864	.898	.932
.9	.965	.998	212.033	212.066	212.100
30.0	212.133	212.167	.201	.234	.268
.1	.301	.335	.368	.410	.436
.2	.469	.502	.536	.569	.603
.3	.636	.669	.702	.736	.769
.4	.802	.836	.869	.902	.935
.5	.968	213.002	213.035	213.068	213.101
.6	213.134	.167	.200	.233	.267
.7	.299	.332	.365	.398	.431
.8	.464	.497	.530	.563	.596
.9	.628	.661	.693	.727	.760

$$t_p = 212 + 50.422 \left(\frac{p}{p_0} - 1\right) - 20.95 \left(\frac{p}{p_0} - 1\right)^2 \text{°F}$$

$$p_0 = 29.921''\text{Hg}$$

equation was used to define temperature from the ice point to the antimony point. The latter is a secondary fixed point standing at about 630°C.)

It was soon realized that the Callendar equation could not be extrapolated far below the ice point without seriously failing to conform to the thermodynamic scale. In 1925, M. S. Van Dusen proposed [7] the use of an additional term with Callendar's equation such that

$$t = \left\{\frac{R_t - R_0}{R_{100} - R_0}\right\} 100 + \delta\left(\frac{t}{100} - t\right)\left(\frac{t}{100}\right) + \beta\left(\frac{t}{100} - 1\right)\left(\frac{t}{100}\right)^3, \quad (4.2)$$

where β is another characteristic constant of the particular thermometer, defined at the oxygen point.

Until 1968, this equation was used to define temperature from the oxygen point to the ice point [8], [9]. From the antimony point to the gold point, a standard platinum/10% rhodium versus platinum thermocouple is currently used as the interpolation instrument. The equation defining temperature in this range is of the form

$$E = a + bt + ct^2, \quad (4.3)$$

where the constants a, b, and c are determined at the antimony, silver, and gold points, respectively. E is the net emf of the thermocouple, with one junction at the ice point and the other at temperature t.

Above the gold point, temperature on the international Celsius scale is currently defined by Planck's radiation formula, used with an optical pyrometer, in the form:

$$\frac{J_t}{J_{Au}} = \frac{\exp\left[\dfrac{1.4388}{\lambda(t_{Au} + 273.15)}\right] - 1}{\exp\left[\dfrac{1.4388}{\lambda(t + 273.15)}\right] - 1}, \quad (4.4)$$

where J_t and J_{Au} are the radiant energies per unit wavelength interval at wavelength λ, emitted per unit time, per unit solid angle, per unit area of a blackbody at the temperature t and at the gold point, respectively; the constant 1.4388 is measured in centimeter-kelvins; λ, the wavelength of the visible spectrum, is measured in centimeters; t is measured in degrees Celsius.

4.3 A Standard Empirical Temperature Scale

As mentioned in the Introduction to Part I, to gain an understanding of any subject we must know at least the outline of its history. This is especially true in the case of the standard temperature scale. We next review some of the steps leading to IPTS-68 in order to understand and appreciate it.

4.3.1 *International Temperature Scale of 1927 (ITS-27)*

This scale [2], [3], adopted by the Seventh General Conference on Weights and Measures in 1927, was designed to provide a practical temperature scale that was easily reproducible and coincided as nearly as possible with the thermodynamic temperature scale. The ITS-27 was based on six fixed and reproducible equilibrium states to which were assigned numerical values corresponding to the best determinations of the respective thermodynamic temperatures given in terms of degrees centigrade, (see Table 4.2). The scale was divided into four ranges based on the means available for interpolation (see Table 4.3).

Table **4.2** *Summary of Fixed-point Values*

Defining fixed points	ITS-27 °C	ITS-48 °C	Scale IPTS-48 °C	IPTS-68 °C	IPTS-68 K
t.p. hydrogen				−259.34	13.81
b.p. hydrogen, 25/76 Atmos.				−256.108	17.042
b.p. hydrogen				−252.87	20.28
b.p. neon				−246.048	27.102
t.p. oxygen				−218.789	54.361
b.p. oxygen	−182.97	−182.970	−182.97	−182.962	90.188
f.p. water	0.000	0			
t.p. water			+0.01	+0.01	273.16
b.p. water	100.000	100	100	100	373.15
f.p. zinc				419.58	692.73
b.p. sulfur	444.60	444.600	444.6		
f.p. silver	960.5	960.8	960.8	961.93	1235.08
f.p. gold	1063.0	1063.0	1063	1064.43	1337.58
Some secondary fixed points					
f.p. tin		231.9	231.91	231.9681	505.1181
f.p. lead		327.3	327.3	327.502	600.652
f.p. zinc		419.5	419.505		
b.f. sulfur				444.674	717.824
f.p. antimony		630.5	630.5	630.74	903.89
f.p. aluminum		660.1	660.1	660.37	933.52

t.p., triple point; b.p., boiling point; f.p., freezing point.

Table **4.3** *Comparison of Temperature Ranges and Interpolating Methods*

Item of Comparison	Scale			
	ITS-27	ITS-48	IPTS-48	IPTS-68
Range 1 Interpolating instrument is platinum resistance thermometer				
Temperature limits	−190 to 0°C	−182.970 to 0°C	−182.97 to 0°C	13.81 to 273.15 K
Interpolating relation	Callendar-VanDusen Quartic	Callendar-VanDusen Quartic	Callendar-VanDusen Quartic	Reference function Eq. (4.9)
Range 2 Interpolating instrument is platinum resistance thermometer				
Temperature limits	0 to 660°C	0 to 630.5°C	0 to 630.5°C	0 to 630.74°C
Interpolating relation	Callendar Parabola	Callendar Parabola	Callendar Parabola	Modified Callendar, Eq. (4.11)
Range 3 Interpolating instrument is platinum, 10% rhodium and platinum thermocouple				
Temperature limits	660 to 1063.0°C	630.5 to 1063.0°C	630.5 to 1063°C	630.74 to 1064.43°C
Interpolating relation	Parabola	Parabola	Parabola	Parabola
Range 4 Interpolating instrument is optical pyrometer				
Temperature limits	Above 1063.0°C	Above 1063.0°C	Above 1063°C	Above 1064.43°C
Interpolating relation	Wien's law	Planck's law	Planck's law	Planck's law

Range 1—From −190°C to 0°C temperature was defined by the resistance of a standard platinum thermometer and the unique quartic passed through the temperature values assigned to the oxygen, ice, steam, and sulfur points.

Range 2—From 0°C to 660°C temperature was defined by the resistance of a standard platinum thermometer and the unique parabola passed through the temperature values assigned to the ice, steam, and sulfur points.

Range 3—From 660°C to 1063°C temperature was defined by the emf of a standard platinum, 10% rhodium and platinum thermocouple, and the unique parabola passed through the temperature values assigned to the antimony, silver, and gold points.

Range 4—Above the gold point temperature was defined by a brightness match achieved by a standard optical pyrometer (by means of which the radiant energy of a black body at a given wavelength and at an unknown temperature could be compared with that of a black body at a closely similar wavelength at the gold point) and Wien's law; that is

$$J_{b,\lambda} = \frac{c_1 \lambda^{-5}}{\exp[c_2/\lambda T]}. \tag{4.5}$$

Table 4.4 Comparison of Interpolating Instruments

Item of Comparison	Scale			
	ITS-27	ITS-48	IPTS-48	IPTS-68
Standard platinum resistance thermometers R_{100}/R_0 (higher value = increased purity)	≥1.390	≥1.3910	≥1.3920	≥1.3925
Standard platinum, 10% rhodium and platinum thermocouple				
R_{100}/R_0	≥1.390	≥1.3910	≥1.3920	≥1.3920
E_{Au}		10,300±50 μV	10,300±50 μV	10,300±50 μV
$E_{Au}-E_{Ag}$		1185+0.158($E_{Au}-$ 10,310)±3 μV	1183+0.158($E_{Au}-$ 10,300)±4 μV	1183+0.158($E_{Au}-$ 10,300)±4 μV
$E_{Au}-E_{Sb}$		4776+0.631($E_{Au}-$ 10,310)±5 μV	4766+0.631($E_{Au}-$ 10,300)±8 μV	4766+0.631($E_{Au}-$ 10,300)±8 μV
Standard optical pyrometer Second radiation constant, c_2 (in meter kelvins)	0.01432	0.01438	0.01438	0.014388

Equation 4.5 was used to extrapolate temperatures in this range (see Table 4.4).

It was the feeling in 1927 that there was not enough evidence to establish definite differences between ITS-27 and thermodynamic temperatures. It was further believed at this time that differences between empirical and thermodynamic temperatures did not exceed 0.05°C between the ice and sulfur points.

4.3.2 International Temperature Scale of 1948 (ITS-48)

In October 1948, the International Committee on Weights and Measures and the Ninth General Conferences adopted nine significant changes in the international empirical temperature scale [2], [3], as recommended by the Advisory Committee of Thermometry. The changes incorporated in ITS-48 were:

(a) In Range 1, temperature was no longer extrapolated below the oxygen point.

(b) The triple point of water was stated to be susceptible of being a more precise thermometric reference point than the melting point of ice.

(c) Zero of ITS-48 was defined as being 0.0100°C below the triple point of water.

(d) In Range 2, temperature was no longer extrapolated above the antimony point.

(e) The ratio, R_{100}/R_0, which indicates the purity of platinum used in standard resistance and thermoelectric thermometers, was tightened from 1.390 to 1.3910.

(f) In Range 3, the freezing point of silver was changed from 960.5°C to 960.8°C.

(g) In Range 4, temperatures above the gold point were extrapolated using Planck's law; i.e.,

$$J_{b,\lambda} = \frac{c_1\lambda^{-5}}{\exp[c_2/\lambda T]-1},$$ (4.6)

in place of Wien's law (see Table 4.3).

(h) The second radiation constant, c_2, used in optical pyrometry (as in (4.5) and (4.6) was changed from 0.01432 meter kelvins to 0.01438 meter kelvins (see Table 4.4).

(i) Degree of temperature was designated "degree Celsius" in place of "degree, centigrade."

Otherwise, the same fixed points and the same interpolating methods were used to define ITS-48 as were used to define ITS-27.

In the NBS text of the 1948 scale, it was recognized that significant differences existed between ITS-48 and the thermodynamic temperature scale. For example, by gas thermometry, the sulfur point was determined to be at 444.74±0.05°C (thermo), whereas the ITS-48 value was 444.600°C. It was suggested that the relation

$$t_{\text{thermo}} - t_{\text{ITS-48}} = \frac{t}{100}\left(\frac{t}{100}-1\right)(0.04217 - 7.481\times10^{-5}t),\,°C \quad (4.7)$$

predicted the differences that existed at that time in Range 2 (i.e., between the ice and sulfur points). Differences in these scales between the oxygen and ice points (i.e., in Range 1) were still felt to be less than 0.05°C.

While differences in ITS-48 and thermodynamic temperatures were thought to be small, the same could not be said for temperature differences between ITS-48 and ITS-27. The shift in the silver point changed all temperatures in Range 3; for example, a maximum difference of 0.4°C was reached near 800°C. (This change in the silver point, however, had the beneficial effect of making the interpolated temperatures of Ranges 2, 3, and 4 join more smoothly.) Adjusting the value of "c_2" changed all temperatures above the gold point, and the use

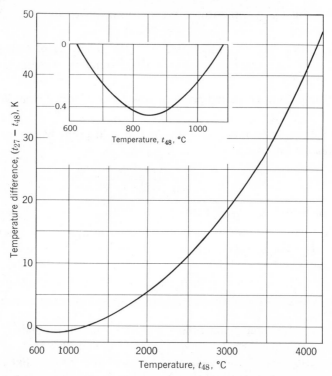

Figure 4.1   Differences between the International Temperature Scales of 1927 and 1948 in Ranges 3 and 4.

of Planck's equation (in place of Wien's) affected the very high temperatures. Figure 4.1 summarizes differences between these two empirical scales.

4.3.3   *International Practical Temperature Scale of* 1948 (*IPTS-48*)

In October 1960, the International Committee on Weights and Measures and the Eleventh General Conference adopted six additional changes in the international empirical temperature scale [5], as recommended by the Advisory Committee on Thermometry. The changes incorporated in IPTS-48 (amended edition of 1960) were:

(a) The ITS-48 was renamed, because of text revisions, the IPTS-48. The 48 was retained in the title because values of temperature remained the same on these two scales.

(b) The thermodynamic scale was redefined in terms of a single fixed point, the triple point of water, with an assigned value of 273.16 K, rather than in terms of the difference between the ice and steam points.

(c) The triple point of water became one of the six defining fixed points of IPTS-48. (This meant that the triple point of water had a common value on the thermodynamic and empirical scales.)

(d) Zero on IPTS-48 was redefined as being 0.01°C below the triple point of water.

(e) The ratio, R_{100}/R_0, was again tightened for both resistance and thermoelectric platinum standards from 1.3910 to 1.3920 (see Table 4.4).

(f) The emfs of standard thermocouples in Range 3 were changed from $E_{Au} - E_{Ag} = 1185 + 0.158$ $(E_{Au} - 10,310) \pm 3\mu V$ to $E_{Au} - E_{Ag} = 1183 + 0.158$ $(E_{Au} - 10,300) \pm 4\mu V$ and from $E_{Au} - E_{Sb} = 4776 + 0.631$ $(E_{Au} - 10,310) \pm 5\mu V$ to $E_{Au} - E_{Sb} = 4766 + 0.631$ $(E_{Au} - 10,300) \pm 8\mu V$ (see Table 4.4).

Again, it was recognized that significant differences existed between IPTS-48 and the thermodynamic temperature scale. On the basis of changes (b) and (c) above, in the new IPTS, the relation

$$t_{thermo} - t_{IPTS-48} = \frac{t}{100}\left[\left(\frac{t}{100} - 1\right)(0.04106 - 7.363 \times 10^{-5}t) - 0.0060\right], °C$$

(4.8)

was given to predict the differences between the empirical and thermodynamic scales in Range 2 [10], [11]. It was noted that average values from gas thermometry in this period compared with IPTS-48 values as follows:

Thermodynamic Scale (°C)	Fixed Point	International Scale (°C)
−182.982	b.p. oxygen	−182.97
444.66	b.p. sulfur	444.6
961.55	f.p. silver	960.8
1064.24	f.p. gold	1063

But, in the interest of avoiding too frequent changes in the international scale, the values assigned to the fixed points of IPTS-48 were not changed essentially from those of ITS-48 at that time.

4.3.4 International Practical Temperature Scale of 1968 (IPTS-68)

In October 1968, the International Committee on Weights and Measures, as empowered by the Thirteenth General Conference, adopted eight major changes in the international empirical temperature scale [12], as recommended by the Advisory Committee on Thermometry. The changes incorporated in IPTS-68 are:

(a) The name kelvin and the symbol K are taken to designate the unit of thermodynamic temperature.

(b) All values assigned to the defining fixed points are changed (except for that of the triple point of water) to conform as closely as possible to the corresponding thermodynamic temperatures (see Table 4.2).

(c) The lower limit of Range 1 is changed from the boiling point of oxygen to the triple point of hydrogen (see Table 4.3).

(d) The standard instrument to be used in Ranges 1 and 2 is the platinum resistance thermometer with R_{100}/R_0 tightened from 1.3920 to 1.3925 (see Table 4.4).

(e) Range 1 is now divided into four parts: Part 1 extends from the triple point to the boiling point of hydrogen. Part 2 extends from the boiling point of hydrogen to the triple point of oxygen. Part 3 extends from the triple point to the boiling point of oxygen. Part 4 extends from the boiling point of oxygen to the freezing point of water,

Table 4.5   Constants and Equations for Use With the New Reference Function for Interpolation in Range 1

Coefficients (A_i) for use in equation 4.9

i	A_i
1	0.2508462096788033 D 03
2	0.1350998699649997 D 03
3	0.5278567590085172 D 02
4	0.2767685488541052 D 02
5	0.3910532053766837 D 02
6	0.6556132305780693 D 02
7	0.8080358685598667 D 02
8	0.7052421182340520 D 02
9	0.4478475896389657 D 02
10	0.2125256535560578 D 02
11	0.7679763581708458 D 01
12	0.2136894593828500 D 01
13	0.4598433489280693 D 00
14	0.7636146292316480 D–01
15	0.9693286203731213 D–02
16	0.9230691540070075 D–03
17	0.6381165909526538 D–04
18	0.3022932378746192 D–05
19	0.8775513913037602 D–07
20	0.1177026131254774 D–08

(f) In Range 1, the Callendar–VanDusen equation is no longer used, but interpolation is by a new reference function*

$$T_{68} = \left[273.15 + \sum_{i=1}^{20} A_i (\ln W_{ref})^i \right], K \tag{4.9}$$

where the coefficients, A_i, are given in Table 4.5. The reference resistance ratio represented by W_{ref} is defined by

$$W_{ref} = W_M - \Delta W \tag{4.10}$$

where W_M is the measured resistance ratio, R_t/R_0, and ΔW is a deviation defined by a specific polynomial-interpolation equation, one being given for each part of Range 1 (see Table 4.6).

**Table 4.6   Polynomial Interpolation Equations for Deviation ΔW,
to be Used in Equation 4.10**

Part 1

$$\Delta W_1 = A_1 + B_1 T + C_1 T^2 + D_1 T^3 \tag{4.13}$$

(with constants to be determined at the triple point of hydrogen, boiling point of hydrogen, and from $d(\Delta W_2)/dT$).

Part 2

$$\Delta W_2 = A_2 + B_2 T + C_2 T^2 + D_2 T^3 \tag{4.14}$$

(with constants to be determined at the boiling point of hydrogen, boiling point of neon, triple point of oxygen, and from $d(\Delta W_3)/dT$).

Part 3

$$\Delta W_3 = A_3 + B_3 T + C_3 T^2 \tag{4.15}$$

(with constants to be determined at the triple point and boiling point of oxygen, and from $d(\Delta W_4)/dT$).

Part 4

$$\Delta W_4 = A_4 t + C_4 t^3 (t_{68} - 100) \tag{4.16}$$

(with constants to be determined at the boiling point of oxygen and the boiling point of water). Assigned values of W_{ref} at the fixed points:

Boiling point of oxygen	0.24379909
Freezing point of water	1
Boiling point of water	1.39259668

---

*In Spring 1976, the IPTS-68 was amended slightly without in anyway changing the measured temperatures [18].

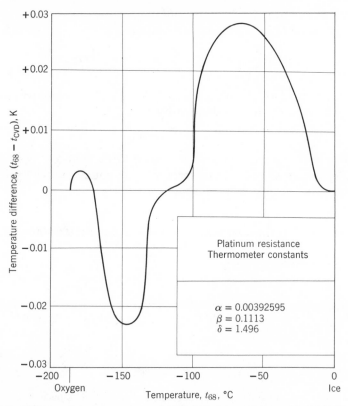

Figure 4.2 Differences in temperature obtained by interpolation with IPTS-68 and with Callendar-Van Dusen equation in Range 1, Part 4.

Figure 4.2 shows, for Range 1, Part 4, the differences in temperature obtained by interpolating with the new reference function and with the old Callendar–VanDusen equation for typical resistance thermometer constants. The effect on the temperature measurement of using the new interpolation method is seen to be less than ± 0.03 K, an insignificant amount to industrial users.

(g) In Range 2, the Callendar equation is modified by a correction term so that interpolated values of temperature will conform more closely with thermodynamic temperatures; that is,

$$t_{68} = t' + \Delta t \qquad (4.11)$$

where t' is temperature by the Callendar equation, and Δt is a correction term given by

$$\Delta t = 0.045 \left(\frac{t'}{100}\right)\left(\frac{t'}{100} - 1\right)\left(\frac{t'}{419.58} - 1\right)\left(\frac{t'}{630.74} - 1\right). \qquad (4.12)$$

Figure 4.3 shows, for Range 2, the differences is temperature introduced by this correction term. That is, Δt is to be recognized as the difference between a parabola (the Callendar equation) and IPTS-68. This difference is seen to be less than ± 0.05 K.

(h) In Range 4, the second radiation constant, c_2, is changed from 0.01438 meter kelvins to 0.014388 meter kelvins (see Table 4.4).

By design, IPTS-68 has been chosen in such a way that temperatures

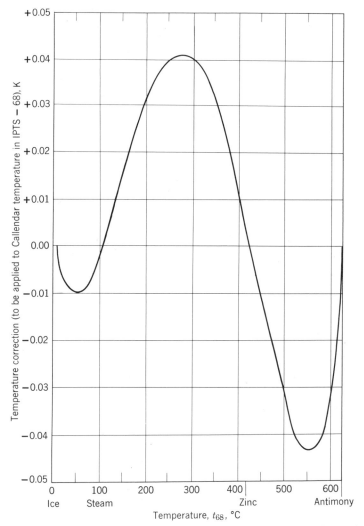

Figure 4.3 Differences between the Callendar equation and IPTS-68 in Range 2.

measured on it closely approximate thermodynamic temperatures; that is, differences are within the limits of the present accuracy of measurement.

Adjustments in values assigned to the defining fixed points naturally cause changes in the international empirical temperature scale. Differences in interpolation methods in Ranges 1 and 2 introduce further shifts in temperature, and changing the value of c_2 affects all temperatures in Range 4. Figure 4.4 summarizes total differences between the empirical scales of IPTS-48 and IPTS-68. Briefly, in English

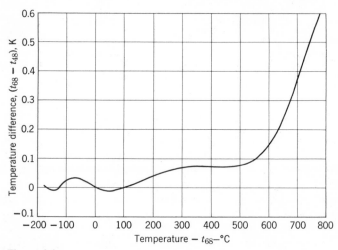

Figure 4.4*a*

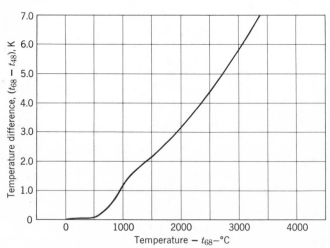

Figure 4.4*b*   Differences between the International Practical Temperature Scales of 1948 and 1968.

units, the temperature differences between the latest two scales amount to:

from $-150°F$ to $0°F$, a change of less than $0.05°F$,
from $0°F$ to $1000°F$, a change of less than $0.2°F$,
from $1000°F$ to $2000°F$, a change of less than $3°F$,
and from $2000°F$ to $3000°F$, the change is less than $5°F$.

In addition to reference 12, many other works are available that document the development of the new international practical temperature scale. Among these are [13]–[15]. The evolution of IPTS-68 is summarized in [16].

4.3.5 *The Engineering Approximation*

Although designed to conform as closely as possible to the thermodynamic Celsius scale, the International Practical Temperature Scale must be considered as nothing more than a practical, standard, empirical temperature scale. One should therefore accept with reservations the usual relation

$$T(K) = t(°C) + 273.15, \tag{4.13}$$

realizing that no such simple relation exists between the practical, empirical Celsius scale and the absolute thermodynamic Kelvin scale. However, equation 4.13 is an extremely useful engineering approximation.

In a similar vein, several other useful engineering approximations can be given:

$$T(°R) = t(°F) + 459.67, \tag{4.14}$$

and, from the relation between Celsius and Fahrenheit scales.

$$T(K) = \frac{5}{9}\left[t(°F) + 459.67\right] = \frac{5}{9} T(°R), \tag{4.15}$$

where $T(°R)$ signifies absolute temperatures on the Rankine scale.

4.4 Prognostications

To date, there exists no continuous thermodynamic temperature scale. The empirical temperature scale that is currently the law of the land must be expected to be changed in time. There are at least three areas in the scale in which changes undoubtedly will occur.

4.4.1 *Fixed Points*

Values assigned to the fixed points are traditionally determined by gas thermometry in spite of the great experimental difficulties involved. These values have been and are subject to refinements and changes over the years as improved techniques are developed. Also, as commercially available fixed-point standards become more common, the characteristics of these fixed points will be studied more closely by many workers using standard interpolating instruments. Those fixed points which are most desirable from the users' point of view will be more readily discovered and, we may hope, will find their way into the IPTS. Thus, although the fixed-point baths are truly gifts of nature, they do not come with numbers preassigned, and we must expect today's values to be changed as time goes on.

4.4.2 *Interpolating Instruments*

A resistance thermometer, a thermocouple and an optical pyrometer fill the role as interpolating instruments on IPTS-68. It would be quite desirable, however, to define temperatures from the helium point all the way up to the gold point by a single interpolating instrument. Among other reasons, smoother transitions would be accomplished in the scale when passing from one range to another. The resistance thermometer would best serve this purpose because of its inherent sensitivity and precision. To date, the platinum resistance thermometer has been used as a standard only up to the antimony point because of temperature limitations. For example: the Pyrex tube yields to stresses above 1000°F; the water of crystallization in the mica insulators may distort the mica above 1000°F, and the platinum wire may become sensibly thinner (by sublimation) at the higher temperatures, and this would cause an increase in resistance.

One important project at the National Bureau of Standards involves work with industry to extend the platinum resistance thermometer range to the gold point [17]. Since the gold point is not only the terminal point of Range 3, but is also the primary fixed point of Range 4, where optical pyrometry takes over, it is always obtained under black-body conditions. Thus the resistance thermometer must not only withstand the high temperature of the gold point, but it must at the same time be small enough to allow its insertion into a black-body enclosure, which, itself, must be kept small to hold the necessary uniform temperature. The advantages outweigh the difficulties, and the prognostication is to expect a single interpolation instrument up to the gold point in the next IPTS.

4.4.3 *Interpolating Methods*

Better methods for interpolation, mathematically speaking, are bound to be forthcoming. It is interesting to note a change in direction in going from IPTS-48 to IPTS-68. Previously, very low degree interpolating equations were the rule. In fact, the fortuitous agreement between simple parabolas and thermodynamic temperatures, from the ice point to the gold point, has long, and thankfully, been noted. Now we are faced with a 20th degree interpolating equation [see (4.9)] for Range 1, with coefficients having 16 significant figures (see Table 4.5). Mathematical niceties aside, the uncertainties inherent in the fixed-point temperature measurements should strongly influence determinations of degree and number of significant figures called for in the interpolating equations of the IPTS. Thus we foresee in future scales the use of lower degree, less complex interpolation methods.

4.5 References

[1] G. K. Burgess, "The International Temperature Scale," *J. Res. NBS*, July–December 1928, RP 22, p. 635.

[2] H. F. Stimson, "The International Temperature Scale of 1948," *J. Res. Nat. Bur. Std.*, March 1949, RP 1926, p. 209.

[3] J. A. Hall, "The International Temperature Scale," in *Temperature*, Vol. 2, Reinhold, New York, 1955, p. 116.

[4] H. F. Stimson, "Heat Units and Temperature Scale for Calorimetry," *Am. J. Phys.*, **23,** No. 9, December 1955, p. 614.

[5] H. F. Stimson, "International Temperature Scale of 1948, Text Revision of 1960," *Nat. Bur. Std. Monograph 37*, September 8, 1961.

[6] H. L. Callendar, "On the Practical Measurement of Temperature," *Phil. Trans. Roy. Soc. London*, **178,** 1887, p. 160.

[7] M. S. VanDusen, *J. Am. Chem. Soc.*, **47,** 1925, p. 3326.

[8] E. F. Mueller, "Precision Resistance Thermometry," in *Temperature*, Vol. 1, Reinhold, New York, 1941, p. 162.

[9] H. F. Stimson, "Precision Resistance Thermometry and Fixed Points," in *Temperature*, Vol. 2, Reinhold, New York, 1955, p. 141.

[10] J. A. Beattie, "Gas Thermometry," in *Temperature*, Vol. 2, Reinhold, New York, 1955, p. 93.

[11] J. A. Beattie, M. Benedict, B. E. Blaisdell, and J. Kaye, "An Experimental Study of the Absolute Temperature Scale. XI. Deviation of the International Practical from the Kelvin Temperature Scale in the Range 0 to 444.6°C," M.I.T. Preprint 474, 1964.

[12] "The International Practical Temperature Scale of 1968," A Committee Report, English version appeared in *Metrologia*, **5,** 2, April 1969.

[13] R. P. Benedict, "International Practical Temperature Scale of 1968," *Leeds and Northrup Technical Journal*, Spring 1969.

[14] H. Preston-Thomas, "The Origin and Present Status of the IPTS-68," in *Temperature*, Vol. 4, Instrument Society of America, Pittsburgh, 1972.

[15] J. G. Hust, "A Compilation and Historical Review of Temperature Scale Differences," *Cryogenics*, **9**, 6, December 1969.

[16] *Evolution of the International Practical Temperature Scale of* 1968, ASTM STP 565, 1974.

[17] D. J. Curtis and G. J. Thomas, "Long Term Stability and Performance of Platinum Resistance Thermometers for Use to 1063 C," *Metrologia*, **4**, 4, October 1968, pp. 184–190.

[18] "The International Practical Temperature Scale of 1968–Amended Edition of 1975" A Committee Report, English version appeared in *Metrologia*, **12**, 1976, pp. 7–17.

Nomenclature

Roman

A_i	coefficients
A, B, C	constants
a, b, c	constants
exp	base of natural logarithms, e, raised to exponent in [ ]
E	net emf of thermocouple
J	radiant energy per unit wavelength per unit time per unit area
R	resistance
t	empirical temperature
t'	temperature by the Callendar equation
T	absolute temperature
W_{ref}	reference resistance ratio
W_M	measured resistance ratio

Greek

α	fundamental coefficient of resistance thermometer
β	resistance thermometer constant determined at oxygen point
Δ	finite difference
δ	resistance thermometer constant determined at sulfur point
λ	wavelength

Subscripts

Ag	at silver point temperature
Au	at gold point temperature
Sb	at antimony point temperature
t	at temperature t
0	at ice point temperature
100	at steam point temperature

Problems

1. Estimate by equation the deviation between the Callendar temperature of tin and t_{68}.

   Ans. $\Delta t \sim 0.03894°C$ for first try (actually, $\Delta t = 0.038937°C$).

2. Estimate graphically the deviation of Problem 1.

   Ans. $\Delta t \sim 0.04°C$.

3. Determine the variation in the temperature of the boiling point of water if the barometric pressure varies from 29.56 to 30.72 inches of mercury.

   Ans. $\Delta T_{b.p.} = 1.943°F$.

Chapter 5

LIQUID-IN-GLASS THERMOMETERS

"... *the degree of their scales agree with one another, and their variations are within fixed limits...*"

Daniel Gabriel Fahrenheit (1706)

5.1 Principles and Definitions

A liquid-in-glass thermometer [1], [2] is a temperature-measuring instrument consisting of a thin-walled glass *bulb* (the reservoir for the thermometer liquid) attached to a glass *stem* (the capillary tube through which the meniscus of the liquid moves with a change in temperature), with the bulb and stem system sealed against its environment. The portion of the bulb–stem space that is not occupied with the thermometer liquid usually is filled with a dry inert gas under sufficient pressure to prevent separation of the thermometer liquid. A *scale* is provided to indicate the height to which the liquid column rises in the stem, and this reading is made to indicate closely the temperature of the bulb. A *contraction chamber* (an enlargement of the capillary) often is provided below the main capillary to avoid the need for a long length of capillary or to prevent contraction of the liquid column into the bulb. An *expansion chamber* (an enlargement of the capillary) often is provided above the main capillary to protect the thermometer in case of overheating. A *reference point* (usually marking the height of the thermometer fluid in the stem when the bulb is at the ice point) should be provided as a check against changes in the bulb volume.

The operation of a liquid-in-glass thermometer thus depends on the coefficient of expansion of the liquid being greater than that of the containing glass bulb. Any increase in the bulb temperature causes the liquid to expand and rise in the stem, with the difference in volume between the bulb and the stem serving to magnify the change in volume of the liquid.

In establishing a thermometer scale, certain conditions prevailed during the calibration. Specifically, there was some definite amount of the thermometer liquid exposed to the calibration environment, and some definite ambient temperature prevailed during the test. Variations in these important conditions are distinguished by the following three definitions (see also Figure 5.1).

PARTIAL IMMERSION. A liquid-in-glass thermometer designed to indicate temperatures correctly when the bulb and a specified portion of the stem are exposed to the temperature being measured.

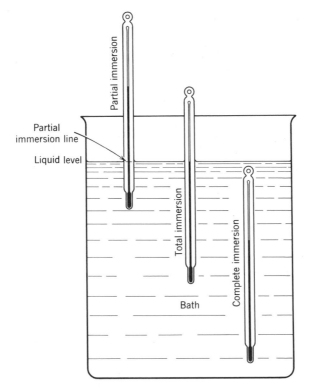

Figure 5.1 Partial, total, and complete immersion thermometer. (After ASME PTC 19.3, 1974, p. 44.)

TOTAL IMMERSION. A liquid-in-glass thermometer designed to indicate temperatures correctly when just that portion of the thermometer containing the liquid is exposed to the temperature being measured.

COMPLETE IMMERSION. A liquid-in-glass thermometer designed to indicate temperatures correctly when the entire thermometer is exposed to the temperature being measured.

Whenever the calibration conditions are not duplicated in the application of a thermometer, an emergent-stem correction must be applied. These corrections are discussed next.

5.2   Stem Corrections

Most liquid-in-glass thermometers in use are of the total immersion type. This is because ambient temperature variations can introduce large uncertainties in the indications of partial immersion thermometers. Countering this advantage of the total immersion type, however, is the difficulty necessarily experienced in reading temperatures when the thermometer liquid column is totally immersed in the environment of the bulb; for example, in measuring a bath temperature, a short length of the thermometer liquid in the stem is usually left emergent from the bath so the meniscus will be visible. But this practice may expose the stem to an appreciable temperature gradient near the bath surface. This will, of course, cause the reading of the thermometer to be either too high or too low, depending on the ambient temperature level with respect to that of the bath. The equation usually used to account for this effect is [3]

$$C_s = KN(t_1 - t_2),\tag{5.1}$$

where C_s = stem correction in degrees, to be added algebraically to the indicated temperature,

K = differential expansion coefficient of thermometer liquid with respect to thermometer glass (for mercury thermometers the values usually recommended are 0.00016 for Celsius scales and 0.00009 for Fahrenheit scales),

N = length of thermometer liquid column emergent from bath, expressed as number of degrees on the thermometer scale,

t_1 = temperature of bulb (for first try, use indicated temperature; then adjust by C_{s1} and try again for a better approximation to C_s),

t_2 = average temperature of emergent liquid column.

See also Figure 5.2.

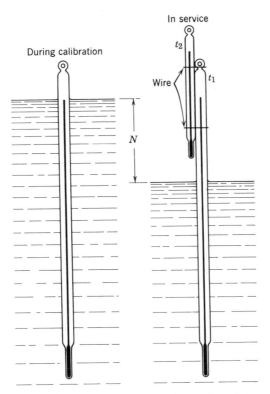

Figure 5.2 Thermometer calibrated for total immersion and used for partial immersion. (After ASME PTC 19.3, 1974, p. 50.)

Example 1. A total immersion thermometer indicated 780°F when the mercury column was immersed to the 200 degree mark on the scale. The mean temperature of the emergent column was 170°F. What was the corrected temperature of the bulb?

$$C_{s1} = 0.00009 \times (780 - 200) \times (780 - 170) = 32°F.$$
$$C_{s2} = 0.00009 \times (780 - 200) \times (780 + 32 - 170) = 34°F.$$
$$t_{corrected} = 780 + 34 = 814°F.$$

Example 2. A total immersion thermometer indicated 100.00°C when the mercury column was immersed to the 80 degree mark on the scale. The mean temperature of the emergent column was 60°C. What was the corrected temperature of the bulb?

$$C_s = 0.00016 \times (100 - 80) \times (100 - 60) = 0.13°C,$$
$$t_{corrected} = 100.00 + 0.13 = 100.13°C.$$

A partial immersion thermometer also can be used at conditions other than specified. The emergent stem correction in such cases is given by

$$C_s = KN(t_{sp} - t_2), \qquad (5.2)$$

where C_s, K, and t_2 are as defined previously,

> $N =$ length of emergent column from immersion mark to top of
> thermometer liquid, expressed as number of degrees on the
> thermometer scale (note that even ungraduated scale lengths
> must be considered),
>
> $t_{sp} =$ specified mean temperature of emergent stem.

Example 3. A partial immersion thermometer indicated 250.0°F when used improperly as
a total immersion thermometer. The equivalent length of the submerged column was 110°F
(taken from the correct immersion mark to the top of the mercury column). The specified
mean temperature of the emergent stem was 75°F, whereas the approximate mean temperature of the emergent stem was 250°F. What was the corrected temperature of the bulb?

$$C_s = 0.00009 \times 110 \times (75 - 250) = -1.7°F.$$
$$t_{corrected} = 250.0 - 1.7 = 248.3°F.$$

A general stem correction curve that can be used for either total or
partial immersion thermometers is given in Figure 5.3.

A final question concerns the immersion effect for shielded thermometers. Sometimes a metal case is used to protect the more-or-less fragile
thermometer, and in such situations the stem corrections differ significantly from those of (5.1) and (5.2). This points to the fact that every
temperature-sensing element must be treated as an individual. Some wear
metal jackets, others are clothed in ceramics, and still others face the
environments bare. Whenever deviations from specified conditions are
encountered, individual tests can provide the only answer.

5.3   Note on Glass Characteristics

Glass flows when subjected to temperature stresses, and the progressive
change in bulb volume causes a corresponding change in the scale
calibration. According to J. Busse [4] of the NBS, the bulb of a
Fahrenheit mercury thermometer has a volume equivalent to about
10,000 degrees of the stem scale. Thus a change in bulb volume of only 1
part in 100,000 will introduce a 0.1°F change in the stem scale, and such
changes are to be expected. To note the effect of a change in bulb volume
on the temperature measurement, the ice point indication should be
observed with respect to the original etched mark. Since the change in
bore volume is relatively small compared with the change in bulb volume,
it is assumed that all points on the thermometer scale change by the same
amount at that noted as the ice point.

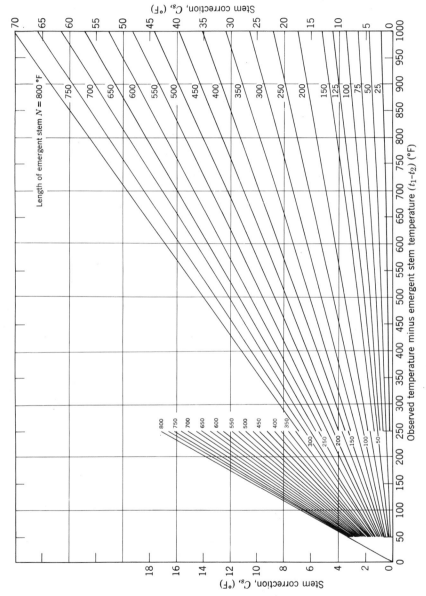

Figure 5.3 Emergent-stem corrections for liquid-in-glass thermometers. (After ASME PTC 19.3, 1974, p. 51.)

5.4 References

[1] J. F. Swindells, "Calibration of Liquid-In-Glass Thermometers," *Nat. Bur. Std.* Monograph 90, February 1965.
[2] *Liquid-In-Glass Thermometry*, Nat. Bur. Std. Monograph 150, 1975.
[3] "Liquid-In-Glass Thermometers," Chapter 5 of *Temperature Measurement*, Supplement to ASME Performance Test Codes, PTC 19.3, 1974.
[4] J. Busse, "Liquid-In-Glass Thermometers," in *Temperature*, Vol. 1, Reinhold, New York, 1941, p. 228.

Nomenclature

Roman

C_s stem correction term
K differential expansion coefficient, liquid to glass
N number of degrees improperly exposed
t_1 temperature of bulb
t_2 average temperature of emergent column
t_{sp} specified mean temperature of emergent column

Problems

1. A total immersion thermometer indicated 600°F when the mercury column was immersed to the 400 degree mark on the scale. The mean temperature of the emergent column was estimated at 150°F. What was the actual bulb temperature?

   Ans. $t_{bulb} = 608.2°F$.

2. A total immersion thermometer indicated 250°C when the mean column was immersed to the 100 degree mark on the scale. The mean temperature of the emergent column was estimated at 150°C. What was the actual bulb temperature?

   Ans. $t_{bulb} = 252.5°C$.

3. A partial immersion thermometer indicated 500°F when used improperly as a total immersion thermometer. The equivalent length of the submerged column was 300°F. The specified mean temperature of the emergent stem was 100°F. What was the actual bulb temperature?

   Ans. $t_{bulb} = 489.5°F$.

4. Confirm the stem corrections for Problems 1 and 3 by Figure 5.3.

Chapter 6

RESISTANCE THERMOMETRY

"... the essential elements required in precise resistance thermometry (are) a resistor properly mounted and protected, a means for measuring its resistance, and a relation between resistance and temperature ..."

E. F. Mueller (1941)

6.1 Principles

We cannot improve on Mueller's concise description given above. A resistance thermometer is a temperature-measuring instrument consisting of a *sensor,* an electrical circuit element whose resistance varies with temperature; a *framework* on which to support the sensor; a *sheath* by which the sensor is protected; and wires by which the sensor is connected to a *measuring instrument* (usually a *bridge*), which is used to indicate the effect of variations in the sensor resistance.

Sir Humphry Davy had noted as early as 1821 that the conductivity of various metals was always lower in some inverse ratio as the temperature was higher. Sir William Siemens, in 1871, first outlined the method of temperature measurement by means of a platinum resistance thermometer. It was not until 1887, however, that precision resistance thermometry as we know it today began. This was when Hugh Callendar published his famous paper on resistance thermometers [1]. Callendar's equation describing the variation in platinum resistance with temperature is still pertinent; it is now used in part to define temperature on the IPTS (see Chapter 4) from the ice point to the antimony point.

Resistance thermometers provide absolute temperatures in the sense that no reference junctions are involved, and no special extension wires are needed to connect the sensor with the measuring instrument. The three *s*s that characterize resistance thermometers are *s*implicity of circuits, *s*ensitivity of measurements, and *s*tability of sensors.

6.2 Sensors

The sensors [2], [3] can be divided conveniently into two basic groups: resistance temperature detectors (RTDs) and thermistors. RTDs are electrical circuit elements formed of solid conductors (usually in wire form) that are characterized by a positive coefficient of resistivity. It is interesting to note in this regard that there is a direct relation between the temperature coefficient of resistivity of metals and the coefficient of expansion of gases. The RTDs of general usage are of platinum, nickel, and copper.

Platinum Sensors

Platinum, being a noble metal, is used exclusively for precision resistance thermometers. Platinum is stable (i.e., it is relatively indifferent to its environment, it resists corrosion and chemical attack, and it is not readily oxidized), is easily workable (i.e., it can be drawn into fine wires), has a high melting point (i.e., it shows little volatilization below 1000°C), and can be obtained to a high degree of purity (i.e., it has reproducible electrical and chemical characteristics). All this is evidenced by a simple and stable resistance-temperature (R-T) relationship that characterizes the platinum sensor over a wide range of temperatures. These desirable and necessary features, however, do not come without effort. Platinum is actually extremely sensitive to minute contaminating impurities and to strains, and these both alter the simple R-T relationship. The sensing wire must be mounted so as to be almost free of strains (to avoid the strain gage effect whereby extraneous changes in resistance occur), and the thermometer must be manufactured in a manner that ensures freedom from contaminants. These two effects are guarded against by elaborate precautions.

Since the time of Callendar, the sensors have been constructed by forming a coil of very pure platinum wire around two mica strips that are joined to form a cross. Meyers [4], at the National Bureau of Standards, first wound the wire in a fine helical coil around a steel wire mandrel, and then wound this coil helically around a mica cross framework (see Figure 6.1). Such procedures leave the wire relatively free from mechanical constraints as the wire experiences differential thermal expansions. The sensor is also annealed to relieve winding strains. The diameter of the platinum wire is on the order of 0.1 mm, and it takes a length of about 2 m to yield a satisfactory 25.5 Ω resistance at the ice point. The resulting precision platinum sensors are usually about 1 in. long.

The sensors are then assembled in a protecting sheath (usually a tube of

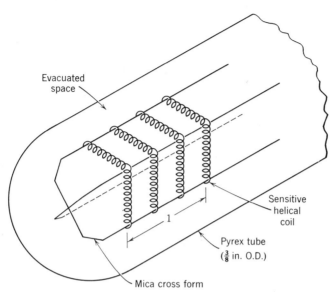

Figure 6.1 Construction detail of a platinum resistance thermometer tip.

pyrex or quartz) about 18 in. long with four gold wires joining the sensor at the bottom of the tube to four copper connecting wires outside the tube. The tube assembly is evacuated and filled with dry air at about one-half atmosphere and hermetically sealed to exclude contaminants. Mica disks separate the internal connecting wires and also serve to break up any convection currents in the tube [5].

Construction is such that δ (the characteristic coefficient of the particular thermometer as determined at the zinc point) is between 1.488 and 1.498. The purity of the platinum wire is assured by testing at the ice and steam points, where R_{100}/R_0 must be greater than 1.3925, where the higher this value, the greater the purity of the platinum.

The reasons for the temperature limitation of present-day platinum resistance thermometers are threefold:

1. The pyrex tube yields to stresses above 1000°F.
2. The water of crystallization in the mica insulators may distort the mica above 1000°F.
3. The platinum wire may become sensibly thinner (by evaporation) at the higher temperatures, and this would cause an increase in resistance.

The National Bureau of Standards hopes to extend the range of the platinum resistance thermometer standard from the antimony to the gold point, in accord with proposals of H. Moser [6] of Germany, by use of special sheaths and special insulators. Several papers describing studies on long-term stability and performance of gold point resistance.thermometers have been published [7], [8]. Their conclusion is that high temperature resistance thermometers can lead to a better practical temperature scale up to the gold point, just as prognosticated in Section 4.4.2. Meantime, good precision platinum resistance thermometers are commercially available with NBS certificates up to 1000°F so that construction details need not be discussed further here.

Nickel and Copper Sensors

These are much cheaper than the platinum sensors and are useful in many special applications. Nickel is appreciably nonlinear, however, and has an upper temperature limit of about 600°F. Copper is quite linear, but is limited to about 250°F, and has such a low resistance that very accurate measurements are required. All in all, the platinum RTDs continue to dominate the field in present resistance thermometry.

Thermistors

Temperature-sensitive resistor materials (such as metallic oxides) have been known to science for some time. However, the chief stumbling block to the widespread use of these materials was their lack of electrical and chemical stability. In 1946, Becker, Green, and Pearson, at the Bell Laboratories, reported [9] on success in producing metallic oxides that had high negative coefficients of resistivity and exhibited the stable characteristics necessary for manufacturing to reproducible specifications. Thus thermistors (a contraction for "thermally sensitive resistors") are electrical circuit elements formed of solid semiconducting materials that are characterized by a high negative coefficient of resistivity. At any fixed temperature, a thermistor acts exactly as any ohmic conductor. If its temperature is permitted to change, however, either as a result of a change in ambient conditions or because of a dissipation of electrical power in it, the resistance of the thermistor is a definite, reproducible function of its temperature. Typical thermistor resistance variations are from $50,000 \, \Omega$ at 100°F to $200 \, \Omega$ at 500°F. Their characteristic temperature-resistance relationship can be approximated by a power function of the form

$$R = ae^{b/T}, \tag{6.1}$$

where R is the thermistor resistance at its absolute ambient temperature T, and a and b are constants for the particular thermistor under consideration, with typical values of 0.06 and 8000, respectively. Note that the logarithm of R plots as a straight line of intercept log a and slope b, when charted versus $(1/T)$.

Their use for temperature measurement is based on the direct or indirect determination of the resistance of a thermistor immersed in the environment whose temperature is to be measured. Because we are dealing with an ohmic circuit, ordinary copper wires suffice throughout the thermistor circuit; thus special extension wires, reference junctions, and bothersome thermal emfs are avoided. Contrary to common belief, thermistors are quite stable when they are properly aged before use (less than 0.1 % drift in resistance during periods of months). Thermistors exhibit great temperature sensitivity (up to ten times the sensitivity of the usual base-metal thermocouples), whereas thermistor response can be in the order of milliseconds. The practical range for which thermistors are useful is from the ice point to about 600°F [10]. Routine temperature measurements to 0.1°F are not unusual if the current through a thermistor is limited to a value that does not increase the thermistor temperature (by $I^2 R$ heating) above ambient by an amount greater than that consistent with the measuring accuracy required.

6.3 Circuits and Bridges

The conventional Wheatstone bridge circuit (Figure 6.2a) is not very satisfactory for platinum resistance thermometry because (a) the slide-wire contact resistance is a variable depending on the condition of the slide; (b) the resistance of the extension wires to the sensor is a variable depending on the temperature gradient along them; and (c) the supply current itself causes variable self-heating depending on the resistance of the sensor [11]. The three extension-wire Callendar–Griffiths bridge circuit, Figure 6.2b, while alleviating some of the faults of the Wheatstone bridge, still is not entirely satisfactory. The four-extension-wire Mueller bridge circuit, Figure 6.2c, is almost always used to determine the platinum sensor resistance. Most of the important measuring resistors are protected from ambient temperature changes by incorporating them in a thermostatically controlled constant-temperature chamber. The range of resistance for which these bridges are intended is from 0.0 to 422.1111 Ω. The temperature rise caused by the bridge current must be kept small so that no reduction in the current produces an observable change in the indication of resistance. In direct opposition to this requirement of low self-heating current is the fact that for maximum bridge sensitivity the bridge current should be as large as possible.

Since the time of his 1939 paper [12], Mueller has designed a new bridge for precision resistance thermometry. H. F. Stimson, at the NBS, gave a brief description of this bridge somewhat as follows [13]. The intention of the design was to make measurements possible within an

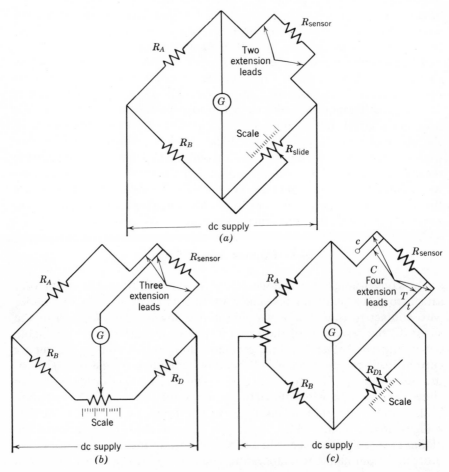

Figure 6.2 (a) Conventional Wheatstone bridge circuit. When G is zeroed by moving R_{slide}, $I_{\text{sensor}}R_{\text{sensor}} = I_{\text{slide}}R_{\text{slide}}$ and $I_{\text{sensor}}R_A = I_{\text{slide}}R_B$. Dividing one by the other, $R_{\text{sensor}} = (R_A/R_B)R_{\text{slide}}$; therefore R_{sensor} is easily determined by the fixed ratio (R_A/R_B) and the variable R_{slide} which is obtained from the calibrated scale. (b) The Callendar-Griffiths bridge circuit, in which slide-wire contact has no effect, the effect of extension-lead resistance variation is reduced, and the effect of self-heating of the sensor is reduced if $R_A \approx R_{\text{sensor}}$. (c) The Mueller bridge circuit. With the switch in one position (as shown), $R_{D1} + R_T = R_{\text{sensor}} + R_C$. With the switch in the other position, extension leads c and t and C and T are interchanged, respectively, and $R_{D2} + R_C = R_{\text{sensor}} + R_T$. Then, by addition, $R_{\text{sensor}} = (R_{D1} + R_{D2})/2$ and the effects of the extension leads have been eliminated.

uncertainty not exceeding 2 or 3 $\mu\Omega$. This bridge has a seventh decade that makes one step in the last decade the equivalent of 10 $\mu\Omega$. One uncertainty that was recognized was that of the contact resistance in the dial switches of the decades. Experience shows that with care this uncertainty can be kept down to the order of 0.0001 Ω.

In Mueller's design, the ends of the equal-resistance ratio arms are at the dial-switch contacts on the 1-Ω and the 10-Ω decades. He has made the resistances of the ratio arms 3000 Ω; thus the uncertainty of contact resistance, 0.0001 Ω, produces an uncertainty of only one part in 30,000,000. With 30 Ω, for example, in each of the other arms of the bridge, these two ratio-arm contacts should produce an uncertainty of resistance measurement of 1.41 $\mu\Omega$.

The 10-Ω decade has mercury contacts and the commutator switch has large mercury contacts. One new feature in the commutator switch is the addition of an extra pair of mercury-contact links that serve to reverse the ratio coils simultaneously with the thermometer leads. This feature automatically makes the combined error of the normal and reverse readings only one in a million, for example, when the ratio arms are unequal by as much as one part in a thousand, and thus almost eliminates any error from lack of balance in the ratio arms. It does not eliminate any systematic error in the contact resistance at the end of the ratio arms on the decades, however, because the mercury reversing switch must be in the arms.

The commutators on these bridges are made so as to open the battery circuit before breaking contacts in the resistance leads to the thermometer. This makes it possible to lift the commutator, rotate it, and set it down in the reverse position in less than a second and thus not disturb the galvanometer or the steady heating of the resistor by a very significant amount. In balancing the bridge, snap switches are used to reverse the current so the heating is interrupted only a small fraction of a second. This practice of reversing the current has proved very valuable. It not only keeps the current flowing almost continuously in the resistor but also gives double the signal of the bridge unbalances.

Basic thermistor circuits are given in Figure 6.3*a* and *b*. One is the familiar Wheatstone bridge, and the other is a simple series circuit that is quite effective in determining thermistor temperature as a function of the emf drop across a fixed resistor.

6.4 Equations and Solutions

In Chapter 4 the Callendar interpolating equation for the platinum resistance thermometer, which was used in part to define temperature on

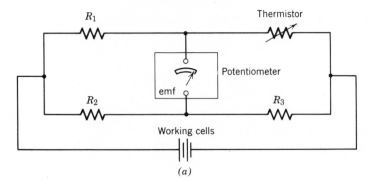

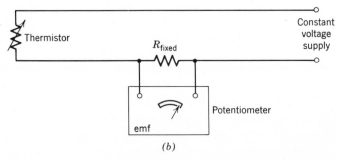

Figure 6.3 Basic thermistor circuits: (*a*) bridge circuit and (*b*) series circuit.

the IPTS between the ice and antimony points, was given by (4.1) as

$$t = \left(\frac{R_t - R_0}{R_{100} - R_0}\right)100 + \delta\left(\frac{t}{100} - 1\right)\left(\frac{t}{100}\right). \tag{6.2}$$

Similarly, between the oxygen and ice points, the Callendar–VanDusen equation was given by (4.2). The point here is that it is one thing to measure a resistance precisely (no mean feat in itself, as indicated in Section 6.3) and quite another to convert the measurement readily to a temperature.

Hand calculations, involving quadratics and cubics, are practical if only a few conversions are required. Eggenberger [14] has presented a solution to (6.2), based in part on tables and in part on graphs. With the advent of the computer, machine solutions make resistance-to-temperature conversions routine, precise, and nearly instantaneous [15].

Note that (6.2), for example, also can be expressed as

$$t^2 - \left(\frac{10^4}{\delta} + 10^2\right)t + \frac{10^6}{\delta}\left(\frac{R_t - R_0}{R_{100} - R_0}\right) = 0. \tag{6.3}$$

Solving the quadratic yields

$$t = A - B\sqrt{C - R_t},\qquad(6.4)$$

where
$$A = \frac{50}{\delta}(\delta + 10^2)$$

$$B = \frac{50}{\delta}\left(\frac{400\delta}{R_{100} - R_0}\right)^{1/2}$$

$$C = (R_{100} - R_0)\left(\frac{\delta}{400} + \frac{1}{2} + \frac{25}{\delta}\right) + R_0.$$

Equation 6.3 can be solved readily by hand, whereas (6.4) may be more amenable to machine solution. However, (6.4) may be manipulated still further to obtain Eggenberger's tabular-graphical solution; that is, to avoid solution of (6.4) for each measurement, the temperature also can be expressed as

$$t = t_i + \Delta t + \text{correction},\qquad(6.5)$$

where $t =$ the empirical temperature corresponding to the total resistance measurement;

  $t_i =$ that part of the temperature measurement corresponding to the integral part of the resistance measurement;

  $\Delta t =$ the approximate part of the temperature measurement corresponding to the decimal part of the resistance measurement (the approximation is that the slope $\Delta t / \Delta R$ is 10°C/Ω);

correction = the adjustment necessary to account for the approximation of a constant temperature-resistance slope.

Equation 6.5 is synthesized as follows. First, the resistance (R_t) of (6.4) is expressed in parts. Thus

$$t = A - B\sqrt{(C - R_i) - \Delta R},\qquad(6.6)$$

where R_i represents the integral part of the resistance, and ΔR the decimal part. The temperature corresponding to the integral part of the resistance is then

$$t_i = A - B\sqrt{C - R_i}.\qquad(6.7)$$

This expression can be tabulated for the individual resistance thermometer (e.g., see Table 6.1). The approximate temperature, including the integral part of the resistance plus the approximate effect of the decimal part of the resistance is

$$t_a = A - B\sqrt{S} + 10\,\Delta R,\qquad(6.8)$$

Table 6.1   Typical Platinum Resistance Thermometer Chart
(absolute ohms versus degrees Fahrenheit)

Resistance (Ω)	0	1	2	3	4
			t_i (°F)		
25	22.852	40.552	58.303	76.106	93.961
30	111.869	129.830	147.845	165.915	184.039
35	202.218	220.453	238.744	257.092	275.498
40	293.961	312.482	331.063	349.703	368.403
45	387.164	405.986	424.870	443.816	462.826
50	481.899	501.037	520.240	539.508	558.843
55	578.245	597.715	617.253	636.861	656.538
60	676.287	696.106	715.998	735.963	756.003
65	776.116	796.306	816.571	836.914	857.336
70	877.836	898.416	919.078	939.821	960.647
75	981.557	1002.552	1023.633	1044.802	1066.057

$$R_0 = 25.517 \qquad R_{100} = 35.538 \qquad \delta = 1.493$$

where $S = (C - R_i)$, and the slope, $\Delta t / \Delta R$, is assumed to be 10°C/Ω (or 18°F/Ω). The complete expression for the temperature is then

$$t = t_a + [A - B\sqrt{S - \Delta R} - A + B\sqrt{S} - 10\,\Delta R]. \qquad (6.9)$$

Expressed as a correction, (6.9) yields

$$t - t_a = BS^{1/2}\left[1 - \left(1 - \frac{\Delta R}{S}\right)^{1/2}\right] - 10\,\Delta R, \qquad (6.10)$$

or

$$t - t_a = \left(\frac{B}{2S^{1/2}} - 10\right)\Delta R + \left(\frac{B}{8S^{3/2}}\right)\overline{\Delta R^2}. \qquad (6.11)$$

This correction can be plotted for the individual resistance thermometer (e.g., see Figure 6.4).

Example 1. A platinum resistance thermometer having the constants $R_0 = 25.517\,\Omega$, $R_{100}/R_0 = 1.392719$, and $\delta = 1.493$, indicates $R_t = 28.5547\,\Omega$. What is the corresponding Callendar temperature on the Fahrenheit scale?

A.   By hand calculation of (6.3)

$$t^2 - \left(\frac{10^4}{1.493} + 10^2\right)t + \frac{10^6}{1.493}\left(\frac{28.5547 - 25.517}{35.538 - 25.517}\right) = 0,$$

$$t^2 - 6797.924t + 203036.449 = 0,$$

$$t = 29.998°C = 85.996°F.$$

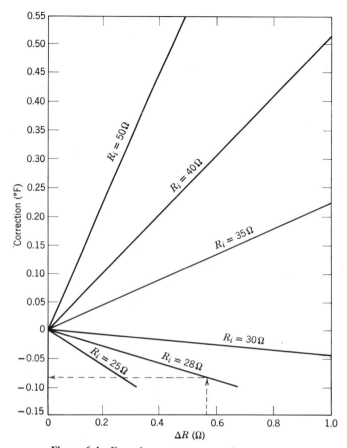

Figure 6.4 Eggenberger-type correction curve.

B. By machine calculation of (6.4)

$$t = 30°C = 86°F \quad \text{(see Table 6.2)}.$$

C. By Eggenberger's method of (6.5)

$$R_i = 28 \ \Omega, \qquad t_i = 76.106°F \text{ (see Table 6.1)}.$$
$$\Delta R = 0.5547 \ \Omega, \qquad \Delta t = 9.985°F \text{ (i.e., } 18 \ \Delta R).$$

Correction $= -0.084°F$ (see Figure 6.4).

Summing: $t = 76.106 + 9.985 - 0.084 = 86.007°F$.

Example 2. A copper resistance thermometer has the constants $R_0 = 12 \ \Omega$, $R_{100}/R_0 = 1.431$, and $R_{200}/R_0 = 1.862$ (see Table 6.3). Find its resistance at 100°F.

A. 100°F $= 37.78°C$, thus $R_{37.78}$ is required.
B. $R_{100} = 1.431 \times 12 = 17.172 \ \Omega$,
    $R_{200} = 1.862 \times 12 = 22.344 \ \Omega$.

Table 6.2 Typical Standard Resistance Thermometer Table[a]

Temperature (°F)	0	1	2	3	4	5	6	7	8	9
						R, (Ω)					
0	23.7042	23.7610	23.8177	23.8745	23.9313	23.9880	24.0448	24.1015	24.1583	24.2150
10	24.2717	24.3284	24.3851	24.4418	24.4984	24.5551	24.6118	24.6684	24.7250	24.7817
20	24.8383	24.8949	24.9515	25.0081	25.0647	25.1212	25.1778	25.2344	25.2909	25.3474
30	25.4040	25.4605	25.5170	25.5735	25.6300	25.6865	25.7429	25.7994	25.8559	25.9123
40	25.9687	26.0252	26.0816	26.1380	26.1944	26.2508	26.3071	26.3635	26.4199	26.4762
50	26.5326	26.5889	26.6452	26.7015	26.7578	26.8141	26.8704	26.9267	26.9830	27.0392
60	27.0955	27.1517	27.2079	27.2642	27.3204	27.3766	27.4328	27.4890	27.5451	27.6013
70	27.6575	27.7136	27.7698	27.8259	27.8820	27.9381	27.9942	28.0503	28.1064	28.1625
80	28.2185	28.2746	28.3306	28.3867	28.4427	28.4987	28.5547	28.6107	28.6667	28.7227
90	28.7787	28.8346	28.8906	28.9465	29.0025	29.0584	29.1143	29.1702	29.2261	29.2820
100	29.3379	29.3938	29.4496	29.5055	29.5613	29.6171	29.6730	29.7288	29.7846	29.8404
110	29.8962	29.9520	30.0077	30.0635	30.1192	30.1750	30.2307	30.2864	30.3421	30.3979

[a] Computer solution of (6.4). $R_0 = 25.517$; $(R_{100} - R_0)/100R_0 = 0.00392719$; $\delta = 1.493$

Table 6.3 Resistance Versus Temperature of Various Metals[a]

Metal	Resistivity (microhm-cm)	Relative Resistance (R_t/R_0) at °C											
		−200	−100	0	+100	200	300	400	500	600	700	800	900
Alumel	28.1	—	—	1.000	1.239	1.428	1.537	1.637	1.726	1.814	1.899	1.982	2.066
Copper	1.56	0.117	0.557	1.000	1.431	1.862	2.299	2.747	3.210	3.695	4.208	4.752	5.334
Iron	8.57	—	—	1.000	1.650	2.464	3.485	4.716	6.162	7.839	9.790	12.009	12.790
Manganin	48.2	—	—	1.000	1.002	0.996	0.991	0.983	—	—	—	—	—
Nickel	6.38	—	—	1.000	1.663	2.501	3.611	4.847	5.398	5.882	6.327	6.751	7.156
Platinum	9.83	0.177	0.599	1.000	1.392	1.773	2.142	2.499	2.844	3.178	3.500	3.810	4.109
Pt 90, Rh 10.	18.4	—	—	1.000	1.166	1.330	1.490	1.646	1.798	1.947	2.093	2.234	2.370
Silver	1.50	0.176	0.596	1.000	1.408	1.827	2.256	2.698	3.150	3.616	4.094	4.586	5.091

[a] Data from: *Temperature, Its Measurement and Control in Science and Industry, Reinhold, New York, 1941, p. 1312.* Data for temperatures below zero not given for same sample as data for temperatures above zero.

C. To find $R_{37.78}$ a linear relationship between R and t is first assumed; that is

$$R_t = R_0[1+at],$$

with t in °C. Then

$$\frac{R_{100}}{R_0} = 1 + a(100),$$

and $a = 0.00431$. Using this coefficient,

$$\frac{R_{200}}{R_0} = 1 + 0.00431(200) = 1.862.$$

Since this checks with the thermometer constant given, the assumption of linearity is valid. Thus

$$R_{37.78°C} = R_{100°F} = 12[1+0.00431(37.78)] = 13.954 \ \Omega.$$

6.5   The Callendar Coefficients

In this section, we consider a practical method for determining the required resistance thermometer interpolating equations [16], [17].

According to (4.11), namely,

$$t_{68} = t' + \Delta t \tag{6.12}$$

we are assured that the Callendar temperature, t', as defined by (6.2), is still very important in determining t_{68}.

IPTS-68 calls for the platinum resistance thermometer to be calibrated at the triple point of water (0.01°C), the steam point (100°C), and the zinc point (419.58°C). However, it is common practice to use the tin point (231.9681°C) in place of the steam point. This complicates the computation of the Callendar coefficients in that R_{100} may not be available experimentally.

Hence, to evaluate the Callendar interpolation equation used to obtain t_{68}, one must determine δ and R_{100} analytically. A detailed description for doing this is given next.

Example 3.   A certain resistance thermometer exhibits the following resistances at the fixed points of interest:

Resistance R, Ω	Fixed Point	Temperature t_{68}, °C
25.56600	Triple	0.01
48.39142	Tin	231.9681
65.67700	Zinc	419.58

These resistances represent the means of many experimental determinations.

DETERMINATION OF CALLENDAR TEMPERATURE OF TIN.   Although the Callendar temperatures of the triple and zinc points are identical with

t_{68}s, since Δts at these points are zero (see equation 4.12), this cannot be said of the tin point. The temperature correction at the tin point is $\Delta t_{SN} = 0.038937°C$ as determined by a short iterative procedure. First, t is assumed equal to t_{68} and Δt is calculated from (4.12). Then, t is adjusted for this Δt according to (6.12). The process is repeated until t does not change by more than some acceptable value, say 1 in the fifth decimal place. For the Δt so determined, $t_{SN} = 231.92916°C$.

DETERMINATION OF CALLENDAR COEFFICIENTS, b AND c.   The Callendar equation can be written in the entirely equivalent form

$$R_t = R_0 + bt + ct^2 \qquad (6.13)$$

where the two unknown coefficients can be determined from the algebraic relations

$$c = \frac{R_0(t_{ZN} - t_{SN}) - R_{SN}t_{ZN} + R_{ZN}t_{SN}}{t_{SN}t_{ZN}(t_{ZN} - t_{SN})} \qquad (6.14)$$

and

$$b = \frac{R_{ZN} - R_0 - ct_{ZN}^2}{t_{ZN}} \qquad (6.15)$$

where t_{SN} = Callendar temperature at the tin point
t_{ZN} = Callendar temperature at the zinc point
R_{SN} = Measured resistance at the tin point
R_{ZN} = Measured resistance at the zinc point
R_0 = Measured resistance at the ice point (this should be about 0.001 Ω less than the triple point resistance).

For the values given in this example,

$$b = 1.019045 \times 10^{-1}$$
$$c = -1.502488 \times 10^{-5}$$

DETERMINATION OF R_{100}.   Although R_{100} is no longer a measured quantity, it can be determined readily from (6.13) written as

$$R_{100} = R_0 + b(10^2) + c(10^4) \qquad (6.16)$$

For the values given, $R_{100} = 35.6052$ Ω and $R_{100}/R_0 = 1.3927323$.

DETERMINATION OF δ.   As indicated earlier, the characteristic constant of the thermometer can be determined from any fixed point data other than the ice and steam points. For example:

$$\delta = \frac{t_{ZN} - pt_{ZN}}{\left(\dfrac{t_{ZN}}{100} - 1\right)\left(\dfrac{t_{ZN}}{100}\right)} = \frac{t_{SN} - pt_{SN}}{\left(\dfrac{t_{SN}}{100} - 1\right)\left(\dfrac{t_{SN}}{100}\right)} \qquad (6.17)$$

where $$pt = 100(R_t - R_0)/(R_{100} - R_0) \qquad (6.18)$$

and pt is known as the platinum temperature. For the values given, $\delta = 1.496472$, and $pt_{SN} = 227.35022°C$ and $pt_{ZN} = 399.51390°C$. Note that the platinum temperatures and the resistances measured at the fixed points are independent of the temperature scale being used.

GENERAL OBSERVATIONS. (1) Since the IPTS-68 Δt is zero at the ice, steam, zinc, and antimony points, it follows that the $R - t_{68}$ points representing these baths are all on the 68-Callendar parabola. (2) It further follows that the tin and sulfur points are not. Δt_{SN} has already been given in this example. By the same method, $\Delta t_{SU} = -0.012174$ (See Figure 6.5). (3) Given the Callendar constants, b and c, and the 68-

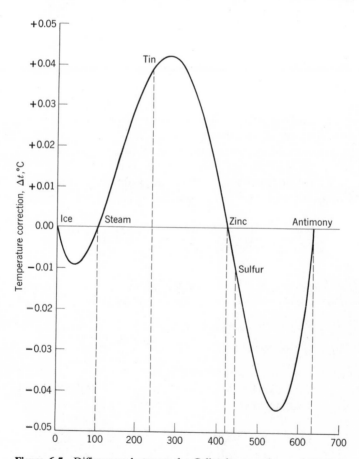

Figure 6.5 Differences between the Callendar equation and IPTS-68.

Callendar temperatures at the sulfur and antimony points, the corresponding resistances can be determined from (6.13) as: $R_{SU} = 67.9094\ \Omega$ and $R_{SB} = 83.8628\ \Omega$.

Example 4.   A certain resistance thermometer yields the following resistances at the fixed points of interest:

Resistance R, Ω	Fixed Point	Temperature t_{68}, °C
6.2268	oxygen	−182.962
25.5500	ice	0
35.5808	steam	100
65.6246	zinc	419.58

RESISTANCE THERMOMETER COEFFICIENTS.   Platinum temperature at the zinc point:

$$pt_{ZN} = \frac{100(R_{ZN} - R_0)}{R_{100} - R_0} = \frac{4007.46}{10.0308} = 399.5158°C.$$

Fundamental coefficient:

$$\alpha = \frac{R_{100} - R_0}{100\ R_0} = \frac{10.0308}{2555.00} = 0.0039259491.$$

Callendar constant:

$$\delta = \frac{t_{ZN} - pt_{ZN}}{\left(\frac{t_{ZN}}{100} - 1\right)\left(\frac{t_{ZN}}{100}\right)} = \frac{419.58 - 399.5158}{(3.1958)(4.1958)} = 1.496334.$$

VanDusen constant:

$$\beta = \frac{t_{OX} - \frac{100(R_{OX} - R_0)}{(R_{100} - R_0)} - \delta\left(\frac{t_{OX}}{100} - 1\right)\left(\frac{t_{OX}}{100}\right)}{\left(\frac{t_{OX}}{100} - 1\right)\left(\frac{t_{OX}}{100}\right)^3} = 0.1113$$

where t_{OX} = Callendar–VanDusen temperature at the oxygen point.
     R_{OX} = measured resistance at the oxygen point.

Measured resistance ratios:

$$(W_M)_{OX} = \frac{R_{OX}}{R_0} = \frac{6.2268}{25.5500} = 0.24371037,$$

$$(W_M)_{100} = \frac{R_{100}}{R_0} = \frac{35.5808}{25.5500} = 1.39259491.$$

By equation 4.10 and Table 4.6:

$$(\Delta W_4)_{OX} = 0.24371037 - 0.24379909 = -0.00008872,$$

$$(\Delta W_4)_{100} = 1.39259491 - 1.39259668 = -0.00000177.$$

Also, by Table 4.6, the interpolating polynomial for determining deviations in Range 1, Part 4 is:

$$\Delta W_4 = A_4 t + C_4 t^3 (t - 100). \tag{6.18}$$

By simultaneous solution at the oxygen and steam points, the constants, A_4 and C_4, are determined to be:

$$A_4 = -0.177 \times 10^{-7},$$
$$C_4 = -0.5306 \times 10^{-13}.$$

With this information about this resistance thermometer available, we can ask for example: If the measured resistance ratio is $W_M = 0.5$, what is t_{68}?

To determine ΔW_4 by (6.18), one must estimate t and then show by iteration (trial and error) that this estimate is close enough to satisfy accuracy requirements. A useful estimate of t is given by the Callendar–VanDusen equation (4.2), which can be rewritten as

$$t_{CVD} = \frac{1}{\alpha}(W_M - 1) + \delta\left(\frac{t}{100} - 1\right)\left(\frac{t}{100}\right) + \beta\left(\frac{t}{100} - 1\right)\left(\frac{t}{100}\right)^3.$$

With the thermometer constants given, and by standard calculation procedures

$$t_{CVD} = -122.8038°C.$$

Using this value of temperature, ΔW_4 is determined to be

$$\Delta W_4 = -0.00001973.$$

By (4.10), the reference resistance ratio is

$$W_{ref} = 0.5 + 0.00001973 = 0.50001973.$$

By (4.9), an estimate of t_{68} is found by computer calculation to be

$$t_{68} = -122.8128°C.$$

Since ΔW does not change significantly for the small temperature difference $(t_{68} - t_{CVD})$ of -0.0090 K, this estimate of t_{68} can serve as t_{68}.

6.6 References

[1] H. L. Callendar, "On the Practical Measurement of Temperature," *Phil. Trans. Roy. Soc. London*, **178**, 1887, p. 160.

[2] "Resistance Thermometers," Chapter 4 of *Temperature Measurement*, Supplement to ASME Performance Test Codes, PTC 19.3, 1974.

[3] D. P. Eckman, *Industrial Instrumentation*, Wiley, New York, 1950, p. 105.

[4] C. H. Meyers, "Coiled Filament Resistance Thermometers," *Nat. Bur. Std. Research Paper 508*, 1932.

[5] D. A. Robertson and K. A. Walch, "Calibration Techniques for Precision Platinum Resistance Thermometers," *Leeds and Northrup Tech. Pub.* A1.2101, 1961.

[6] H. Moser, "Uber die Temperaturmessung mit dem Platinwiderstandsthermometer bis 1100," *Ann. D. Physik*, **5**, 6, 1930, p. 852.

[7] D. J. Curtis and G. J. Thomas, "Long Term Stability and Performance of Platinum Resistance Thermometry for Use to 1063°C," *Metrologia*, **4**, 4, October 1968, pp. 184–190.

[8] J. P. Evans and S. D. Wood, "An Intercomparison of High Temperature Platinum Resistance Thermometers and Standard Thermocouples," *Metrologia*, **7**, 3, July 1971, pp. 108–130.

[9] J. A. Becker, C. B. Green, and G. L. Pearson, "Properties and Uses of Thermistors," *Trans. AIEE*, **65**, November 1946, p. 711.

[10] R. P. Benedict, "Thermistors versus Thermocouples for Temperature Measurements," *Elec. Mfg.*, August 1954, p. 120.

[11] R. P. Benedict, "Temperature and Its Measurement," *Electro-Technol.*, July 1963, p. 71.

[12] E. F. Mueller, "Precision Resistance Thermometry," in *Temperature*, Vol. 1, Reinhold, New York, 1941, p. 162.

[13] H. F. Stimson, "Precision Resistance Thermometry and Fixed Points," in *Temperature*, Vol. 2, Reinhold, New York, 1955, p. 141.

[14] D. N. Eggenberger, "Converting Platinum Resistance to Temperature," *A. Chem.*, **22**, No. 10, October 1950, 1335.

[15] D. R. Stull, "Machine Computation of Relation Between Resistance and Temperature of a Resistance Thermometer," *A. Chem.*, **22**, No. 9, September 1950, p. 1172.

[16] R. P. Benedict and R. J. Russo, "Calibration and Application Techniques for Platinum Resistance Thermometers," *Trans. ASME, J. Basic Eng.*, June 1972, pp. 381–386.

[17] R. P. Benedict, "International Practical Temperature Scale of 1968," *Leeds and Northrup Technical Journal*, Spring 1969.

Nomenclature

Roman

A, B, C	constants
I	electric current
R	resistance
S	the constant $(C - R)$
t	empirical temperature

Greek

δ resistance thermometer constant determined at zinc point
Δ finite difference

Subscripts

i integral part
t at temperature t
0 at ice point temperature
100 at steam point temperature
OX   oxygen
SB   antimony
SN   tin
SU   sulfur
ZN   zinc

Problems

1. Find the Fahrenheit temperature indicated by a thermistor characterized by (6.1), with $a = 0.06\,\Omega$ and $b = 8000°$Rankine, if the measured resistance of the thermistor is $10,000\,\Omega$.
   Ans. $t = 205.68°$F.

2. A platinum resistance thermometer exhibits a resistance ratio $R_{100}/R_0 = 1.3926$ and a characteristic constant $\delta = 1.495$. When placed in an ice bath, the thermometer indicates $25.550\,\Omega$. When exposed to an environment whose temperature is required, it indicates $35.550\,\Omega$. What is the Callendar temperature on the Celsius scale?
   Ans. $t = 99.687°$C.

3. A nickel resistance thermometer, as described by Table 6.3, indicates $20\,\Omega$ at the ice point. What temperature in degrees Celsius is indicated by $R_t = 45\,\Omega$?
   Ans. $t = 172.146°$C.

4. A platinum resistance thermometer yields the following resistance values at the ice, tin, and zinc points respectively: $25.5440\,\Omega$, $48.3434\,\Omega$, $65.6091\,\Omega$.
   a. Find the Callendar coefficients, b, c, δ.
   b. If $R_t = 50.0521\,\Omega$, find the Callendar temperature.
   Ans. $b = 0.10178216$, $c = -0.0000149997$, $t = 250°$C.

5. Find the Callendar–VanDusen coefficients, δ and β, for a platinum resistance thermometer that yields the following resistance values at the oxygen, ice, steam, and zinc points, respectively: $6.230\,\Omega$, $25.540\,\Omega$, $35.590\,\Omega$, and $65.650\,\Omega$.
   Ans. $\delta = 1.527$, $\beta = 0.073385$.

Chapter 7

THERMOELECTRIC THERMOMETRY

"... I sought a method which above all would be rapid and simple, and decided on the use of thermo-electric couples ..."

Henri Le Chatelier (1885)

7.1 Historical Development of Basic Relations

Thomas Johann Seebeck, the German physicist, discovered in 1821 the existence of thermoelectric currents while observing electromagnetic effects associated with bismuth-copper and bismuth-antimony circuits [1], [2]. His experiments showed that, when the junctions of two dissimilar metals forming a closed circuit are exposed to different temperatures, a net thermal electromotive force is generated that induces a continuous electric current.

The Seebeck effect concerns the net conversion of thermal energy into electrical energy with the appearance of an electric current. The Seebeck voltage refers to the net thermal electromotive force set up in a thermocouple under zero-current conditions.

The direction and magnitude of the Seebeck voltage E_s depend on the temperatures of the junctions and on the materials making up the thermocouple. For a particular combination of materials, A and B, and for a small temperature difference, dT, the Seebeck voltage may be written

$$dE_s = \alpha_{A,B} \, dt, \tag{7.1}$$

where $\alpha_{A,B}$ is a coefficient of proportionality called the Seebeck coefficient. (This is commonly called the thermoelectric power.) Thus the

Seebeck coefficient represents, for a given material combination, the net change in thermal emf caused by a unit temperature difference; that is,

$$\alpha_{A,B} = \lim_{\Delta T \to 0} \frac{\Delta E_s}{\Delta t} = \frac{dE_s}{dt}. \tag{7.2}$$

Thus if $E = at + \frac{1}{2}bt^2$ is determined by calibration, then $\alpha = a + bt$. Note that, based on the validity of the experimental relation,

$$E_s = \int_{t_2}^{t} \alpha \, dT = \int_{t_1}^{t} \alpha \, dT - \int_{t_1}^{t_2} \alpha \, dt, \tag{7.3}$$

where $t_1 < t_2 < t$, it follows that α is entirely independent of the reference temperature employed. In other words, for a given combination of materials, the Seebeck coefficient is a function of temperature level only.

Jean Charles Athanase Peltier, the French physicist, discovered in 1834 peculiar thermal effects when he introduced small, external electric currents in Seebeck's bismuth-antimony thermocouple [3], [4], [5]. His experiments show that when a small electric current is passed across the junction of two dissimilar metals in one direction, the junction is cooled (i.e., it acts as a heat sink) and thus absorbs heat from its surroundings. When the direction of the current is reversed, the junction is heated (i.e., it acts as a heat source) and thus releases heat to its surroundings.

The Peltier effect concerns the reversible evolution, or absorption, of heat that usually takes place when an electric current crosses a junction between two dissimilar metals. (In certain combinations of metals, at certain temperatures, there are thermoelectric neutral points where no Peltier effect is apparent.) This Peltier effect takes place whether the current is introduced externally or is induced by the thermocouple itself. The Peltier heat was early found to be proportional to the current and may be written

$$dQ_P = \pm \pi I \, d\theta, \tag{7.4}$$

where π is a coefficient of proportionality known as the Peltier coefficient or the Peltier voltage. Note that π represents the reversible heat that is absorbed, or evolved, at the junction when unit current passes across the junction in unit time, and that it has the dimensions of voltage. The direction and magnitude of the Peltier voltage depend on the temperature of the junction and on the materials making up the junction; however, π at one junction is independent of the temperature of the other junction.

External heating, or cooling, of the junctions results in the converse of the Peltier effect. Even in the absence of all other thermoelectric effects, when the temperature of one junction (the reference junction) is held constant and the temperature of the other junction is increased by external heating, a net electric current will be induced in one direction. If the temperature of the latter junction is reduced below the reference junction temperature by external cooling, the direction of the electric current will be reversed. Thus the Peltier effect is seen to be closely related to the Seebeck effect. Peltier himself observed that, for a given electric current, the rate of absorption, or liberation, of heat at a thermoelectric junction depends on the Seebeck coefficient α of the two materials. Note that the Peltier thermal effects build up a potential difference that opposes the thermoelectric current, thus negating the perpetual-motion question.

William Thomson (Lord Kelvin), the English physicist, came to the remarkable conclusion in 1851 that an electric current produces different thermal effects, depending on the direction of its passage from hot to cold, or from cold to hot, in the same metal. By applying the (then) new principles of thermodynamics to the thermocouple and by disregarding (with tongue-in-cheek) the irreversible I^2R and conduction-heating processes, Thomson reasoned that, if an electric current produced only the reversible Peltier heating effects, the net Peltier voltage would equal the Seebeck voltage and would be linearly proportional to the temperature difference at the junctions of the thermocouple [6]–[9].

This reasoning led to requirements at variance with observed characteristics (i.e., $dE_s/dt \neq$ constant). Therefore, Thomson concluded that the net Peltier voltage was not the only source of emf in a thermocouple circuit but that the single conductor itself, whenever it is exposed to a longitudinal temperature gradient, must also be a seat of emf. A. C. Becquerel had at that time already discovered a thermoelectric neutral point, that is, $E_s = 0$, for an iron-copper couple at $\approx 280°C$. Thomson agreed with Becquerel's conclusion and started his thermodynamic reasoning from there.

The Thomson effect concerns the reversible evolution, or absorption, of heat occurring whenever an electric current traverses a single homogeneous conductor, across which a temperature gradient is maintained, regardless of external introduction of the current or its induction by the thermocouple itself. The Thomson heat absorbed, or generated, in a unit volume of a conductor is proportional to the temperature difference and to the current; that is,

$$dQ_T = \pm \left[\int_{T_1}^{T_2} \sigma \, dT \right] I \, d\theta, \qquad (7.5)$$

where σ is a coefficient of proportionality called the Thomson coefficient.

Thomson refers to this as the specific heat of electricity because of an apparent analogy between σ and the usual specific heat c of thermodynamics. Note that σ represents the rate at which heat is absorbed, or evolved, per unit temperature difference per unit current, whereas c represents the heat transfer per unit temperature difference per unit mass. The Thomson coefficient is also seen to represent an emf per unit difference in temperature. Thus the total Thomson voltage set up in a single conductor may be expressed as

$$E_T = \int_{T_1}^{T_2} \sigma \, dT, \tag{7.6}$$

where its direction and magnitude depend on temperature level, temperature difference, and material considered. Note that the Thomson voltage alone cannot sustain a current in a single homogeneous conductor forming a closed circuit, since equal and opposite emfs will be set up in the two paths from heated to cooled parts.

Soon after his heuristic reasoning, Thomson succeeded in demonstrating indirectly the existence of the predicted Thomson emfs (see Figure 7.1). He sent an external electric current through a closed circuit formed of a single homogeneous conductor that was subjected to a temperature gradient and found the $I^2 R$ heat to be slightly augmented or diminished

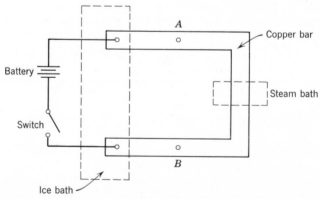

Figure 7.1 Thomson experiment to demonstrate existence of reversible emfs in a single conductor. (1) Points A and B can be found, when the switch is open, having the same temperature. (2) When an electric current is passed through the copper bar, point A becomes cooler while point B becomes hotter. (3) That is, Joule heating ($I^2 R$) is slightly affected by the reversible Thomson emf (E_T), which either opposes or adds to the external emf when the copper bar is preferentially heated. (4) Thomson concluded that hot copper was electrically positive with respect to cold copper.

by the reversible Thomson heat in the paths from cold to hot or from hot to cold, depending on the direction of the current and the material under test.

In summary, thermoelectric currents may exist whenever the junctions of a circuit formed of at least two dissimilar materials are exposed to different temperatures. This temperature difference is always accompanied by irreversible Fourier heat conduction; the passage of electric currents is always accompanied by irreversible Joule heating effects. At the same time, the passage of electric currents is always accompanied by reversible Peltier heating or cooling effects at the junctions of the dissimilar metals; the combined temperature difference and passage of electric current is always accompanied by reversible Thomson heating or cooling effects along the conductors. The two reversible heating-cooling effects are manifestations of four distinct emfs that make up the net Seebeck emf:

$$E_s = \pi_{A,B}|t_2 - \pi_{A,B}|t_1 + \int_{T_1}^{T_2} \sigma_A \, dT - \int_{T_1}^{T_2} \sigma_B \, dT = \int_{t_1}^{t_2} \alpha_{A,B} \, dt, \qquad (7.7)$$

where the three coefficients, α, π, σ, are related by the Kelvin relations.

7.2 The Kelvin Relations

Assuming that the irreversible $I^2 R$ and heat-conduction effects can be completely disregarded (actually, they can only be minimized, since, if thermal conductivity is decreased, electrical resistivity usually is increased, and vice versa, see Figure 7.2), the net rate of absorption of heat required by the thermocouple to maintain equilibrium in the presence of an electric current is

$$q = \frac{Q_{net}}{\Delta \theta} = \left[\pi_2 - \pi_1 + \int_1^2 (\sigma_A - \sigma_B) \, dT \right] I = E_s I. \qquad (7.8)$$

This is in accord with the first law of thermodynamics, according to which heat and work are mutually convertible. Thus the net heat absorbed must equal the electric work accomplished or, in terms of a unit charge of electricity, must equal the Seebeck emf E_s, which may be expressed in the differential form

$$dE_s = d\pi + (\sigma_A - \sigma_B) \, dT. \qquad (7.9)$$

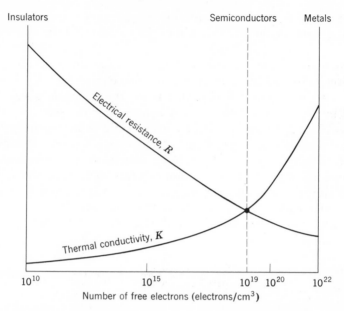

Figure 7.2 Relation between electrical resistance and thermal conductivity of a material.

The second law of thermodynamics may also be applied to the thermocouple cycle (the unit charge of electricity again being considered) as

$$\Delta S_{\text{reversible}} = \sum \frac{\Delta Q}{T_{\text{absolute}}} = 0, \tag{7.10}$$

where ΔQ represents the various components of the net heat absorbed (i.e., the components of E_s), and T_{absolute} is the temperature at which the heat is transferred across the system boundaries. Equation 7.10 can be expressed in the differential form

$$dS_{\text{reversible}} = d\left(\frac{\pi}{T}\right) + \frac{(\sigma_A - \sigma_B)}{T}\, dT = 0. \tag{7.11}$$

Combining the differential expressions for the first and second laws of thermodynamics, we obtain the Kelvin relations

$$\pi_{A,B} = T_{\text{absolute}}\left(\frac{dE_s}{dT}\right) = T_{\text{absolute}}\alpha_{A,B}, \tag{7.12}$$

and

$$(\sigma_A - \sigma_B) = -T_{\text{absolute}}\left(\frac{d^2 E_s}{dT^2}\right), \tag{7.13}$$

from which we can determine α, π, and $\Delta\sigma$, when E_s is obtained as a function of T. Thus if

$$E_s = at + \tfrac{1}{2}bt^2 + \cdots \qquad (7.14)$$

is taken to represent the thermoelectric characteristic of a thermocouple whose reference junction is maintained at 0°C and in which the coefficients a and b are obtained, for example, by the curve fitting of calibration data, then

$$\alpha = (a + bt + \cdots), \qquad (7.15)$$

$$\pi = T_{\text{absolute}}(a + bt + \cdots), \qquad (7.16)$$

$$\Delta\sigma = -T_{\text{absolute}}(b + \cdots). \qquad (7.17)$$

An example of the use of these coefficients is given in Figure 7.3.

Given the two constants, a and b, as determined with respect to platinum,

Metal	a, μV/°C	b, μV/(°C)^2
Iron (I)	+16.7	−0.0297
Copper (Cu)	+2.7	+0.0079
Constantan (C)	−34.6	−0.0558

By way of illustration, consider the following combinations of materials: iron-copper and iron-constantan, with their measuring junctions at 200°C and their reference junctions at 0°C:

Iron-copper

$$a_{\text{I-Cu}} = a_\text{I} - a_\text{Cu} = 16.7 - 2.7 = 14 \; \mu\text{V/°C}$$

$$b_{\text{I-Cu}} = b_\text{I} - b_\text{Cu} = -0.0297 - 0.0079$$

$$b_{\text{I-Cu}} = -0.0376 \; \mu\text{V/(°C)}^2$$

Iron-constantan

$$a_{\text{I-C}} = a_\text{I} - a_\text{C} = 16.7 - (-34.6) = 51.3 \; \mu\text{V/°C}$$

$$b_{\text{I-C}} = b_\text{I} - b_\text{C} = -0.0297 - (-0.0558)$$

$$b_{\text{I-C}} = 0.0261 \; \mu\text{V/(°C)}^2$$

Since Seebeck voltage $E_S = at + \tfrac{1}{2}bt^2$,

Iron-copper

$$E_S = 14(200) + \tfrac{1}{2}(-0.0376)(200)^2$$

$$E_S = 2048 \; \mu\text{V}$$

Figure 7.3 Determination of various thermoelectric quantities applied to thermocouples.

Iron-constantan

$$E_S = 51.3(200) + \tfrac{1}{2}(0.0261)(200)^2$$
$$E_S = 10{,}782 \ \mu\text{V}$$

Note how different combinations of materials give widely different thermal emfs.

Now we proceed to write expressions for α, π, and $\Delta\sigma$, to note how the separate emfs combine to give the (net) Seebeck emf. Since $\alpha_{A,B} = a_{A,B} + b_{A,B}T = $ Seebeck coefficient

Iron-copper

$$\alpha_0 = 14 + (-0.0376)(0) = 14 \ \mu\text{V/}^\circ\text{C}$$
$$\alpha_{200} = 14 + (-0.0376)(200) = 6.48 \ \mu\text{V/}^\circ\text{C}$$

Iron-constantan

$$\alpha_o = 51.3 + 0.0261(0) = 51.3 \ \mu\text{V/}^\circ\text{C}$$
$$\alpha_{200} = 51.3 + 0.0261(200) = 56.52 \ \mu\text{V/}^\circ\text{C}$$

Note that it is the great difference in Seebeck coefficients (thermoelectric powers) for the two combinations that accounts for the difference in thermal emfs:

$$E_S = \int_{T_R}^{T} \alpha_{A,B} \, dT$$

Since $\pi_{A,B} = T_{\text{absolute}} \alpha_{A,B} = $ Peltier coefficient $ = $ Peltier voltage

Iron-copper

$$\pi_0 = 273(14) = 3822 \ \mu\text{V}$$
$$\pi_{200} = 473(6.48) = 3065 \ \mu\text{V}$$

Iron-constantan

$$\pi_0 = 273(51.3) = 14{,}005 \ \mu\text{V}$$
$$\pi_{200} = 473(56.52) = 26{,}734 \ \mu\text{V}$$

Note that, in the case of the iron-copper (I-Cu) couple, $\pi_{\text{cold}} > \pi_{\text{hot}}$, whereas in the more usual I-C couple, $\pi_{\text{hot}} > \pi_{\text{cold}}$.

Figure 7.3 (*continued*)

Since $\Delta\sigma_{A,B} = -b_{A,B}T_{\text{absolute}} = $ Thomson coefficient, and

$$E_T = \int_{T_R}^{T} \Delta\sigma\, dT = \tfrac{1}{2}b_{A,B}(T_R^2 - T^2)$$

$$= \text{Thomson voltage}$$

Iron-copper

$$E_T = -\frac{0.0376}{2}(273^2 - 473^2)$$

$$E_T = 2805\ \mu V$$

Iron-constantan

$$E_T = \frac{0.0261}{2}(273^2 - 473^2)$$

$$E_T = -1947\ \mu V$$

We sum the various components

$$E_s = \pi_2 - \pi_1 + \int_{1}^{2}\Delta\sigma\, dT = \text{Seebeck voltage}$$

Iron-copper

$$E_S = 3065 - 3822 + 2805$$

$$E_S = 2048\ \mu V$$

Iron-constantan

$$E_S = 26{,}734 - 14{,}005 - 1947$$

$$E_S = 10{,}782\ \mu V$$

These figures, of course, check with the original calculations. Note that, in the I-Cu case, the net Thomson emf far outweighs in importance the net Peltier emf, whereas in the I-C case, the converse is true.

Figure 7.3 (*concluded*)

7.3   Microscopic Viewpoint of Thermoelectricity

What gives rise to these thermoelectric effects [10], [11]? Recall from your chemistry that when the loosely bound outer (valence) electrons of a material absorb enough energy from external sources, they may become essentially free from the influence of their nuclei. Once free, these electrons can absorb any amount of energy supplied to them, and it is usual to suppose that the free electrons in a metal act collectively as an idealized gas (see Chapter 2).

Even at a common temperature, however, the energies and densities of the free electrons in different materials need not be the same. Thus when two different materials in thermal equilibrium with each other are brought in contact, there usually will be a tendency for electrons to diffuse across the interface. The electric potential of the material accepting electrons would become more negative at the interface, whereas that of the material emitting electrons would become more positive. In other words, an electric field would be set up by the electron displacement that opposes the osmotic process. When the difference in potential across the interface just balances the thermoelectric (diffusion) force, equilibrium with respect to a transfer of electrons would be established.

If two different homogeneous materials are formed into a closed circuit and the two junctions are maintained at the same temperature, the resultant electric fields would exactly oppose each other, and there would be no net electron flow.

However, if these two junctions are maintained at different temperatures, a net diffusion current will be induced in coincidence with a net electric field (the random motion of the free electrons is on the average in the direction of the net potential gradient, and this gives rise to the electric current).

From conservation principles, the power to drive this electric current could only come from a net absorption of thermal energy from the surroundings by the free electrons of the materials, since there is no observable change in the nature or composition of the thermoelectric materials. But what accounts for this net absorption of heat? Why does the thermocouple act as a heat engine (i.e., a device that makes available as work some portion of the thermal energy acquired from a source)?

Consider first a closed circuit of a single material (thus avoiding for a time the thermoelectric effects). Under the influence of a temperature gradient alone, the material will conduct a thermal current. But all the thermal energy absorbed by the circuit at one zone is rejected by the circuit at another zone. Thus the Fourier effect exhibits no accumulation of heat in the steady state, and cannot account for a thermally induced electric current in a thermocouple circuit. Again, under the influence of a voltage gradient alone, this same single material closed circuit will conduct an electric current. But the only thermal effect associated with this current is the inevitable I^2R heat generation. Thus the Joule effect exhibits no absorption of heat from outside the circuit and cannot account for a thermally induced electric current in a thermocouple circuit. Evidently, when examining the mechanism by which a closed circuit formed of two unlike materials acts as a heat engine, we need concern ourselves only with the reversible thermoelectric effects. (Perhaps similar reasoning guided Thomson when he successfully obtained the valid Thomson relations by intuitively disregarding the irreversible Joule and Fourier processes.)

From this viewpoint, the resultant electric current in the thermocouple circuit having its junctions maintained at different temperatures agrees in direction with the natural (Peltier) potential gradient at the hot junction, and thus tends to wipe out (do away with the need for) this field. At the cold junction, this same thermally induced current must cross the interface against the natural potential gradient, and this tends to build up a stronger field of opposition there. In terms of the idealized gas, the free electrons expand isothermally across the hot junction interface because the flow is in the direction of the Peltier potential gradient there. This expansion tends to cool the hot junction but, in face of the isothermal restriction and in accord with the first law of thermodynamics, the hot junction absorbs just enough heat from its surroundings to maintain its temperature.

These Peltier (junction) effects, although usually predominating, do not tell the complete story. As Thomson first pointed out, there will be thermoelectric heating effects along each of the materials making up the circuit whenever an electric current and a temperature gradient exist. The Thomson effect may be visualized as follows. If a single conductor is preferentially heated, electrons usually tend to leave the hot end more frequently than the cold. Just as we found at the junctions, an opposing electric field would be set up along the single conductor by this diffusion of electrons. Thus whenever an electric current occurs in a closed thermocouple circuit, it must either agree with or oppose these Thomson (material) potential gradients, and this accounts for heating and cooling effects in addition to the Peltier (junction) effects discussed previously.

Hence the thermocouple in a temperature gradient does qualify as a heat engine in that there will always be a net absorption of heat from its surroundings per unit time, and we see once more that it is just this excess of thermal power arising entirely from reversible thermoelectric effects that sustains the thermoelectric current in the circuit.

7.4 Macroscopic Viewpoint of Thermoelectricity

The historical viewpoint presented thus far has avoided the real irreversible I^2R and heat conduction effects in order to arrive at the useful and experimentally confirmed Kelvin relations. We now discuss how the present-day irreversible thermodynamic viewpoint removes this flaw in our reasoning.

Basically, we judge whether a given process is reversible or irreversible by noting the change in entropy accompanying a given change in the thermodynamic state. Thus, if $dS > \delta Q_q / T_{\text{absolute}}$, we say the process is irreversible; or, stated in a more useful manner,

$$dS_{\text{system}} = dS_{\substack{\text{across} \\ \text{boundary}}} + dS_{\substack{\text{produced,} \\ \text{inside}}} \tag{7.18}$$

or

$$dS_s = dS_0 + dS_i = \frac{\delta Q_q}{T_{\text{absolute}}} + \frac{\delta F}{T_{\text{absolute}}}. \tag{7.19}$$

Hence only in the absence of entropy within the system boundaries do we have the reversible case, $dS_{\text{reversible}} = \delta Q_q / T_{\text{absolute}}$, which may be handled adequately by classical thermodynamics in the steady and quasi-steady states. Evidently the rate of production of entropy per unit volume ξ is an important quantity in irreversible thermodynamics. It may be expressed as

$$\xi = \left(\frac{1}{A\,dx}\right)\frac{dS_i}{d\theta} = \left(\frac{1}{A\,dx}\right)\frac{\delta F}{T_{\text{absolute}}\,d\theta}, \tag{7.20}$$

where $A\,dx$ is the area times the differential length.

Another significant quantity, the product $T_{\text{absolute}}\xi$ (called the dissipation), can always be split either into two terms or into a sum of two terms, one associated with a flow J, and the other associated with a force X. Furthermore, in many simple cases a linear relation is found (by experiment) to exist between the flow and force terms so defined; for example, in the one-dimensional, isothermal, steady flow of electric charges, $\delta Q_e / d\theta$, across a potential gradient $-dE/dx$, it may be shown that

$$T_{\text{absolute}}\xi = \left(\frac{I}{A}\right)\left(-\frac{dE}{dx}\right) = J_e X_e, \tag{7.21}$$

where J_e and X_e represent, respectively, the electric flow and force terms, as defined by the entropy production method. (D. G. Miller gives an excellent review of the thermodynamic theory of irreversible processes [12]. We follow his development closely.) The term J_e represents the electric current density and the term X_e the electric field strength or the electromotive force. These are, of course, related by the linear Ohm's law (i.e., $J_e = L_e X_e$, where L_e represents the electrical conductivity). Again, in the one-dimensional, steady flow of thermal charges $\delta Q_q / d\theta$ across a temperature gradient, $-dT/dx$, it may be shown that

$$T_{\text{absolute}}\xi = \left(\frac{Q}{A}\right)\left(\frac{1}{T_{\text{absolute}}}\frac{dT}{dx}\right) = J_q X_q, \qquad (7.22)$$

where J_q and X_q represent, respectively, the thermal flow and force, as defined by the entropy production method. The term J_q represents the thermal current density, and the term X_q represents the thermomotive force. These are, of course, related by the linear Fourier's law (i.e., $J_q = L_q X_q$, where L_q represents the product of the thermal conductivity and the absolute temperature). It has been found that even in complex situations it may always be stated that

$$T_{\text{absolute}}\xi = \sum J_k X_k. \qquad (7.23)$$

When several irreversible transport processes occur simultaneously (e.g., the electric and thermal conduction in a thermocouple), they will usually interfere with each other. Therefore, the linear relations must be generalized to include the various possible interaction terms. Thus for the combined electric and thermal effects we would write

$$J_e = L_{ee}X_e + L_{eq}X_q, \qquad (7.24)$$

$$J_q = L_{qe}X_e + L_{qq}X_q, \qquad (7.25)$$

or, in general, $\qquad J_i = \sum L_{ij}X_j. \qquad (7.26)$

We have just seen that an entropy production necessarily accompanies both the $I^2 R$ and heat conduction effects (i.e., they are irreversible); therefore, the Kelvin relations could not follow from reversible thermodynamic theory without certain intuitive assumptions. By reasoning that the electric and thermal currents were independent, Thomson tacitly assumed that $L_{eq} = L_{qe}$ as we shall show subsequently. Experimentally, this reciprocal relationship was often found to be true. The American chemist Lars Onsager proved, in 1931, from a statistical-mechanics viewpoint, that the assumption

$$L_{ij} = L_{ji} \qquad (7.27)$$

is always true when the linear relations between flows J_k and forces X_k are valid. (For this work Onsager subsequently was awarded a Nobel Prize.) The Onsager reciprocal relation forms the basis of irreversible thermodynamics [13]. By applying these concepts to the processes involved in the thermocouple, we are led rationally and unambiguously to the Kelvin relations [14], [15], [16]. Thus whenever the junctions of a thermocouple are maintained at different temperatures, we expect that an electric potential difference, an electric current, and a thermal current will be present. The dissipation for this thermoelectric process is simply the sum of the electric and thermal terms previously given. That is

$$T_{absolute}\xi = \frac{I}{A}\left(-\frac{dE}{dT}\right) + \frac{Q}{A}\left(\frac{1}{T_{absolute}}\frac{dT}{dx}\right). \tag{7.28}$$

The generalized linear laws for this case have also been given as

$$J_e = L_{ee}\left(-\frac{dE}{dT}\right) + \frac{L_{eq}}{T_{absolute}}\left(\frac{dT}{dx}\right), \tag{7.29}$$

$$J_q = L_{qe}\left(-\frac{dE}{dT}\right) + \frac{L_{qq}}{T_{absolute}}\left(\frac{dT}{dx}\right). \tag{7.30}$$

Recalling that the Seebeck emf is determined under conditions of zero electric current, the Seebeck coefficient α may be expressed in terms of the Onsager coefficients as

$$\alpha = \left(\frac{dE_s}{dT}\right)_{I=0} = \frac{L_{eq}}{L_{ee}T_{absolute}}. \tag{7.31}$$

Recalling that the Peltier coefficient π represents the heat absorbed or evolved with the passage of an electric current across an isothermal junction, this too may be expressed in terms of the Onsager coefficients as

$$\pi = \left(\frac{J_q}{J_e}\right)_{dT=0} = \frac{L_{qe}}{L_{ee}}. \tag{7.32}$$

Finally we recall that Thomson found experimentally (and expressed in the Kelvin relations) that the Seebeck and Peltier coefficients are related, as shown in (7.12). Thus

$$\pi = T_{absolute}\left(\frac{dE_s}{dT}\right)_{I=0}. \tag{7.33}$$

In terms of the Onsager coefficients, this requires that

$$\frac{L_{qe}}{L_{ee}} = T_{absolute}\left(\frac{L_{eq}}{L_{ee}T_{absolute}}\right), \tag{7.34}$$

or
$$L_{qe} = L_{eq}, \tag{7.35}$$

which indicates that the experimental results agree with those which are predicted by the entropy-production linear-law Onsager reciprocal relation approach; in other words, by irreversible thermodynamics, without using any intuitive assumption. The Kelvin relations, also in accord with experiment, must follow.

7.5 Laws of Thermoelectric Circuits

We have seen that, in 1821, T. J. Seebeck discovered "thermomagnetism" when the junctions between two different metals forming a closed circle were subjected to a temperature difference.

Just a few years later, in 1826, A. C. Becquerel read a paper to the Royal Academy of Sciences in Paris describing the first recorded suggestion to make use of Seebeck's discovery as a means of measuring temperature.

It was not until 1885, however, that the problem of measuring high temperatures was successfully attacked by Henry LeChatelier. He determined experimentally that the use of a platinum wire against a platinum-rhodium alloy wire gave the most satisfactory "thermo-electric" results [17], [18], [19]. LeChatelier's work is still the basis of modern standard thermoelectric thermometry.

Numerous investigations of thermoelectric circuits [20], [21], [22] in which accurate measurements were made of the current, resistance, and electromotive force have resulted in the establishment of several basic laws. These laws have been established experimentally beyond a reasonable doubt and may be accepted in spite of any lack of a theoretical development.

Law of Homogeneous Materials

A thermoelectric current cannot be sustained in a circuit of a single homogeneous material however it varies in cross section by the application of heat alone.

A consequence of this law is that two different materials are required for any thermocouple circuit. Experiments have been reported suggesting that a nonsymmetrical temperature gradient in a homogeneous wire gives rise to a measurable thermoelectric emf. A preponderance of evidence, however, indicates that any emf observed in such a circuit arises from the effects of local inhomogeneities. Furthermore, any current detected in such a circuit when the wire is heated in any way whatever is taken as evidence that the wire is inhomogeneous.

Law of Intermediate Materials

The algebraic sum of the thermoelectromotive forces in a circuit composed of any number of dissimilar materials is zero if all of the circuit is at a uniform temperature.

A consequence of this law is that a third homogeneous material can always be added in a circuit with no effect on the net emf of the circuit as long as its extremities are at the same temperature. Therefore, it is evident that a device for measuring the thermoelectromotive force may be introduced into a circuit at any point without affecting the resultant

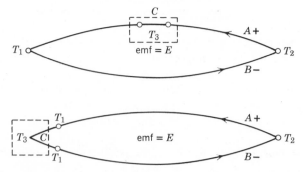

Figure 7.4 The emf is unaffected by the third material C.

emf, provided all of the junctions that are added to the circuit by introducing the device are all at the same temperature. It also follows that any junction whose temperature is uniform and that makes good electrical contact does not affect the emf of the thermoelectric circuit regardless of the method employed in forming the junction (see Figure 7.4).

Another consequence of this law may be stated as follows. If the thermal emfs of any two metals with respect to a reference metal (such as C) are known, the emf of the combination of the two metals is the algebraic sum of their emfs against the reference metal (see Figure 7.5).

Law of Successive or Intermediate Temperature

If two dissimilar homogeneous materials produce a thermal emf of E_1 when the junctions are at temperatures T_1 and T_2, and a thermal emf of E_2 when the junctions are at T_2 and T_3, the emf generated when the junctions are at T_1 and T_3 will be $E_1 + E_2$.

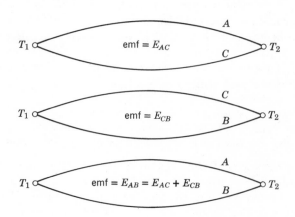

Figure 7.5 Emfs are additive for materials.

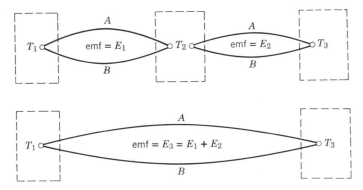

Figure 7.6 Emfs are additive for temperature intervals.

One consequence of this law permits a thermocouple calibrated for a given reference temperature to be used with any other reference temperature through the use of a suitable correction (see Figure 7.6).

Another consequence of this law is that extension wires having the same thermoelectric characteristics as those of the thermocouple wires can be introduced in the thermocouple circuit (say from region T_2 to region T_3 in Figure 7.6) without affecting the net emf of the thermocouple.

7.6 Elementary Thermoelectric Circuits

Two continuous, dissimilar thermocouple wires extending from the measuring junction to the reference junction, when used with copper connecting wires and a potentiometer connected as shown in Figure 7.7, make up the basic thermocouple circuit.

The ideal circuit given in Figure 7.8 is for use when more than one thermocouple is involved. Note that each thermocouple is made up of two continuous wires between the measuring and reference junctions. This circuit should be used for all precise work in thermoelectric thermometry.

A circuit that requires only one reference junction is known in Bureau of Standards publications as a *zone-box circuit*. A diagram of one is shown in Figure 7.9. Such a circuit should not be used when uncertainties under $\pm 1°F$ are required.

The usual thermocouple circuit, however, includes measuring junctions, thermocouple extension wires, reference junctions, copper connecting wires, a selector switch, and a potentiometer, as indicated in Figure 7.10. The uncertainties introduced by this circuit should be expected to be greater than $\pm 1°F$.

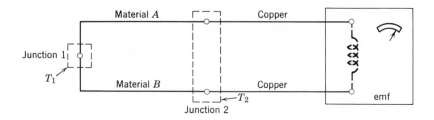

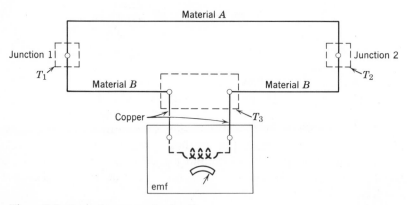

Figure 7.7 Basic thermocouple circuits.

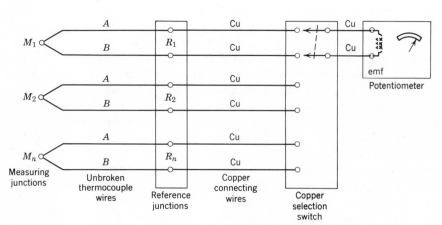

Figure 7.8 Ideal circuit when more than one thermocouple is involved.

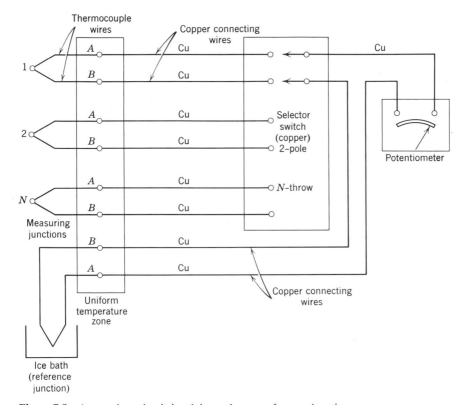

Figure 7.9 A zone-box circuit involving only one reference junction.

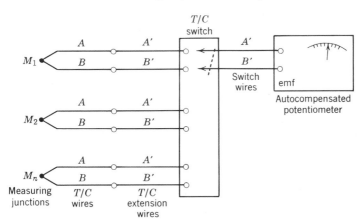

Figure 7.10 Typical industrial thermocouple circuit. Note: the primed quantities (as A' and B') denote materials of the same nominal composition as the unprimed quantities (as A and B), but of different T/E composition in fact.

Many different circuit arrangements of these components also are acceptable, depending on given circumstances, and some are shown in Figure 7.11*a*, *b*, and *c*. These circuits can be characterized briefly as follows:

a. Thermopile, for sensing average temperature level. *Advantages:* magnifies thermoelectric power by the number of couples in series, enabling the detection of small changes in temperature level; one observation gives, by calibration, the arithmetic mean of temperatures sensed by *N* individual measuring junctions, when total emf is divided by *N*.

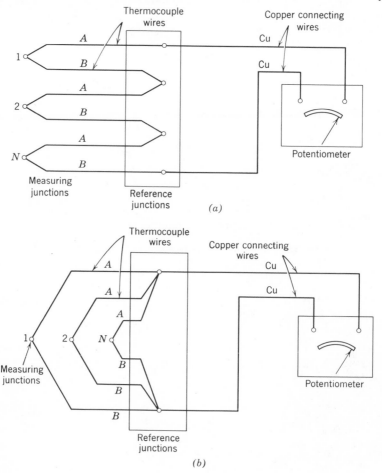

Figure 7.11 Various thermocouple circuits. (*a*) Thermopile for sensing average temperature level. (*b*) Parallel arrangement for sensing average temperature level. (*c*) Thermopile for sensing average temperature differences.

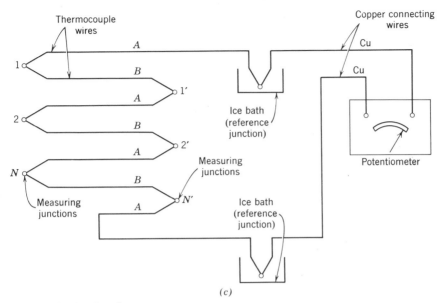

(c)

Figure 7.11 Continued

Precautions: the emf may be too great for the indicator; a short circuit may escape detection (giving erroneous readings); this thermopile gives no indication of temperature distribution; and the temperature/emf calibrations of all couples must be identical.

b. Parallel arrangement, for sensing average temperature level. *Advantage:* one observation gives, by calibration, the arithmetic mean of the temperatures sensed by N individual measuring junctions. *Precautions:* gives no indication of temperature distribution; the temperature/emf calibrations of all couples must be identical and linear; and all couples must have equal resistances.

c. Thermopile, for sensing average temperature differences. *Advantages:* the signal is magnified by the number of couples in series (permits detection of very small temperature differences); and one observation, by calibration, gives the arithmetic mean of the temperature differences sensed by two groups of N measuring junctions when the total emf is divided by N. *Precautions:* temperature level is not indicated; a short circuit may escape detection; and the temperature/emf calibrations of all couples must be identical.

A typical ice bath is shown in Figure 7.12. Various circuits for using potentiometers, compensated to indicate temperature directly and uncompensated to indicate voltages directly, are shown in Figure 7.13.

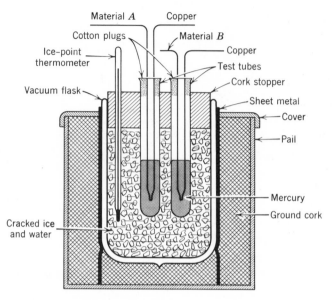

Figure 7.12 A typical ice bath for reference junctions.

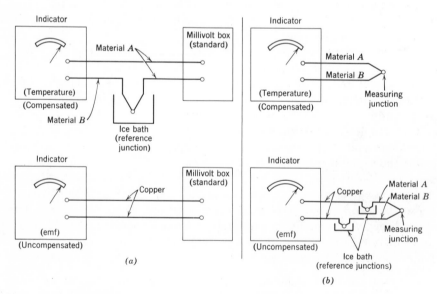

Figure 7.13 Potentiometer circuits: (a) for the calibration of indicators and (b) for indicating temperatures.

Grounded Thermocouple Circuits

Relatively little attention has been paid to electrical effects that are extraneous to the basic thermocouple circuit. Prime examples of external effects not always considered are: effects of electrical and magnetic fields, cross-talk effects, and effects connected with common mode voltage rejection. An excellent discussion of these noise effects in general instrument circuits and of methods for reducing them is given by Klipec [23]. (On common mode voltage rejection see also [24].)

A brief review of these effects follows [25]:

1. Electric fields radiated from voltage sources are capacitively coupled to thermocouple extension wires and cause an alternating noise signal to be superimposed on the desired signal. Electric noise is minimized by *shielding* the thermocouple extension wire and *grounding* the shield.

2. Magnetic fields radiated from current-carrying conductors produce noise current and hence noise voltage in the thermocouple circuit. Magnetic noise is minimized by *twisting* the thermocouple extension wires.

3. Adjacent pairs of a multipair cable tend to pick up noise when pulsating d.c. signals are transmitted. Cross-talk noise is minimized by shielding the *individual* pairs of thermocouple extension wires.

4. Electrical connections made between a thermocouple and a grounded instrument may introduce common mode noise if different ground potentials exist along the wire path. Common mode noise is minimized by grounding the thermocouple and its shielding at a *single point* as *close* as practical to the measuring junction.

Several arrangements of thermocouple/extension wire/shield/ground combinations, acceptable from the noise viewpoint, are shown in Figure 7.14.

It is evident from the foregoing that the grounding of thermocouple circuits is an important consideration. It is a fact that grounding is often done improperly. It is shown in Section 7.9, under circuit analysis, that the introduction of multiple grounds in a thermocouple circuit can lead to serious temperature errors, errors that are over and above those associated with thermoelectro-motive force generation, with static and dynamic calibrations, with heat transfer effects, and so on.

Circuits for Rotating Parts

Often in engineering the temperature of a rotating part must be obtained. There are several methods for accomplishing this task. A

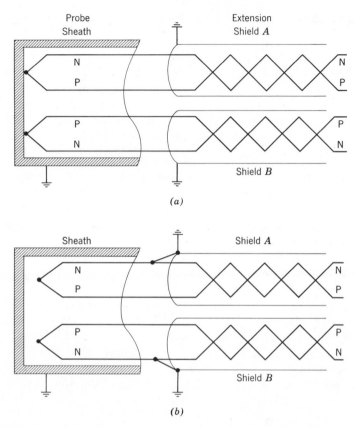

Figure 7.14 Several acceptable grounded thermocouple arrangements. (*a*) Properly grounded circuit when measuring junctions are grounded to sheath. (*b*) Alternative grounded circuit to be used with ungrounded measuring junctions.

rotating transformer may be used wherein one coil remains stationary while the other rotates, and hence the signal is passed electrically from rotating to stationary parts. Radio telemetry may be employed wherein a rotating transmitter sends the temperature signal from a rotating sensor to a stationary indicating device by means of a radio frequency signal, antenna, and receiver system. However, the most usual method is by a slip ring arrangement wherein the electrical signal from a rotating thermocouple is passed to the stationary world by means of rotating slip rings and fixed brushes. These three methods are illustrated in Figure 7.15.

The slip ring circuit that has proven to be useful for the highest accuracy temperature measurement of a rotating part is shown in Figure 7.16*a*. This makes use of a separate rotating thermocouple to sense the temperature in the zone where the necessary transition takes place between the thermocouple wires from the measuring junctions and the copper wires of the slip ring assembly.

In Figure 7.16*b*, the accuracy of the basic slip ring circuit is compromised through the built-in assumption that a common temperature exists at both the rotating and the fixed terminals of the slip ring assembly. If this

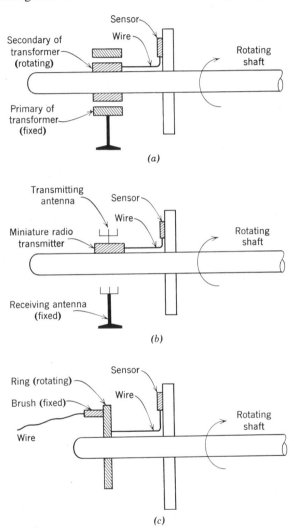

Figure 7.15 Several methods for obtaining temperature measurements of rotating parts. (*a*) By a rotating transformer arrangement. (*b*) By radio telemetry. (*c*) By a slip ring arrangement.

can be safely posited, or if the temperature difference across the slip ring can be tolerated in the overall measurement accuracy (e.g., a 2° Δt across the slip ring will introduce a 2° error in the measurement), the simplification of circuit *b* over *a* may be warranted.

A third arrangement, wherein the rotating zone thermocouple of Figure 7.16*a* is replaced by a fixed zone thermocouple extending into the slip ring assembly may sometimes be used if it is felt that such a fixed zone thermocouple adequately proclaims the temperature in the rotating portion of the slip ring assembly.

A fourth arrangement, wherein a rotating resistance thermometer (RTD, see Chapter 6) is used in place of any of the above rotating or fixed zone thermocouple junctions, can be used to advantage if it is important that copper wires be used throughout the slip ring assembly.

A slip ring circuit is analyzed, thermoelectrically, in Section 7.9.

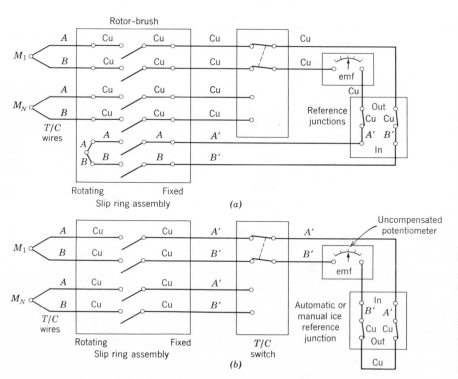

Figure 7.16 Several acceptable circuits for measuring the temperature of rotating parts by thermocouples and slip rings. (*a*) Circuit making use of a rotating zone thermocouple for highest accuracy. (*b*) Circuit that presumes that rotating and fixed terminals of slip ring are at a common temperature.

7.7   Circuit Component Uncertainties

Typical uncertainties that can accompany the various circuit components are discussed briefly below [26], [27], [28].

Measuring Junction

The junction of a thermocouple that is at the temperature to be measured is referred to as the measuring junction. Each measuring junction has its own peculiar characteristics depending on such factors as materials, methods of joining, age, and history of junction. Yet the measuring uncertainties introduced by this junction are small; they are accounted for in the overall circuit calibration (to be discussed), and do not introduce extraneous insidious uncertainties in a temperature measurement.

Thermocouple Extension Wires

Extension wires are thermoelements inserted between the measuring junction and the reference junction that have approximately the same thermoelectric properties as the thermocouple wire with which they are used (the A' and B' wires of Figure 7.10). The use of extension wires introduces at their extremities at least four extraneous thermoelectric junctions in each thermocouple circuit. Even with normal usage, the resulting uncertainties can amount to as much as ±4°F, depending on the temperatures at the ends of the extension wires, and on the particular wires used. Such large uncertainties can be minimized by calibrating with given extension wires and then using the same wires in the field, maintaining (as nearly as possible) the same temperatures at their extremities in both circumstances. These uncertainties can be entirely avoided by not introducing any extension wires in the thermocouple proper.

Selector Switch

When placed in any circuit other than the basic circuit the switch can introduce uncertainties of ±1.5°F. It is not enough to keep a switch at a uniform temperature, nor even at a constant temperature. To minimize uncertainties, the switch must be kept at the same temperature in the field as it was during the calibration. The uncertainties can be avoided entirely by placing the switch in the copper extension wires leading from the reference junctions to the potentiometer.

Reference Junctions

The junction of a thermocouple that is maintained at a known temperature is referred to as the reference junction. In any but the basic circuit, reference junctions can introduce uncertainties of ±1°F. These can be minimized by using the same reference junction in the field as was used in the calibration. The uncertainties can be avoided entirely by using a proper thermocouple, that is, one for which the measuring and reference junctions are simply the extremities of the thermocouple wires.

Copper Connecting Wires

Copper wires used to connect the reference junctions of a thermocouple to the switch or potentiometer are called copper connecting wires. These wires cause no uncertainty when they are used between the reference junctions and the potentiometer.

Potentiometer

A potentiometer is used to measure the emf generated by the thermocouple. A standardized current is passed through a wire of fixed resistance. The wire is calibrated by marking the corresponding voltage drop along its length. Thus any unknown external emf (such as that encountered in the thermocouple circuit) can be compared with the known voltage drop along the calibrated wire. A sliding contact on the calibrated wire facilitates the comparison, whereas a galvanometer indicates the balance (null) condition by absence of needle deflection. With reference to Figure 7.17, operation of the potentiometer is as follows. The switch is placed on (1), and the current from the working cells is

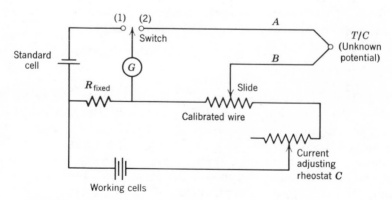

Figure 7.17 Basic potentiometer circuit.

adjusted by rheostat C until the galvanometer indicates the null. Under this condition the current through the calibrated wire is "standardized." The switch is then placed on (2), and the slide on the calibrated wire is adjusted until the galvanometer again indicates the balance condition. The unknown voltage is now determined by the position of the slide on the calibrated wire. Since no current flows in the external circuit, the resistance of that circuit introduces no error in the voltage measurement. An increase in external circuit resistance, however, does decrease the sensitivity of the potential measurement. This component introduces no uncertainty that is not accounted for in the overall calibration.

7.8   Thermoelectric Reference Tables

Only a few of the large number of possible combinations of materials are actually used in thermoelectric thermometry. These are chosen on the basis of their standing in the *Thermoelectric Series,* their *Seebeck coefficients* (i.e., thermoelectric powers), and on their stability and reproducibility as evidenced by the establishment of *Thermocouple Reference Tables.* In the following section, these three tabulations are presented.

Thermoelectric Series

The various conductors have been tabulated in an order such that, at a specified temperature, each material in the list is thermoelectrically negative with respect to all above it and positive with respect to all below it (see Table 7.1). Although this tabulation resembles the electrochemical

Table 7.1   *Thermoelectric Series for Selected Metals and Alloys*

100°C	500°C	900°C
Antimony	Chromel	Chromel
Chromel	Nichrome	Nichrome
Iron	Copper	Silver
Nichrome	Silver	Gold
Copper	Gold	Iron
Silver	Iron	$Pt_{90}Rh_{10}$
$Pt_{90}Rh_{10}$	$Pt_{90}Rh_{10}$	Platinum
Platinum	Platinum	Cobalt
Palladium	Cobalt	Alumel
Cobalt	Palladium	Nickel
Alumel	Alumel	Palladium
Nickel	Nickel	Constantan (Adams)
Constantan (Adams)	Constantan (Adams)	
Copel	Copel	
Bismuth		

electromotive force series, the position of a material in one series bears no relation to that in the other.

Seebeck Coefficients

Nominal thermoelectric powers of various thermoelements with respect to Platinum 67 (a standard maintained at the NBS), and of various common thermocouple types, are presented in Tables 7.2 and 7.3. These are quite useful in the analysis of circuits such as those dealt with in Section 7.9. Curves to present this same information graphically are given in Figures 7.18 and 7.19.

Thermocouple Reference Tables

Thermocouple reference values represent in both form and magnitude the nominal characteristics of particular thermocouple types. Thermocouple wires are carefully selected and matched to conform with these tables within prescribed limits. The printed scales of all direct temperature-indicating potentiometers are based on such tables. Calibration curves, in the form of voltage- or temperature-difference plots, are obtained by comparing calibration data with reference values.

Table 7.2 **_Nominal Seebeck Coefficients (T/E Power) of Various Thermoelements with Respect to Platinum 67_**[a]

Temperature, °C	Thermoelement Type (Seebeck Coefficient, μV/°C)					
	JP	JN	TP	TN, EN	KP, EP	KN
0	17.9	32.5	5.9	32.9	25.8	13.6
100	17.2	37.2	9.4	37.4	30.1	11.2
200	14.6	40.9	11.9	41.3	32.8	7.2
300	11.7	43.7	14.3	43.8	34.1	7.3
400	9.7	45.4	16.3	45.5	34.5	7.7
500	9.6	46.4	—	46.6	34.3	8.3
600	11.7	46.8	—	46.9	33.7	8.8
700	15.4	46.9	—	46.8	33.0	8.8
800	—	—	—	46.3	32.2	8.8
900	—	—	—	45.3	31.4	8.5
1000	—	—	—	44.2	30.8	8.2

[a] From NBS Monograph 125, March 1974.
E = Chromel(+)   Constantan(−)
J = Iron(+)   Constantan(−)
K = Chromel(+)   Alumel(−)
T = Copper(+)   Constantan(−)
P = Positive(+)
N = Negative(−)

Table 7.3 **Nominal Seebeck Coefficients (T/E Power) of Various Thermocouple Types**[a]

| Temperature, °C | Thermocouple Type (Seebeck Coefficient, $\mu V/°C$) | | | | | |
	E	J	K	R	S	T
−200	25.1	21.9	15.3	—	—	15.7
−100	45.2	41.1	30.5	—	—	28.4
0	58.7	50.4	39.5	5.3	5.4	38.7
100	67.5	54.3	41.4	7.5	7.3	46.8
200	74.0	55.5	40.0	8.8	8.5	53.1
300	77.9	55.4	41.4	9.7	9.1	58.1
400	80.0	55.1	42.2	10.4	9.6	61.8
500	80.9	56.0	42.6	10.9	9.9	—
600	80.7	58.5	42.5	11.3	10.2	—
700	79.8	62.2	41.9	11.8	10.5	—
800	78.4	—	41.0	12.3	10.9	—
900	76.7	—	40.0	12.8	11.2	—
1000	74.9	—	38.9	13.2	11.5	—

[a] From NBS Monograph 125, March 1974.

E = Chromel(+)	Constantan(−)
J = Iron(+)	Constantan(−)
K = Chromel(+)	Alumel(−)
R = Platinum/13% Rhodium(+)	Platinum(−)
S = Platinum/10% Rhodium(+)	Platinum(−)
T = Copper(+)	Constantan(−)

A satisfactory table of reference values for use with thermocouples must be capable of easy and unique generation for a number of reasons: (1) the values are used by many people; (2) their storage in computer applications must not present a problem; and (3) since interpolation must be used, unique values at all arbitrary points must be available. The last requirement means that a mathematically continuous functional relation must exist between the temperatures and voltages with which the table is concerned. Satisfactory reference values must also agree closely with the characteristic of the thermocouple type being considered so that differences in values change slowly and smoothly.

The established reference tables [29], [30], [31] adequately fulfill all these requirements. Unique polynomials are available for generating tables for each of the seven thermocouples now in common use. These are given in Tables 7.4 and 7.5 for easy reference. A short discussion of each thermocouple type follows [32].

TYPE T (copper/constantan). This type can be represented by an eighth-degree equation as in Table 7.4a. This is used to generate exact

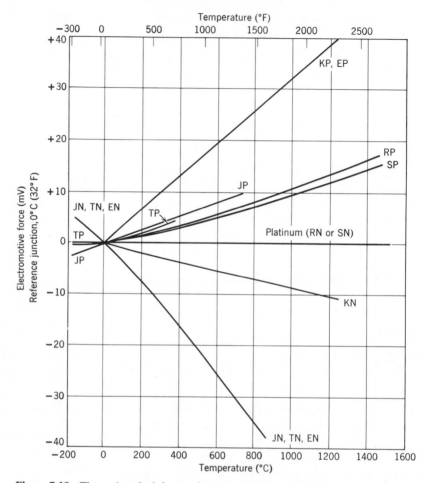

Figure 7.18 Thermal emf of thermoelements relative to platinum.

values of voltage as a function of temperature from 0° to 400°C. The inverse relation, for temperature as a function of voltage, is generated over the same temperature range by another polynomial also given in the table. This procedure is necessarily inexact, but will yield reference temperatures to within known uncertainty bands. For type T, temperatures are given to ±0.2°C from 0° to 400°C.

TYPE J (iron/constantan). Output of this type can be determined as a function of temperature from 0° to 760°C by a seventh-degree equation as shown in Table 7.4*b*. The resulting values are exact by definition. This table also shows the inverse relation for temperature as a function of

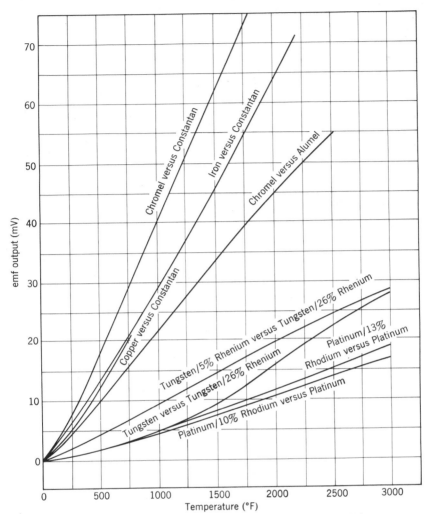

Figure 7.19 Temperature versus emf curves. (Reference junction, 32°F.)

voltage over the same temperature range. For type J, temperatures are given to ±0.1°C from 0° to 760°C.

TYPE E (chromel/constantan). Thermocouples of this type can be represented by a ninth-degree polynomial as in Table 7.4c. Voltage generated as a function of temperature from 0° to 1000°C can be calculated, as can temperature as a function of voltage using the inverse relationship. For the latter case, temperatures are given to ±0.1°C from 0° to 1000°C.

Table 7.4 Power Series Expansions for Base Metal Thermocouples

Temperature Range	Exact Reference Voltage in Millivolts	Reference Temperature within ±0.2°C
0° to 400°C	$E = (+3{,}8740773840 \times 10 \times T$ $+3.3190198092 \times 10^{-2} \times T^2$ $+2.0714183645 \times 10^{-4} \times T^3$ $-2.1945834823 \times 10^{-6} \times T^4$ $+1.1031900550 \times 10^{-8} \times T^5$ $-3.0027581898 \times 10^{-11} \times T^6$ $+4.5653337165 \times 10^{-14} \times T^7$ $-2.7616878040 \times 10^{-17} \times T^8) \times 10^{-3}$	$T = +2.5661297 \times 10 \times E$ $-6.1954869 \times 10^{-1} \times E^2$ $+2.2181644 \times 10^{-2} \times E^3$ $-3.550000 \times 10^{-4} \times E^4$

(a) Type T Thermocouples

Temperature Range	Exact Reference Voltage in Millivolts	Reference Temperature within ±0.1°C
0° to 400°C	$E = (+5.0372753027 \times 10 \times T$ $+3.0425491284 \times 10^{-2} \times T^2$ $-8.5669750464 \times 10^{-5} \times T^3$ $+1.3348825735 \times 10^{-7} \times T^4$ $-1.7022405966 \times 10^{-10} \times T^5$ $+1.9416091001 \times 10^{-13} \times T^6$ $-9.6391844859 \times 10^{-17} \times T^7) \times 10^{-3}$	$T = +1.9750953 \times 10 \times E$ $-1.8542600 \times 10^{-1} \times E^2$ $+8.3683958 \times 10^{-3} \times E^3$ $-1.3280568 \times 10^{-4} \times E^4$ $-1.3280568 \times 10^{-4} \times E^4$
400° to 760°C	(applies to both ranges of temperature)	$T = +9.2808351 \times 10$ $+5.4463817 \times E$ $+6.5254537 \times 10^{-1} \times E^2$ $-1.3987013 \times 10^{-2} \times E^3$ $+9.9364476 \times 10^{-5} \times E^4$

(b) Type J Thermocouples

Temperature Range	Exact Reference Voltage in Millivolts	Reference Temperature within ±0.1°C
0° to 400°C	$E = (+5.8695857799 \times 10 \times T$ $+4.3110945462 \times 10^{-2} \times T^2$ $+5.7220358202 \times 10^{-5} \times T^3$ $-5.4020668085 \times 10^{-7} \times T^4$ $+1.5425922111 \times 10^{-9} \times T^5$ $-2.4850089136 \times 10^{-12} \times T^6$ $+2.3389721459 \times 10^{-15} \times T^7$ $-1.1946296815 \times 10^{-18} \times T^8$ $+2.5561127497 \times 10^{-22} \times T^9) \times 10^{-3}$	$T = +1.7022525 \times 10 \times E$ $-2.2097240 \times 10^{-1} \times E^2$ $+5.4809314 \times 10^{-3} \times E^3$ $-5.7669892 \times 10^{-5} \times E^4$
400° to 1000°C	(applies to both ranges of temperature)	$T = +2.9347907 \times 10$ $+1.3385134 \times 10 \times E$ $-2.6669218 \times 10^{-2} \times E^2$ $+2.3388779 \times 10^{-4} \times E^3$

(c) Type E Thermocouples

Table 7.4   (Continued)

Temperature Range	Exact References Voltage in Millivolts	Reference Temperature within ±0.02°C
0° to 400°C	$E = (-1.8533063273 \times 10$ $+3.8918344612 \times 10 \times T$ $+1.6645154356 \times 10^{-2} \times T^2$ $-7.8702374448 \times 10^{-5} \times T^3$ $+2.2835785557 \times 10^{-7} \times T^4$ $-3.5700231258 \times 10^{-10} \times T^5$ $+2.8932909136 \times 10^{-13} \times T^6$ $-1.2849848798 \times 10^{-16} \times T^7$ $+2.2239974336 \times 10^{-20} \times T^8$ $+125 \exp\left(-\frac{1}{2}\left\{\frac{T-127}{65}\right\}^2\right)\right) \times 10^{-3}$	No inverse equation available to a reasonable degree of accuracy
400· to 1100°C	(applies to both ranges of temperature)	$T = -2.4707112 \times 10$ $+2.9465633 \times 10 \times E$ $-3.1332620 \times 10^{-1} \times E^2$ $+6.5075717 \times 10^{-3} \times E^3$ $-3.9663834 \times 10^{-5} \times E^4$

(d) Type K Thermocouples

Table 7.5   Power Series Expansions for Noble Metal Thermocouples

Temperature Range	Exact Reference Voltage in Millivolts
0° to 1820°C	$E = (-2.4674601620 \times 10^{-1} \times T$ $+5.9102111169 \times 10^{-3} \times T^2$ $-1.4307123430 \times 10^{-6} \times T^3$ $+2.1509149750 \times 10^{-9} \times T^4$ $-3.1757800720 \times 10^{-12} \times T^5$ $+2.4010367459 \times 10^{-15} \times T^6$ $-9.0928148159 \times 10^{-19} \times T^7$ $+1.3299505137 \times 10^{-22} \times T^8) \times 10^{-3}$

(a) Type B Thermocouples

Temperature Range	Exact Reference Voltage in Millivolts
630° to 1064°C	$E = (-2.6418007025 \times 10^2$ $+8.0468680747 \times 10^0 \times T$ $+2.9892293723 \times 10^{-3} \times T^2$ $-2.6876058617 \times 10^{-7} \times T^3) \times 10^{-3}$
1064° to 1665°C	$E = (+1.4901702702 \times 10^3$ $+2.8639867552 \times 10^0 \times T$ $+8.0823631189 \times 10^{-3} \times T^2$ $-1.9338477638 \times 10^{-6} \times T^3) \times 10^{-3}$

(b) Type R Thermocouples

107

Table 7.5   *(Continued)*

Temperature· Range	Exact Reference Voltage in Millivolts
630° to 1064°C	$E = (-2.9824481615 \times 10^2$ $+8.2375528221 \times 10^0 \times T$ $+1.6453909942 \times 10^{-3} \times T^2) \times 10^{-3}$
1064° to 1665°C	$E = (+1.2766292175 \times 10^3$ $+3.4970908041 \times 10^0 \times T$ $+6.3824648666 \times 10^{-3} \times T^2$ $-1.5722424599 \times 10^{-6} \times T^3) \times 10^{-3}$

(c) Type S Thermocouples

TYPE K (chromel/alumel). Outputs of these thermocouples can be represented by the equation in Table 7.4d. An eighth-degree equation with an exponential term is used to generate values of voltage as a function of temperature from 0° to 1372°C. The inverse relation provides reference temperatures accurate to ±0.02°C from 400° to 1100°C.

TYPE B (platinum-30% rhodium/platinum). An eighth-degree equation, given in Table 7.5a is used to generate these voltages as a function of temperature from 0° to 1820°C.

TYPE R (platinum-13% rhodium/platinum). Two third-degree equations are required to generate voltage as a function of temperature from 630° to 1665°C, as shown in Table 7.5b.

TYPE S (platinum-10% rhodium/platinum). From the antimony point to the gold point, that is, from 630.74° to 1064.43°C, a second-degree equation is used to generate voltages for this type thermocouple. From 1064° to 1665°C, a third-degree equation is used. The required coefficients of these polynomials are given in Table 7.5c.

A table generated by the polynomials given in Tables 7.4 and 7.5, for six of the most commonly used types of thermocouples, is presented as Table 7.6.

Prior to the publication of the NBS polynomials of Tables 7.4 and 7.5, there was no unique, internationally accepted, mathematical generation scheme available that would reproduce the established thermocouple reference table values [33]. One method that found wide acceptance in the ISA [34], ASME [35], ASTM [36], and industry in general [37], [38] was based on the trial-and-error establishment of a few key values of voltage and temperature and a series of equations of specified degree to interpolate voltages between the key points. Today this generation scheme is no longer required and is only of historical interest. NBS

Monograph 125 [29] is the accepted standard for the nominal characteristic for all conventional thermocouple types.

7.9 Thermoelectric Circuit Analysis

An important phase of each temperature measurement by means of thermocouples is the thermoelectric circuit analysis. This operation can be systematized by observing the following set of rules.

1. Draw an equivalent circuit, numbering all thermoelectric junctions, and show an assumed direction of current.
2. Indicate the direction of the potential drop at each junction, guided by the outputs of each material with respect to platinum-67 (see Figure 7.18).
3. Compute the net emf in terms of the potentials at each junction (assigning a plus sign if the potential drop is in the same direction as the assumed current).
4. Combine the voltages algebraically to a form that allows easy reference table look-up (i.e., combine in terms of usual thermocouple combinations).
5. Compute the errors.

The examples that follow illustrate the application of these rules, and have been chosen to represent common and yet serious thermocouple circuit errors.

Example 1. Use of Incorrect Extension Wires. A chromel–alumel (Ch–Al) thermocouple was connected by mistake with iron–constantan (I–C) extension wires to a potentiometer in such a manner that the output emf was on scale. The actual temperatures were: at measuring junction, $T_M = 60°F$; at reference junction, $T_R = 32°F$; and at the junction between the thermocouple and extension wires, $T_J = 130°F$. What was the error in the measurement?

1. Draw an equivalent circuit (see Figure 7.20).
2. Indicate direction of all potential drops (dotted arrows in Figure 7.20).
3, 4. Compute net emf:

$$E_{net} = -E_1 + E_2 - E_3 + E_4$$

where $E_1 = (Ch–Al)_{60°F} = 0.619 \, mV$;

$\qquad E_2 = (Ch–C)_{130°F} = 3.329 \, mV$;

$\qquad E_3 = (I–C)_{32°F} = 0 \, mV$;

$\qquad E_4 = (I–Al)_{130°F} = (Ch–Al)_{130°F} - (Ch–C)_{130°F} + (I–C)_{130°F}$;

$\qquad\qquad = 2.206 - 3.329 + 2.820 \, mV$

$\qquad\qquad = 1.697 \, mV \text{(see Figure 7.21)}.$

$\qquad\qquad E_{net} = -0.619 + 3.329 - 0 + 1.697 = 4.407 \, mV,$

Table 7.6A

MILLIVOLTS, CELSIUS SCALE						DEG	MILLIVOLTS, FAHRENHEIT SCALE					
T	S	R	K	J	E	DEG	E	J	K	R	S	T
0.000	0.000	0.000	.000	0.000	0.000	0	-1.026	-.865	-.692	-.089	-.092	-.874
.391	.055	.054	.397	.507	.591	10	-.709	-.611	-.478	-.063	-.064	-.647
.769	.113	.111	.798	1.019	1.192	20	-.389	-.334	-.262	-.035	-.035	-.256
1.196	.173	.171	1.203	1.536	1.801	30	-.065	-.056	-.044	-.006	-.006	-.043
1.611	.235	.232	1.611	2.058	2.419	40	.262	.224	.176	.024	.024	.173
2.035	.299	.296	2.022	2.585	3.047	50	.591	.507	.397	.054	.055	.391
2.467	.365	.363	2.436	3.115	3.683	60	.924	.791	.619	.086	.087	.611
2.908	.432	.431	2.850	3.649	4.329	70	1.259	1.076	.843	.118	.119	.834
3.357	.502	.501	3.266	4.186	4.983	80	1.597	1.363	1.068	.150	.152	1.060
3.813	.573	.573	3.681	4.725	5.646	90	1.937	1.652	1.294	.184	.186	1.288
4.277	.645	.647	4.095	5.268	6.317	100	2.281	1.942	1.520	.218	.221	1.518
4.749	.719	.723	4.508	5.812	6.996	110	2.627	2.233	1.748	.253	.256	1.752
5.227	.795	.800	4.919	6.359	7.683	120	2.977	2.526	1.977	.289	.291	1.988
5.712	.872	.879	5.327	6.907	8.377	130	3.329	2.820	2.204	.326	.328	2.226

2.667	.365	.363	2.436	3.115	3.663	140	9.078	7.457	5.733	.959	.950	6.204
2.711	.402	.400	2.666	3.411	4.041	150	9.787	8.008	6.137	1.041	1.029	6.702
2.958	.440	.439	2.896	3.708	4.401	160	10.501	8.560	6.539	1.124	1.109	7.207
3.206	.478	.478	3.127	4.006	4.764	170	11.222	9.113	6.939	1.208	1.190	7.718
3.458	.517	.517	3.358	4.305	5.130	180	11.949	9.667	7.338	1.294	1.273	8.235
3.711	.557	.557	3.589	4.605	5.498	190	12.681	10.222	7.737	1.380	1.356	8.757
3.967	.597	.598	3.819	4.906	5.869	200	13.419	10.777	8.137	1.468	1.440	9.286
4.225	.637	.639	4.049	5.207	6.242	210	14.161	11.332	8.537	1.557	1.525	9.820
4.486	.678	.681	4.279	5.509	6.618	220	14.909	11.887	8.938	1.647	1.611	10.360
4.749	.719	.723	4.508	5.812	6.996	230	15.661	12.442	9.341	1.738	1.698	10.905
5.014	.761	.766	4.737	6.116	7.377	240	16.417	12.998	9.745	1.830	1.785	11.456
5.281	.803	.809	4.964	6.420	7.760	250	17.178	13.553	10.151	1.923	1.873	12.011
5.550	.846	.852	5.192	6.724	8.145	260	17.942	14.108	10.560	2.017	1.962	12.572
5.821	.889	.897	5.418	7.029	8.532	270	18.710	14.663	10.969	2.111	2.051	13.137
6.094	.932	.941	5.643	7.335	8.922	280	19.481	15.217	11.381	2.207	2.141	13.707

Table 7.6B

T	S	R	K	J	E	DEG	E	J	K	R	S	T
14,281	2,242	2,303	11,793	15,771	20,256	290	9,314	7,641	5,868	.986	.976	6,369
14,860	2,323	2,400	12,207	16,325	21,033	300	9,708	7,947	6,092	1,032	1,020	6,647
15,443	2,414	2,498	12,623	16,879	21,814	310	10,103	8,253	6,316	1,077	1,064	6,926
16,030	2,506	2,596	13,039	17,432	22,597	320	10,501	8,560	6,539	1,124	1,109	7,207
16,621	2,599	2,695	13,456	17,984	23,383	330	10,901	8,867	6,761	1,170	1,154	7,490
17,217	2,692	2,795	13,874	18,537	24,171	340	11,302	9,175	6,984	1,217	1,199	7,775
17,816	2,786	2,896	14,292	19,089	24,961	350	11,706	9,483	7,205	1,265	1,245	8,062
18,420	2,880	2,997	14,712	19,640	25,754	360	12,111	9,790	7,427	1,313	1,291	8,350
19,027	2,974	3,099	15,132	20,192	26,549	370	12,518	10,098	7,649	1,361	1,337	8,641
19,638	3,069	3,201	15,552	20,743	27,345	380	12,926	10,407	7,870	1,409	1,384	8,933
20,252	3,164	3,304	15,974	21,295	28,143	390	13,336	10,715	8,092	1,458	1,431	9,227
20,869	3,260	3,407	16,395	21,846	28,943	400	13,748	11,023	8,314	1,508	1,478	9,523
	3,356	3,511	16,818	22,397	29,744	410	14,161	11,332	8,537	1,557	1,525	9,820
	3,452	3,616	17,241	22,949	30,546	420	14,576	11,640	8,759	1,607	1,573	10,120
	3,549	3,721	17,664	23,501	31,350	430	14,992	11,949	8,983	1,657	1,620	10,420

				Index						
3,645	3,826	24,054	32,155	440	15,410	12,257	9,206	1,708	1,669	10,723
3,743	3,933	24,607	32,960	450	15,829	12,566	9,430	1,756	1,717	11,027
3,840	4,039	25,161	33,767	460	16,249	12,874	9,655	1,810	1,765	11,333
3,938	4,146	25,716	34,574	470	16,670	13,183	9,880	1,861	1,814	11,640
4,036	4,254	26,272	35,382	480	17,093	13,491	10,106	1,913	1,863	11,949
4,135	4,362	26,829	36,190	490	17,517	13,800	10,333	1,964	1,912	12,260
4,234	4,471	27,388	36,999	500	17,942	14,108	10,560	2,017	1,962	12,572
4,333	4,580	27,949	37,808	510	18,368	14,416	10,787	2,069	2,011	12,885
4,432	4,689	28,511	38,617	520	18,795	14,724	11,015	2,122	2,061	13,200
4,532	4,799	29,075	39,426	530	19,223	15,032	11,243	2,175	2,111	13,516
4,632	4,910	29,642	40,236	540	19,653	15,340	11,472	2,228	2,161	13,834
4,732	5,021	30,210	41,045	550	20,083	15,648	11,702	2,282	2,211	14,153
4,832	5,132	30,782	41,853	560	20,514	15,956	11,931	2,335	2,262	14,474
4,933	5,244	31,356	42,662	570	20,947	16,264	12,161	2,389	2,313	14,795
5,034	5,356	31,933	43,470	580	21,380	16,571	12,392	2,443	2,363	15,118

Table 7.6C

S	R	K	J	E	DEG	E	J	K	R	S	T
5,136	5,469	24,476	32,513	44,278	590	21,814	16,879	12,623	2,498	2,414	15,443
5,237	5,582	24,902	33,096	45,085	600	22,248	17,186	12,854	2,552	2,465	15,769
5,339	5,696	25,327	33,683	45,891	610	22,684	17,493	13,085	2,607	2,517	16,096
5,442	5,810	25,751	34,273	46,697	620	23,120	17,800	13,317	2,662	2,568	16,424
5,544	5,925	26,176	34,867	47,502	630	23,558	18,107	13,549	2,718	2,620	16,753
5,648	6,040	26,599	35,464	48,308	640	23,996	18,414	13,781	2,773	2,672	17,084
5,751	6,155	27,022	36,066	49,109	650	24,434	18,721	14,013	2,829	2,723	17,416
5,855	6,272	27,445	36,671	49,911	660	24,873	19,027	14,246	2,885	2,775	17,750
5,960	6,388	27,867	37,280	50,713	670	25,313	19,334	14,479	2,941	2,828	18,084
6,064	6,505	28,288	37,893	51,513	680	25,754	19,640	14,712	2,997	2,880	18,420
6,169	6,623	28,709	38,510	52,312	690	26,195	19,947	14,945	3,053	2,932	18,757
6,274	6,741	29,128	39,130	53,110	700	26,637	20,253	15,178	3,110	2,985	19,095
6,380	6,860	29,547	39,754	53,907	710	27,079	20,559	15,412	3,167	3,037	19,434
6,486	6,979	29,965	40,382	54,703	720	27,522	20,866	15,646	3,224	3,090	19,774
6,592	7,098	30,383	41,013	55,498	730	27,966	21,172	15,880	3,281	3,143	20,116

6,699	7,218	30,799	41,647	56,291	740	28,009	21,478	16,114	3,338	3,196	20,458
6,805	7,339	31,214	42,283	57,083	750	28,854	21,785	16,349	3,396	3,249	20,801
6,913	7,460	31,629	42,922	57,873	760	29,299	22,091	16,583	3,453	3,302	
7,020	7,582	32,042		58,663	770	29,744	22,397	16,818	3,511	3,356	
7,128	7,703	32,455		59,451	780	30,190	22,704	17,053	3,569	3,409	
7,236	7,826	32,866		60,237	790	30,636	23,010	17,288	3,627	3,463	
7,345	7,949	33,277		61,022	800	31,082	23,317	17,523	3,686	3,516	
7,454	8,072	33,686		61,806	810	31,529	23,624	17,759	3,744	3,570	
7,563	8,196	34,095		62,588	820	31,976	23,931	17,994	3,803	3,624	
7,672	8,320	34,502		63,366	830	32,423	24,238	18,230	3,862	3,678	
7,782	8,445	34,909		64,147	840	32,871	24,546	18,466	3,921	3,732	
7,892	8,570	35,314		64,924	850	33,319	24,853	18,702	3,980	3,786	
8,003	8,696	35,718		65,700	860	33,767	25,161	18,938	4,039	3,840	
8,110	8,822	36,121		66,473	870	34,215	25,469	19,174	4,099	3,895	
8,225	8,949	36,524		67,245	880	34,664	25,778	19,410	4,158	3,949	

Table 7.6D

S	R	K	J	E	DEG	E	K	R	S
4,004	4,218	19,646	26,087	35,113	890	68,015	36,925	9,076	8,336
4,058	4,278	19,883	26,396	35,562	900	68,783	37,325	9,203	8,468
4,113	4,338	20,120	26,705	36,011	910	69,549	37,724	9,331	8,560
4,168	4,398	20,356	27,016	36,460	920	70,313	38,122	9,460	8,673
4,223	4,458	20,593	27,326	36,909	930	71,075	38,519	9,589	8,786
4,278	4,519	20,830	27,637	37,358	940	71,835	38,915	9,718	8,899
4,333	4,580	21,066	27,949	37,808	950	72,593	39,310	9,848	9,012
4,388	4,640	21,303	28,261	38,257	960	73,350	39,703	9,978	9,126
4,443	4,701	21,540	28,573	38,707	970	74,104	40,096	10,109	9,240
4,498	4,762	21,777	28,887	39,157	980	74,857	40,488	10,240	9,359
4,554	4,824	22,014	29,201	39,606	990	75,608	40,879	10,371	9,470
4,609	4,885	22,251	29,515	40,056	1000	76,358	41,269	10,503	9,585
4,665	4,947	22,488	29,831	40,505	1010		41,657	10,636	9,700
4,721	5,008	22,725	30,147	40,955	1020		42,045	10,768	9,816
4,776	5,070	22,961	30,464	41,402	1030		42,432	10,902	9,932

1040	10,048	11,035	42,617	41,855	30,782	23,198	5,132	4,832
1050	10,165	11,170	43,202	42,303	31,100	23,435	5,194	4,888
1060	10,282	11,304	43,585	42,752	31,420	23,672	5,256	4,944
1070	10,400	11,439	43,968	43,201	31,740	23,908	5,319	5,000
1080	10,517	11,574	44,349	43,650	32,061	24,145	5,381	5,057
1090	10,635	11,710	44,729	44,098	32,384	24,382	5,444	5,113
1100	10,754	11,846	45,108	44,547	32,707	24,618	5,507	5,169
1110	10,872	11,983	45,486	44,995	33,031	24,854	5,570	5,226
1120	10,991	12,119	45,863	45,443	33,356	25,091	5,633	5,283
1130	11,110	12,257	46,238	45,891	33,683	25,327	5,696	5,339
1140	11,229	12,394	46,612	46,339	34,010	25,563	5,759	5,396
1150	11,348	12,532	46,985	46,786	34,339	25,799	5,823	5,453
1160	11,467	12,669	47,356	47,234	34,668	26,034	5,886	5,510
1170	11,587	12,808	47,726	47,681	34,999	26,270	5,950	5,567
1180	11,707	12,946	48,095	48,127	35,331	26,505	6,014	5,625

Table 7.6E

S	R	K		DEG	E	J	K	R	S
11,827	13,065	48,862		1190	48,574	35,664	26,740	6,078	5,682
11,947	13,224	48,828		1200	49,020	35,999	26,975	6,143	5,740
12,067	13,363	49,192		1210	49,466	36,334	27,210	6,207	5,797
12,188	13,502	49,555		1220	49,911	36,671	27,445	6,272	5,855
12,308	13,642	49,916		1230	50,357	37,009	27,679	6,336	5,913
12,429	13,782	50,276		1240	50,802	37,348	27,914	6,401	5,971
12,550	13,922	50,633		1250	51,246	37,688	28,148	6,466	6,029
12,671	14,062	50,990		1260	51,691	38,030	28,382	6,532	6,087
12,792	14,202	51,344		1270	52,135	38,372	28,615	6,597	6,146
12,913	14,343	51,697		1280	52,578	38,716	28,849	6,662	6,204
13,034	14,483	52,049		1290	53,022	39,061	29,082	6,728	6,263
13,155	14,624	52,398		1300	53,465	39,407	29,315	6,794	6,321
13,276	14,765	52,747		1310	53,907	39,754	29,547	6,860	6,380
13,397	14,906	53,093		1320	54,349	40,103	29,780	6,926	6,439
13,519	15,047	53,439		1330	54,791	40,452	30,012	6,992	6,498

13,640	15,188	53,782	1340	55,233	40,802	30,244	7,059	6,557
13,761	15,329	54,125	1350	55,674	41,154	30,475	7,125	6,616
13,883	15,470	54,466	1360	56,115	41,506	30,706	7,192	6,675
14,004	15,611	54,807	1370	56,555	41,859	30,937	7,259	6,734
14,125	15,752		1380	56,995	42,212	31,168	7,326	6,794
14,247	15,893		1390	57,434	42,567	31,399	7,393	6,853
14,366	16,035		1400	57,873	42,922	31,629	7,460	6,913
14,489	16,176		1410	58,312		31,859	7,527	6,972
14,610	16,317		1420	58,750		32,088	7,595	7,032
14,731	16,458		1430	59,188		32,317	7,663	7,092
14,852	16,599		1440	59,626		32,546	7,731	7,152
14,973	16,741		1450	60,063		32,775	7,799	7,212
15,094	16,882		1460	60,499		33,003	7,867	7,272
15,215	17,022		1470	60,935		33,231	7,935	7,333
15,336	17,163		1480	61,371		33,459	8,004	7,393

Table 7.6F

S	R	DEG	E	K	R	S
15,456	17,304	1490	61,806	33,686	8,072	7,454
15,576	17,445	1500	62,240	33,913	8,141	7,514
15,697	17,585	1510	62,675	34,140	8,210	7,575
15,817	17,726	1520	63,108	34,366	8,279	7,636
15,937	17,866	1530	63,542	34,593	8,348	7,697
16,057	18,006	1540	63,974	34,818	8,417	7,758
16,176	18,146	1550	64,406	35,044	8,487	7,819
16,296	18,286	1560	64,838	35,269	8,556	7,880
16,415	18,425	1570	65,269	35,494	8,626	7,942
16,534	18,564	1580	65,700	35,718	8,696	8,003
16,653	18,703	1590	66,130	35,942	8,766	8,065
16,771	18,842	1600	66,559	36,166	8,836	8,126
16,890	18,981	1610	66,988	36,390	8,907	8,188
17,008	19,119	1620	67,416	36,613	8,977	8,250
17,125	19,257	1630	67,844	36,836	9,048	8,312

1640	68,271	37,058	9,118	8,374
1650	68,698	37,280	9,189	8,436
1660	69,124	37,502	9,260	8,498
1670	69,549	37,724	9,331	8,560
1680	69,974	37,945	9,403	8,623
1690	70,398	38,166	9,474	8,685
1700	70,821	38,387	9,546	8,748
1710	71,244	38,607	9,617	8,811
1720	71,667	38,827	9,689	8,874
1730	72,088	39,046	9,761	8,937
1740	72,509	39,266	9,833	9,000
1750	72,930	39,485	9,906	9,063
1760	73,350	39,703	9,978	9,126
1770	73,769	39,922	10,050	9,190
1780	74,188	40,140	10,123	9,253

17,243 19,395

17,360 19,533

Table 7.6G

DEG	E	K	R	S
1790	74,606	40,358	10,196	9,317
1800	75,024	40,575	10,269	9,380
1810	75,441	40,792	10,342	9,444
1820	75,858	41,009	10,415	9,508
1830	76,274	41,225	10,488	9,572
1840		41,442	10,562	9,636
1850		41,657	10,636	9,700
1860		41,873	10,709	9,764
1870		42,088	10,783	9,829
1880		42,303	10,857	9,893
1890		42,518	10,931	9,958
1900		42,732	11,006	10,023
1910		42,946	11,080	10,087
1920		43,159	11,155	10,152
1930		43,373	11,229	10,217

Year			
1940	43,585	11,304	10,282
1950	43,798	11,379	10,348
1960	44,010	11,454	10,413
1970	44,222	11,529	10,478
1980	44,434	11,605	10,544
1990	44,645	11,680	10,609
2000	44,856	11,756	10,675
2010	45,066	11,831	10,740
2020	45,276	11,907	10,806
2030	45,486	11,983	10,872
2040	45,695	12,059	10,938
2050	45,904	12,135	11,004
2060	46,113	12,211	11,070
2070	46,321	12,287	11,136
2080	46,529	12,363	11,202

Table 7.6H

DEG	K	R	S
2090	46,737	12,440	11,268
2100	46,944	12,516	11,335
2110	47,150	12,593	11,401
2120	47,356	12,669	11,467
2130	47,562	12,746	11,534
2140	47,767	12,823	11,600
2150	47,972	12,900	11,667
2160	48,177	12,977	11,734
2170	48,381	13,054	11,800
2180	48,584	13,131	11,867
2190	48,787	13,208	11,934
2200	48,990	13,286	12,001
2210	49,192	13,363	12,067
2220	49,394	13,440	12,134
2230	49,595	13,518	12,201

2240	49,796	13,595	12,268
2250	49,996	13,673	12,335
2260	50,196	13,751	12,402
2270	50,395	13,828	12,469
2280	50,594	13,906	12,536
2290	50,792	13,984	12,604
2300	50,990	14,062	12,671
2310	51,187	14,140	12,738
2320	51,384	14,218	12,805
2330	51,580	14,296	12,872
2340	51,776	14,374	12,940
2350	51,971	14,452	13,007
2360	52,165	14,530	13,074
2370	52,360	14,608	13,142
2380	52,553	14,686	13,209

Table 7.6I

DEG	K.	R	S
2390	52,747	14,765	13,278
2400	52,939	14,843	13,344
2410	53,132	14,921	13,411
2420	53,324	15,000	13,478
2430	53,515	15,078	13,546
2440	53,706	15,156	13,613
2450	53,897	15,235	13,681
2460	54,087	15,313	13,748
2470	54,277	15,391	13,815
2480	54,466	15,470	13,883
2490	54,656	15,548	13,950
2500	54,845	15,627	14,018

Example 1:   Equivalent circuit

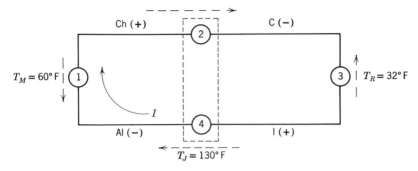

Example 2:   Equivalent circuit

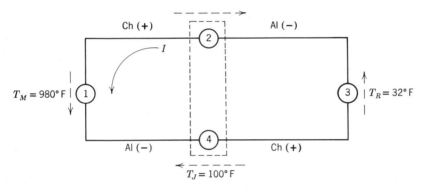

Figure 7.20 Junction method of circuit analysis. Dotted arrows indicate polarity of emf at junctions. Solid arrow indicates direction of net current.

whereas the correct emf should have been:

$$E = (\text{Ch–Al})_{60°\text{F}} - (\text{Ch–Al})_{32°\text{F}} = 0.619 \text{ mV}.$$

5. Compute errors:

net voltage error $= 4.407 - 0.619 = 3.788$ mV,

net temperature error $= 225 - 60 = 165°$F, (on Ch–Al scale),

net temperature error $= 183 - 60 = 123°$F, (on I–C scale).

It goes without saying that, on either thermocouple scale, errors of this magnitude cannot be tolerated.

Example 2. Use of Incorrect Polarity of Extension Wires. The extension wires of a chromel-alumel thermocouple were inadvertently interchanged. The actual temperatures were at measuring junction, $T_M = 980°$F; at reference junction, $T_R = 32°$F; and at the

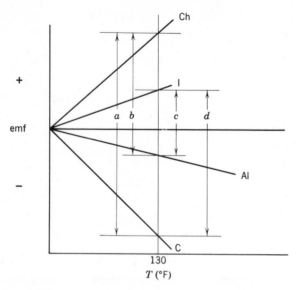

Figure 7.21 Graphical method of circuit analysis for Example 1: Computation of $E_4 = (I\text{-}A1) = (I+A1)$. $E_4 = (I+A1) = (Ch+A1) - (Ch+C) + (1+C)$, $c = b - a + d$, $c = 2.206 - 3.329 + 2.820$, $c = 1.697$ mV.

junction between the thermocouple and extension wires, $T_J = 100°F$. What was the error in the measurement?

1, 2. Draw an equivalent circuit (see Figure 7.20).

3, 4. Compute net emf:

$$E_{net} = E_1 - E_2 + E_3 - E_4$$

where $E_1 = (Ch\text{-}Al)_{980°F} = 21.777$ mV;

$E_2 = (Ch\text{-}Al)_{100°F} = 1.520$ mV;

$E_3 = (Ch\text{-}Al)_{32°F} = 0$ mV;

$E_4 = (Ch\text{-}Al)_{100°F} = 1.520$ mV;

$E_{net} = 21.777 - 1.520 + 0 - 1.520 = 18.737$ mV,

whereas the correct emf should have been:

$$E = (Ch\text{-}Al)_{980°F} - (Ch\text{-}Al)_{32°F} = 21.777 \text{ mV}.$$

5. Compute errors:

net voltage error $= 21.777 - 18.737 = 3.040$ mV.

net temperature error $= 980 - 851.5 = 128.5°F$ (on Ch–Al scale).

Once again, a most common thermocouple circuit error, namely interchanging extension wires, is seen to lead to a very serious temperature measuring error.

Example 3. Use of Improper Grounding. A chromel–constantan thermocouple, grounded at its measuring junction according to Figure 7.14, was inadvertently grounded at its reference junction also. The actual temperatures were: at measuring junction, $T_M = 700°F$; at reference junction, $T_R = 32°F$. What is the predicted error in the measurement?

1. Draw an equivalent circuit:

Figure 7.22 illustrates the multigrounded circuit under consideration. Figure 7.23 depicts the equivalent circuit that represents the multigrounded thermocouple.

2. In this example, the direction of all potential drops is indicated by electrical cell symbols, where the drop is from minus to plus at the junction.

3, 4. Compute net emf:

A voltage summation is written around the outer ground loop to obtain the current in the loop as follows.

$$\sum E = -(E_{pq})_M - iR_g - iR_n + (E_{pn})_M = 0 \qquad (7.36)$$

where $(E_{pg})_M$ signifies the emf set up by the positive thermoelement and the ground wire at the measuring junction temperature, and so on. Thus

$$i = \frac{(E_{pn})_M - (E_{pg})_M}{R_g + R_n} \qquad (7.37)$$

Note that chromel is the positive element with respect to constantan, and that copper is assumed to be the grounding wire.

Given the ground loop current, the net voltage set up by the doubly grounded thermocouple circuit can be determined as follows.

$$E_{net} = iR_g + (E_{pg})_M = -iR_n + (E_{pn})_M \qquad (7.38)$$

The pertinent resistance values are:

for constantan, $R_n = 2.902\ \Omega$

for copper, $R_g = 0.0646\ \Omega$.

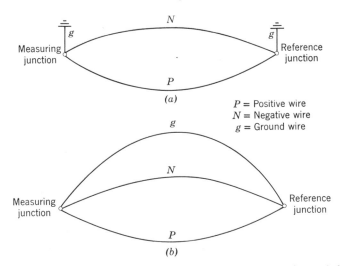

P = Positive wire
N = Negative wire
g = Ground wire

Figure 7.22 Multigrounded thermocouple circuits. (*a*) Schematic multigrounded circuit. (*b*) Experimental multigrounded thermocouple circuit (equivalent to that of (*a*).

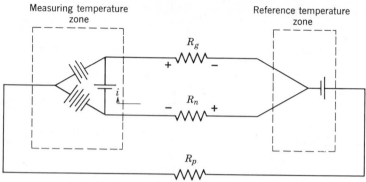

Thermocouple grounded at measuring and reference junctions

Figure 7.23 Equivalent electrical circuit for the multigrounded thermocouples of Figure 7.22. Note: R_n = resistance of negative wire. R_p = resistance of positive wire. R_g = resistance of ground wire.

The pertinent voltage values are:

$$(E_{pn})_M = (E_{Ch-C})_{700°F} = 26.637 \text{ mV}$$

and

$$(E_{pg})_M = (E_{Ch-Cu})_{700°F} = (E_{Ch-C} - E_{Cu-C})_{700°F}$$

or

$$(E_{pg})_M = 26.637 - 19.095 = 7.542 \text{ mV}.$$

By (7.37)

$$i = \frac{26.637 - 7.542}{0.0646 + 2.902} = 6.437 \ mA.$$

By (7.38)

$$E_{net} = iR_g + (E_{pg})_M$$
$$= 6.437(0.0646) + 7.542 = 7.958 \text{ mV}.$$

5. The temperature error for this doubly grounded thermocouple can be estimated as follows.

$$\text{Net voltage error} = 26.637 - 7.958 = 18.679 \text{ mV}.$$

$$\text{Net temperature error} = 700 - 255 = 445°F.$$

This error is obviously much larger than any other error associated with thermocouple circuits, and it is clear that doubly grounded thermocouple circuits are to be avoided at all costs.

Example 4. Rotating Thermocouple Circuit Analysis. A rotating iron-constantan thermocouple is connected through a slip ring to a stationary potentiometer and reference junction. The unknown temperature at the transition between the thermocouple wires and the copper slip ring wires is accounted for by a special rotating iron-constantan thermocouple connected as shown in Figure 7.16a. The actual temperatures were: at measuring junction, $T_M = 500°F$; at reference junction, $T_R = 32°F$; and at the slip ring, $T_J = 150°F$. What was the error in the measurement?

1, 2. Draw an equivalent circuit (see Figure 7.24).

3, 4. Compute net emf:

$$E_{net} = E_1 - E_2 - E_3 + E_4 - E_5 - E_6$$

$$E_1 = (\text{I–C})_{500°F}$$

$$E_2 = (\text{I–Cu})_{150°F}$$

$$E_3 = (\text{Cu–C})_{32°F}$$

$$E_4 = (\text{I–C})_{150°F}$$

$$E_5 = (\text{I–Cu})_{32°F}$$

$$E_6 = (\text{Cu–C})_{150°F}$$

$$E_{net} = (\text{I–C})_{500°F} - \text{I}_{150°F} + \text{Cu}_{150°F} - 0 + (\text{I–C})_{150°F} - 0 - \text{Cu}_{150°F} + \text{C}_{150°F}$$

$$E_{net} = (\text{I–C})_{500°F} - (\text{I–C})_{150°F} + (\text{I–C})_{150°F} = (\text{I–C})_{500°F}$$

Since this is the voltage required, there are no circuit errors in this example.

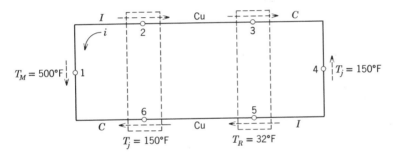

Figure 7.24 Equivalent circuit of Figure 7.16*a* for Example 4.

7.10 References

[1] A. Volta, "On the Electricity Excited by the Mere Contact of Conducting Substances of Different Kinds," *Philosophical Trans.*, 1800, p. 403.

[2] T. J. Seebeck, "Evidence of the Thermal Current of the Combination Bi-Cu by its Action on Magnetic Needle," *Roy. Acad. Sci.*, Berlin, 1822–1823, p. 265.

[3] J. B. J. Fourier, *Analytical Theory of Heat*, Gauthier-Villars, Paris, 1822; English translation by A. Freeman, Cambridge, 1878.

[4] G. S. Ohm, "Determination of the Laws by which Metals Conduct the Contact Electricity, and also a Draft for a Theory of the Voltage Apparatus," *J. Chem. Phys* (Schweigger's Journal), **46**, 1836, p. 137.

[5] J. C. A. Peltier, "Investigation of the Heat Developed by Electric Currents in Homogeneous Materials and at the Junction of Two Different Conductors," *Ann. Chim. Phys.*, **56** (2nd Ser.), 1834, p. 371.

[6] J. P. Joule, "On the Production of Heat by Voltaic Electricity," *Proc. Roy. Soc.*, December 1940.

[7] R. J. E. Clausius, "About the Motive Force of Heat," *Ann. Phys. Chem.*, **79,** 1850, pp. 368, 500.

[8] W. Thomson, "On a Mechanical Theory of Thermo-Electric Currents," *Proc. Roy. Soc. Edinburgh,* December 1851.

[9] W. Thomson, "On the Thermal Effects of Electric Currents in Unequal Heated Conductors," *Proc. Roy. Soc.,* **7,** May 1854.

[10] R. P. Benedict, "Thermoelectric Effects," *Elec. Mfg.,* February 1960, p. 103.

[11] R. Stratton, "On the Elementary Theory of Thermoelectric Phenomena," *Brit. J. Appl. Phys.,* **8,** August 1949, p. 33.

[12] D. G. Miller, "Thermodynamic Theory of Irreversible Processes," *Am. J. Phys.,* May 1956, p. 433.

[13] L. Onsager, "Reciprocal Relations in Irreversible Processes," *Phys. Rev.,* **37,** February 1931, p. 405.

[14] H. B. Callen, "The Application of Onsager's Reciprocal Relations to Thermoelectric, Thermomagnetic, and Galvanomagnetic Effects," *Phys. Rev.* **73,** June 1948, p. 3149.

[15] G. N. Hatsopoulos and J. H. Keenan, "Analyses of the Thermoelectric Effects of Methods of Irreversible Thermodynamics," *J. Appl. Mech.,* **25,** December 1958, p. 428.

[16] R. P. Benedict, "Temperature and Its Measurement," *Electro-Technol.,* July 1963, p. 71.

[17] H. LeChatelier, *Compt. Rend.,* 1886, **102,** p. 819.

[18] H. LeChatelier, *J. de Phys.,* 1887, **6,** p. 23.

[19] L. B. Hunt, "The Early History of the Thermocouples," *Platinum Metals Review,* Vol. 8, No. 1, January 1964 (published by Matthey-Bishop Co.).

[20] D. I. Finch, "General Principles of Thermoelectric Thermometry," in *Temperature,* Vol. 3, Pt. 2, Reinhold, New York, 1962.

[21] W. F. Roeser, "Thermoelectric Circuitry," *J. Appl. Phys.,* **11,** 1940, p. 388.

[22] P. H. Dike, "Thermoelectric Thermometry," Leeds and Northrup Tech. Pub. EN-33A (1a), 3rd ed., April 1955.

[23] B. E. Klipec, "Reducing Electrical Noise in Instrument Circuits," *IEEE Trans. on Industry and General Applications,* Vol. IGA-3, March/April 1967, p. 90.

[24] J. R. Springer, "Common Mode Voltage Rejection," *Instruments and Control Systems,* February 1971, p. 99.

[25] R. P. Benedict and R. J. Russo, "A Note on Grounded Thermocouple Circuits," *Trans. ASME, J. Basic Eng.,* June 1972, p. 377.

[26] B. O. Buckland and S. S. Stack, "Thermocouples for Testing Turbines," in *Temperature,* Vol. 1, Reinhold, New York, 1941, p. 884.

[27] A. I. Dahl, "Stability of Thermocouples," in *Temperature,* Vol. 1, Reinhold, New York, 1941, p. 1238.

[28] J. L. Howard, "Error Accumulation in Thermocouple Thermometry," in *Temperature,* Vol. 4, Pt. 3, Instrument Society of America, Pittsburg, 1972, p. 2017.

[29] R. L. Powell, W. J. Hall, C. H. Hyink, Jr., L. L. Sparks, G. W. Burns, M. G. Scroger, and H. H. Plumb, "Thermocouple Reference Tables Based on IPTS-68," *NBS Monograph 125,* March 1974.

[30] *Temperature-Electromotive Force (EMF) Tables for Thermocouples,* ASTM Standard E-230-73.

[31] *Manual on the Use of Thermocouples in Temperature Measurement,* ASTM STP 470A, March 1974.

[32] R. P. Benedict, "Generating Thermocouple Reference Tables," *Instruments and Control Systems,* January 1974, p. 53.

[33] H. Shenker, J. I. Lawritzer, Jr., R. J. Corruccini, and S. J. Lonberger, "Reference Tables for Thermocouples," *NBS Circular 561*, 1955.
[34] R. P. Benedict and H. F. Ashby, "Improved Reference Tables for Thermocouples," in *Temperature*, Vol. 3, Pt. 2, Reinhold, New York, 1962, p. 51.
[35] R. P. Benedict and H. F. Ashby, "Empirical Determination of Thermocouple Characteristics," *Trans. ASME, J. Eng. Power*, January 1963, p. 9.
[36] *Manual on the Use of Thermocouples in Temperature Measurement*, ASTM STP 470, August 1970.
[37] R. P. Benedict, "The Generation of Thermocouple Reference Tables," *Electro-Technology*, November 1963, p. 80.
[38] "Calibrating and Checking Thermocouples in Plants and Laboratories," *L and N Folder* A1.1121, 1967, p. 13.

Nomenclature

Roman

a, b	constants
A	area
d	exact differential
E	emf of thermocouple
F	friction
I	electric current
J	flow
L	linear coefficient
q	rate of heat transfer
Q	quantity of heat
R	resistance
S	entropy
t	empirical temperature
T	temperature (absolute or empirical as noted in text)
x	length
X	force
x, y	general coordinates
E, J, K, R, S, T	thermocouple type letter designations
Cr, Al, I, C	thermocouple materials
emf	net electromotive force of thermocouple

Greek

α	Seebeck coefficient, also called thermoelectric power
δ	inexact differential
Δ	finite difference
θ	time

ξ rate of entropy production/unit volume
π Peltier coefficient, also called Peltier voltage
σ Thomson coefficient
$\sum$ summation

Subscripts

A, B	materials
e	with respect to electricity
i	inside
o	outside
P	Peltier
q	with respect to heat
S	Seebeck, system
T	Thomson
$0, 1, 2, n$	general subscripts
i, j	general subscripts

Problems

1. For a chromel–alumel thermocouple with its measuring junction at 500°C and its reference junction at 0°C, find the Seebeck voltage, the Seebeck coefficient, the Peltier voltage, and the net Thomson voltage, given the following values:

    $a_{\text{chromel}} = 30.5132 \ \mu\text{V/°C}$, $a_{\text{alumel}} = -9.4332 \ \mu\text{V/°C}$,
    $b_{\text{chromel}} = 7.6296 \times 10^{-3} \ \mu\text{V/(°C)}^2$ and $b_{\text{alumel}} = 2.3032 \times 10^{-3} \ \mu\text{V/(°C)}^2$.

    Ans. $E_S = 20639.6 \ \mu\text{V}$, $\alpha_{500°C} = 42.612 \ \mu\text{V/°C}$, $\pi_{500°C} = 32,945.47 \ \mu\text{V}$, $E_T = -1394.51 \ \mu\text{V}$.

2. Show that the values given in Tables 7.2 and 7.3 agree for each base metal thermocouple type at 400°C.

    Ans. $JP + JN = 9.7 + 45.4 = 55.1 \ \text{mV}$.

3. The measuring junction of a chromel–alumel thermocouple is exposed to 1000°F. At 500°F, extension wires of copper-constantan are mistakenly joined to the chromel–alumel wires, observing proper polarity. What is the net temperature error on the chromel-alumel scale?

    Ans. $\epsilon_T = 85°\text{F}$.

4. Two iron–constantan thermocouples were checked against each other by placing both thermocouples in separate ice baths. However, the

junctions between thermocouple materials and copper wires (leading to the potentiometer) were made at different temperatures, namely, 80°F and 150°F. What would be the resulting emf and how would this be interpreted in terms of a temperature error?

Ans. $\epsilon_T = 72°F$.

5. A chromel–alumel (type K) thermocouple constructed so that its measuring junction was grounded to its sheath was mistakenly grounded at its reference junction also. What would be the expected temperature error in such an arrangement if the measuring junction were at 750°F and the reference junction were at the ice point?

Ans. $\epsilon_T = 356°F$. (Note: use $R_{ground} = 0.0646 \, \Omega$, $R_{alumel} = 4.108 \, \Omega$.)

Chapter 8

OPTICAL PYROMETRY

"... try adding minus one to the denominator ..."

Max Planck (1900)

8.1 Historical Resume

Gustav Kirchoff, in 1860, defined a black body as a surface that neither reflects nor passes radiation [1]. He suggested that such a surface could be realized by heating a hollow enclosure and observing the radiation from a small access hole (e.g., a cylindrical hole of a depth from five to eight times its diameter that radiates from one end closely approximates a black body). Kirchoff defined the emissivity ϵ of a *non*-black body as the ratio of its radiant intensity to that of a similar black body at the same temperature. Note that $\epsilon_{black} = 1$, $\epsilon_{gray} =$ a constant < 1, whereas $\epsilon_{non\text{-}gray}$ varies strongly with wavelength.

Henri LeChatelier introduced, in 1892, the first practical optical pyrometer. It included an oil lamp as the reference light source, a red glass filter to limit the wavelength interval, and an iris diaphragm to achieve a brightness match between the light source and the test body.

Wilhelm Wien, in 1896, derived his law for the distribution of energy in the emission spectrum of a black body as

$$J_{b,\lambda} = \frac{c_1 \lambda^{-5}}{\exp{(c_2/\lambda T)}},\tag{8.1}$$

where J represents the intensity of radiation emitted by a black body at temperature T, and wavelength λ per unit wavelength interval, per unit time, per unit solid angle, per unit area.

Max Planck, to remedy deviations that appeared between (8.1) and the experimental facts at high values of λT, suggested, in 1900, the

136

mathematical expression

$$J_{b,\lambda} = \frac{c_1 \lambda^{-5}}{\exp{(c_2/\lambda T)} - 1},$$ (8.2)

to describe the distribution of radiation among the various wavelengths; that is, he simply added a -1 to the denominator of Wien's equation. However, in attempting to explain the significance of -1 in the denominator, Planck developed the quantum theory (wherein he postulated

Table 8.1 Quantities Involved in equation 8.2

Quantity	S.I. System	U.S. Customary System
$J_{b,\lambda}$ radiant intensity	joule per second per cubic centimetre J/sec · cm³	Btu per hour per cubic foot BTU/hr · ft³
c_1 First radiation constant $(c_1 = 2\pi c_0^2 h)$	3.7413×10^{-12} joule · square centimetre per second J · cm²/sec	13.741×10^{-15} Btu · square foot per hour BTU · ft²/hr
c_2 Second radiation constant $(c_2 = c_0 h/k)$	1.43883 centimetre-kelvin cm · K	0.08497 feet-degree Rankine ft · °R
λ wave length	microns or centimetres μ or cm	feet ft
T absolute temperature	kelvins K	degrees Rankine °R
c_0 speed of light in vacuum	2.99793×10^{10} cm/sec	354.086×10^{10} ft/hr
h Planck's constant	6.6252×10^{-34} J · sec	1.7443×10^{-40} Btu · hr
k Boltzmann's constant	1.38042×10^{-23} J/K	7.26881×10^{-27} Btu/°R
Conversions		
Temperature	$273.15 \text{ K} = 491.67°\text{R}$	
Energy	$1.055056 \times 10^3 \text{ J} = 1 \text{ Btu}$	
Length	$1\mu = 3.28084 \times 10^{-6} \text{ ft}$	

that electromagnetic waves can exist only in the form of certain discrete packages or quanta). Subsequently he received the Nobel Prize for his work.

The numerical values and the units, both in the International System (SI) and in the U.S. Customary System, for all the quantities involved in (8.2), are given in Table 8.1. An example illustrates the use of (8.2).

Example 1. Find the intensity of radiation emitted by a black body at 2200°R at a wavelength of 3μ in both SI and U.S. Customary units, and check.

U.S. Customary

$$\lambda = 3\mu = 3\mu\left[\frac{10^{-4}\,\text{cm}}{\mu}\times\frac{1\,\text{ft}}{30.48\,\text{cm}}\right] = 9.8425\times10^{-6}\,\text{ft}.$$

By (8.2)

$$J_{b,\lambda} = \frac{13.741\times10^{-15}\,\text{BTU.\,ft}^2/\text{hr}\times(9.8425\times10^{-6}\,\text{ft})^{-5}}{\exp\left[0.08497\,\text{ft.°R}/(9.8425\times10^{-6}\,\text{ft}\times2200°\text{R})\right]-1}$$

$$J_{b,\lambda} = \frac{1.48762\times10^{11}}{e^{3.92408}-1} = 2.99884\times10^9\,\frac{\text{BTU}}{\text{hr ft}^3}.$$

SI UNITS

By (4.15)

$$T(K) = \frac{T(°\text{R})}{1.8} = \frac{2200}{1.8} = 1222.22\,K.$$

By (8.2)

$$J_{b,\lambda} = \frac{3.7413\times10^{-12}\,J\cdot\text{cm}^2/\text{sec}\times(3\mu)^{-5}}{\exp\left[1.43883\,\text{cm}\cdot K/(3\mu\times1222.22\,K)\right]-1}$$

$$J_{b,\lambda} = \frac{1.53963\times10^6}{e^{3.92408}-1} = 3.10369\times10^4\,\frac{J}{\text{sec cm}^3}.$$

That the SI answer checks the U.S. answer can readily be seen by applying several conversion constants. That is

$$J_{b,\lambda} = 3.10369\times10^4\,\frac{J}{\text{sec}\cdot\text{cm}^3}\left[\frac{1\,\text{BTU}}{1.055056\times10^3\,J}\times\frac{3600\,\text{sec}}{\text{hr}}\times\frac{(30.48\,\text{cm})^3}{\text{ft}^3}\right]$$

$$= 2.99882\times10^9\,\frac{\text{BTU}}{\text{hr ft}^3}.$$

In 1927 the International Temperature Scale (ITS) was defined. Temperatures above the gold point (then 1063°C) were to be given by Wien's equation (8.1) used in conjunction with a disappearing filament optical pyrometer.

In 1948 the ITS was redefined so that temperatures above the gold point were to be given by Planck's equation (8.2) and a disappearing filament optical pyrometer. This definition of the higher temperatures continues to the present (see Chapter 4). Note that in practice it is the spectral radiance of the test body (at its temperature) relative to that of a black body at the gold point that defines its temperature on the IPTS according to (4.4).

8.2 Principles of Optical Pyrometry

An optical pyrometer [2], [3] consists basically of an optical system and a power supply. The optical system includes a microscope, a calibrated lamp and a narrow band wave filter, all arranged so that the test body and the standard light source can be viewed simultaneously. The power supply provides an adjustable current to the lamp filament (see Figures 8.1 and 8.2).

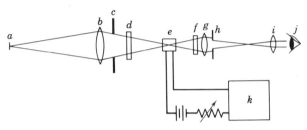

Figure 8.1 Schematic diagram of an optical pyrometer: (*a*) source; (*b*) objective lens; (*c*) objective aperture; (*d*) absorption filter (used for temperatures above 1300°C); (*e*) pyrometer lamp; (*f*) red filter; (*g*) microscope objective lens; (*h*) microscope aperture stop; (*i*) microscope ocular; (*j*) eye; (*k*) current measuring instrument. (After NBS Monograph 41.)

Optical pyrometry is based on the fact that the spectral radiance from an incandescent body is a function of its temperature [4], [5]. For black body radiation, the well-known curves of Planck's equation describe the energy distribution as a function of temperature and wavelength (Figure 8.3). If a non-black body is being viewed, however, its emissivity, which is

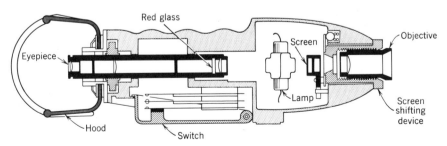

Figure 8.2 Optical pyrometer. (After Leeds and Northrup.)

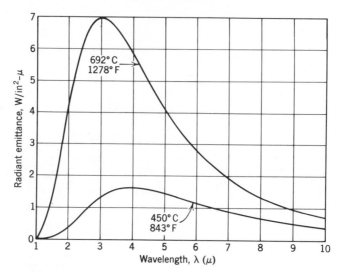

Figure 8.3 Blackbody energy distribution. According to Planck's law, the defining equation is $(c_1/\lambda^5)/[\exp (c_2/\lambda T) - 1]$.

a function of wavelength and temperature, must be taken into consideration (Figure 8.4). In general, to obtain the temperature of a test body, the intensity of its radiation at a *particular wavelength* is compared with that of a standard light source.

Red Filter

The very narrow band of wavelengths required for the comparison just noted is established in part of the use of a red filter in the optical system, and in part by the observer's eye. A red filter exhibits a sharp cutoff at $\lambda = 0.63 \mu$ (where $1 \mu = 1$ micron $= 0.001$ mm). This means that below about 1400°F, the intensity of radiation transmitted by a red filter is too low to give adequate visibility of the test body and the standard filament. On the other hand, a red filter exhibits a high transmission for $\lambda > 0.65 \mu$ (see Figure 8.5), so that it is the diminishing sensitivity of the eye that provides the necessary cutoff at the high end of the band. The particular wavelength that is effective in optical pyrometry is usually taken as 0.655μ.

Brightness Temperature

Several steps are involved in determining temperature by an optical pyrometer [6], [7]. First, the *brightness* of the test body is *matched* against

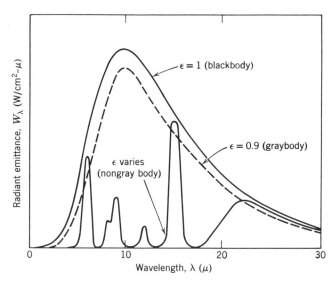

Figure 8.4 Emitted radiation as a function of emissivity for bodies at 300 K. (After King et al. 1963.)

the brightness of the filament of a calibrated lamp at the effective wavelength of 0.655 μ. An optical system (see Figure 8.1) is required such that the filament and the test body can be viewed simultaneously. The match is achieved by adjusting the filament current until the filament image just disappears in the image of the test body (see Figure 8.6). Because the image is nearly monochromatic red, no color difference is seen between the lamp filament and the test body, and thus the filament *seems* to disappear against the background of the target. Of course, matching should be recognized as a null balance procedure. It is the

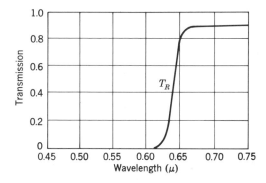

Figure 8.5 Transmission characteristic of a red filter. (After Leeds and Northrup.)

Low   Null   High

Figure 8.6   Schematic response of an optical pyrometer.

optical equivalent of a Wheatstone bridge balance to determine resistance measurements or a potentiometric balance for voltage measurements. Second, the filament current necessary for the brightness match must be measured. For the highest accuracy, this is best achieved by a potentiometric measurement across a fixed resistor in series with the lamp filament. Third, the filament current measurement, obtained at the match condition, must be translated into *brightness temperature* by means of a predetermined calibration relationship (see Section 8.3). This calibration is predicated on the existence of a lamp with a stable, reproducible characteristic with temperature and time. Finally, brightness temperature must be converted to *actual temperature* through applying the emissivity of the test body (see Tables 8.2, 8.3, and 8.4 as well as the examples that follow).

The brightness temperature is defined as that temperature at which a black body would emit the same radiant flux as the test body. This is the temperature as observed with an optical pyrometer. For non-black bodies the brightness temperature is always less than the actual temperature. Thus according to Table 8.2 a black body ($\epsilon = 1$) would appear 1.1 times as bright as carbon ($\epsilon = 0.9$), 2.3 times as bright as tungsten ($\epsilon = 0.43$), and 3.3 times as bright as platinum ($\epsilon = 0.3$), when all are at the same temperature. To say it another way, the actual temperatures of these materials would be 2000°F for the black body, 2016°F for the carbon, 2137°F for the tungsten, and 2198°F for the platinum, when all indicate a brightness temperature of 2000°F (according to Tables 8.2, 8.3, and 8.4).

From Wien's law, (8.1), the relationship between the actual temperature (T) and the brightness temperature (T_B) can be approximated in terms of the pyrometer's effective wavelength of radiation (λ), the second radiation constant (c_2), and the target emissivity (ϵ) by:

$$\frac{1}{T} - \frac{1}{T_B} = \frac{\lambda}{c_2} \ln \epsilon. \qquad (8.3)$$

If the transmission (τ') of the viewing system is not unity, then the effective emissivity (ϵ') should be used in (8.3) in place of the source

Table 8.2 *Spectral Emissivity of Some Materials with Unoxidized Surface at λ = 0.65 μ*[a]

Material	Solid	Liquid
Beryllium	0.61	0.61
Carbon	0.80–0.93	—
Chromium	0.34	0.39
Cobalt	0.36	0.37
Columbium	0.37	0.40
Copper	0.10	0.15
Erbium	0.55	0.38
Gold	0.14	0.22
Iridium	0.30	—
Iron	0.35	0.37
Manganese	0.59	0.59
Molybdenum	0.37	0.40
Nickel	0.36	0.37
Palladium	0.33	0.37
Platinum	0.30	0.38
Rhodium	0.24	0.30
Silver	0.07	0.07
Tantalum	0.49	—
Thorium	0.36	0.40
Titanium	0.63	0.65
Tungsten	0.43	—
Uranium	0.54	0.34
Vanadium	0.35	0.32
Yttrium	0.35	0.35
Zirconium	0.32	0.30
Steel	0.35	0.37
Cast Iron	0.37	0.40
Constantan	0.35	—
Monel	0.37	—
Chromel P (90 Ni-10 Cr)	0.35	—
80 Ni-20 Cr	0.35	—
60 Ni-24 Fe-16 Cr	0.36	—
Alumel (95 Na; Bal. Al, Mn, Si)	0.37	—
90 Pt-10 Rh	0.27	—

[a] From ASME PTC 19.3, 1974.

emittance (ϵ). These quantities are related [8], [9] by

$$\epsilon' = \epsilon \tau' \tag{8.4}$$

Of course, if $\epsilon = 1$ and $\tau' = 1$, the actual temperature will equal the brightness temperature, since, according to (8.3), $\ln 1 = 0$.

Equation 8.3, with (8.4) factored in, is solved graphically in Figure 8.7, and tabularly in Tables 8.3 and 8.4.

Several examples will illustrate the use of these graphs and tables.

Table 8.3   Fahrenheit Corrections to Brightness Temperatures to Obtain Actual Temperatures at $\lambda = 0.655\ \mu^a$

t, °F	1400	1600	1800	2000	2200	2400	2600	2800	3000	3200	3400	3600	3800	4000
T, °R	1860	2060	2260	2460	2660	2860	3060	3260	3460	3660	3860	4060	4260	4460
$\epsilon_\lambda = 0.1$	226	280	342	411	487	571	663	763	872	990	1118	1255	1404	1562
0.2	152	188	229	273	322	376	435	498	566	640	718	803	892	988
0.3	112	138	167	199	234	272	314	359	407	458	513	572	634	700
0.4	84	103	125	149	174	203	233	266	301	339	379	421	466	513
0.5	63	77	93	111	130	151	173	197	223	251	280	311	343	377
0.6	46	57	68	81	95	110	126	143	162	182	202	225	248	272
0.7	32	39	47	56	65	76	87	99	111	125	139	154	170	187
0.8	20	24	29	35	40	47	54	61	69	77	86	95	105	115
0.9	9	11	14	16	19	22	25	28	32	36	40	44	49	53

^{a} From Leeds and Northrup [3].

Table 8.4   Celsius Corrections to Brightness Temperatures to Obtain Actual Temperature at $\lambda = 0.655\ \mu^a$

t, °C	700	800	900	1000	1100	1200	1300	1400	1500	1600	1700	1800	1900	2000	2100	2200
T, K	973	1073	1173	1273	1373	1473	1573	1673	1773	1873	1973	2073	2173	2273	2373	2473
$\epsilon_\lambda = 0.1$	110	136	164	196	231	269	310	355	404	457	514	575	640	710	782	864
0.2	75	91	110	131	153	178	205	233	264	297	333	371	411	453	499	546
0.3	55	67	80	95	112	129	148	169	191	214	239	265	293	323	354	387
0.4	41	50	60	71	83	96	110	125	141	158	177	196	216	238	260	284
0.5	31	38	45	53	62	72	82	93	105	117	131	145	160	175	192	209
0.6	23	27	33	39	45	52	60	68	76	85	95	105	116	127	138	151
0.7	16	19	23	27	31	36	41	47	53	59	65	72	79	87	95	103
0.9	10	12	14	17	19	22	25	29	32	36	40	45	49	54	58	64
0.9	5	6	7	8	9	10	12	13	15	17	19	21	23	25	27	30

^{a} From Leeds and Northrup [3].

Example 1. A target brightness temperature of 1600 K is measured with an optical pyrometer having an effective wavelength of 0.655 μ. At this wavelength the effective emittance of the target is determined to be 0.6. Estimate and check the true target temperature in degrees Celsius and in degrees Fahrenheit.

Graphical Solution. According to Figure 8.7a, at $T_B = 1600$ K, and at $\epsilon' = 0.6$, $\Delta T = T - T_B = 62$°C. Thus $T = T_B + \Delta T = 1662$ K.

Tabular Solutions. According to Table 8.4, at $T_B = 1327$°C, and at $\epsilon' = 0.6$, $\Delta T = 60 \pm \frac{1}{4}(8) = 62$°C, which checks the graphical solution.

According to Table 8.3, at $T_B = \frac{9}{5}(1327) + 32 = 2421$°F, and at $\epsilon' = 0.6$, $\Delta T = 110 \pm \frac{1}{10}(16) = 112$°F. Thus $T = 2421 + 112 = 2533$°F. Or $T = \frac{5}{9}(2533 - 32) = 1389$°C = 1662 K, which checks the graphical solution.

Numerical Solution. For small temperature differences, by Wien's law (8.1)

$$\Delta T \sim \frac{\lambda T_B^2}{c_2} \ln \frac{1}{\epsilon'}. \tag{8.5}$$

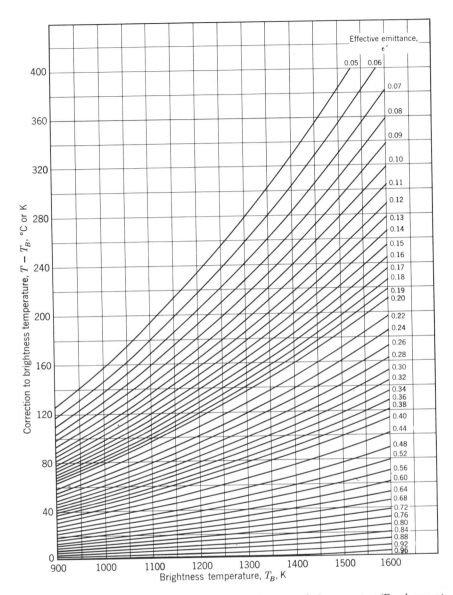

Figure 8.7(a) Brightness temperature correction for an optical pyrometer. (Band pass at 0.653 μ) (After NASA TND-3604.)

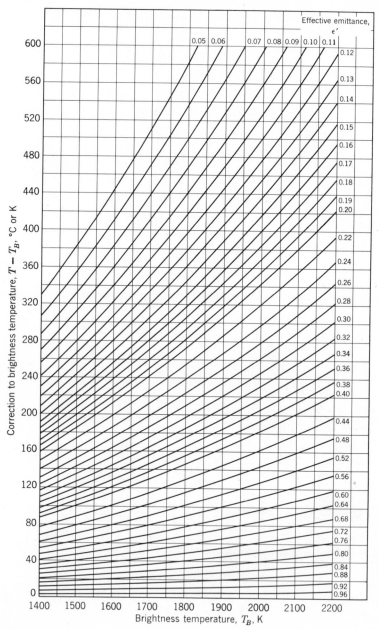

Figure 8.7(b)

But $1\mu = 10^{-6}$ meters $= 10^4$ angstroms, hence

$$\Delta T \sim \frac{0.655 \times 10^{-6} \text{m} \, (1600)^2 \text{K}^2}{0.014388 \text{ m} \cdot \text{K}} \ln\left(\frac{1}{0.6}\right) = 60°\text{C}$$

which once more provides a close check to the graphical and tabular solutions.

Example 2. The clean surface of liquid nickel, when viewed through an optical pyrometer having the conventional effective wavelength of 0.655 μ, yields a brightness temperature of 2600°F. Estimate and check the true nickel temperature.

Tabular Solutions. According to Table 8.3, at $T_B = 2600°\text{F}$, and with an effective emissivity estimated to be 0.37, according to Table 8.1, $\Delta T = 233 \pm 0.3(81) = 257°\text{F}$. Thus $T = T_B + \Delta T = 2857°\text{F}$.

According to Table 8.4, at $T_B = \frac{5}{9}(2600 - 32) = 1427°\text{C}$, and at $\epsilon' = 0.37$, by double interpolation $\Delta T = 143°\text{C}$. Thus $T = 1427 + 143 = 1570°\text{C} = 2858°\text{F}$, which is consistent with the Fahrenheit solution.

Graphical Solution. According to Figure 8.7b, at $T_B = 1427 + 273 = 1700$ K, and at $\epsilon' = 0.37$, $\Delta T = 142°\text{C}$. Thus $T = 1427 + 142 = 1569°\text{C} = 2856°\text{F}$, which checks the tabular solutions.

Numerical Solution. According to (8.5),

$$\Delta T \sim \frac{0.655 \times 10^{-6} \times 1700^2}{0.014388} \ln \frac{1}{0.37} = 131°\text{C}$$

which provides a fair check to the graphical and tubular solutions. Since ΔT is large in this example, the approximation of (8.3) is not as reliable as for smaller ΔTs.

Brightness temperature thus depends on the sensitivity of the eye to differences in brightness, and on a knowledge of the mean effective wavelength of the radiation being viewed. It does not depend on the distance between the test body and the optical pyrometer, however. We must be sure that the radiation observed is that being emitted by the test body rather than reflected radiation, since there is no relationship between the temperature of a surface and the radiation it reflects. Also, smoke or fumes between the optical pyrometer and the target must be avoided as must dust or other deposits on the lenses, screens, and lamp windows.

Pyrometer Lamp

Of all the elements in the optical pyrometer, the pyrometer lamp is the most important, since it provides the reference standard for all radiance measurements. The most stable lamps consist of a pure tungsten filament enclosed in an evacuated glass tube. The vacuum is necessary to minimize convection and conduction heat transfer effects. The tungsten is always annealed before calibration and, once calibrated, can be used typically for 200 hours before recalibration is required. Although theoretically the optical pyrometer has no upper temperature limit, practically, for long-term stability, the lamp filament cannot be operated above a certain current or brightness. This limit corresponds approximately to 1350°C.

Absorbing Glass Filter

Glass filters, which absorb some of the radiation being viewed, are used when temperatures higher than 1350°C are being measured in order to reduce the apparent brightness of the test body to values that the filament can be made to match. Thus it is not necessary to operate the standard lamp filament at as high a temperature as would normally be called for by the brightness of the test body. Such practice adds to the stability and life of the filament. The absorbing glass filters are inserted between the objective lens and the pyrometer lamp as shown in Figures 8.1 and 8.2. Thus a black body at a temperature above 1350°C appears the same through the absorbing glass as another black body would appear in the same pyrometer without a screen at a temperature below 1350°C. In addition, use of the proper glass screen allows closer color matching between the pyrometer lamp and the attenuated source, and this leads to more precise brightness matching.

Black Body

The emissivity of most materials is such a variable quantity and so strongly dependent on surface conditions [10] that it becomes almost mandatory to sight on the test body in a black body furnace. In lieu of laboratory-type testing in the highly favorable conditions of a black body furnace, one can often approximate a black body in field-type applications; for example, if the surface temperature of an incandescent material in a test rig is required, a small hole can be drilled directly into the surface for sighting purposes. The hole depth should be about five times its diameter, as previously mentioned. Regardless of the emissivity of the hole walls, the multiple reflections inside the cavity makes the hole radiate approximately as a black body. Temperatures based on such techniques are almost certain to be more valid than temperatures based on flat surface sightings, as corrected by estimated surface emissivities. In any case, whenever black body conditions prevail, if the brightness match is achieved, and if the pyrometer is properly calibrated, then $T_{observed} = T_{brightness} = T_{IPTS}$.

Hot Gas Measurements

The radiation characteristics of hot gases are not like those of hot solids in that gases do not emit a continuous spectrum of radiation. Instead, the emissivity of a gas exhibits a rapid variation with wavelength. That is, gases radiate or emit strongly only at certain characteristic wavelengths, which correspond to absorption lines (see Figure 8.8). It is clear that principles of radiation pyrometry entirely different from those just described are called for in the temperature measurement of gases, and the appropriate literature should be consulted for further information [11].

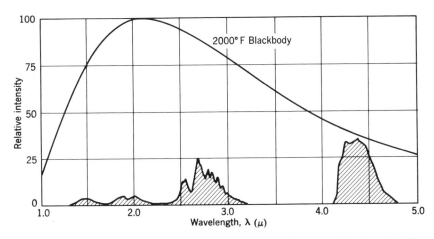

Figure 8.8   Combustion gas radiation at 2000°F. (After Tourin and Tandler, 1959.)

8.3   Calibration

At least three types of calibration are to be distinguished in optical pyrometry.

Primary

This type of calibration is done only at the National Bureau of Standards [2], [12]. The filament current required to balance the standard lamp brightness against pure gold held at the gold point temperature in a black body furnace constitutes the basic point in the calibration. Higher temperature points are determined by a complex method that is detailed in NBS Monograph 41, and is based on the use of tungsten strip lamps and sectored disks. Representative uncertainties in achieving the IPTS by optical pyrometry are: ±4°C at the gold point, ±6°C at 2000°C, and 40°C at ±4000°C.

Secondary

In this type of calibration [3], [4] the output of a primary pyrometer (i.e., one calibrated at the NBS) is compared with that of a secondary pyrometer (i.e., one to be calibrated) when the pyrometers are sighted alternately on a tungsten strip lamp operated at different brightnesses. Such lamps are highly reproducible sources of radiant energy and can be calibrated with respect to brightness temperatures from 800 to 2300°C, with accuracies only slightly less than would be obtained according to the IPTS. Note that in this method the source need not be a black body so long as the pyrometers are optically similar.

Industrial

Two secondary optical pyrometers can be intercompared periodically by sighting them alternately on the same source. Note that here the source need not be a black body and the comparison pyrometers need not have primary calibrations. The method is most useful for indicating the stability of the pyrometers, and thus can indicate the need for a more basic calibration.

8.4   The Two-Color Pyrometer

The accuracy of a temperature determination by the single-color optical pyrometer just discussed is based on black body furnace sightings or on known emissivities. A two-color pyrometer, on the other hand, is used in an attempt to avoid the need for emissivity corrections. The principle of operation is that energy radiated at one color increases with temperature at a different rate from that at another color [13], [14]. The ratio of radiances at two different effective wavelengths is used to deduce the temperature. The two-color temperature will equal the actual temperature whenever the emissivity at the two wavelengths is the same. Unfortunately this is seldom true. All that can be said is that when the emissivity does not change rapidly with wavelength, the two-color temperature may be closer to the actual temperature than the single-color brightness temperature. If the emissivity change with wavelength is large, however, the converse is true. Kostkowski [13] of the NBS indicates that, in any case, the two-color pyrometer is less precise than the single-color optical pyrometer. Typically, when both were sighted on a black body, the optical came within 2°C of the known temperature, whereas the two-color pyrometer was off by 30°C.

8.5 Automatic Optical Pyrometer

Since 1956 the automatic optical pyrometer has dominated the field of accurate high-temperature measurement. In such an instrument, the pyrometer lamp current is adjusted automatically and continuously by means of a photoelectric detector that views alternately the target and the pyrometer lamp. Thus a brightness-temperature balance is achieved by the automatic optical pyrometer which, in principle, has considerably greater sensitivity and precision than a manually adjusted pyrometer that depends on the subjective judgement of the human eye (see Figure 8.9) [3], [9], [15].

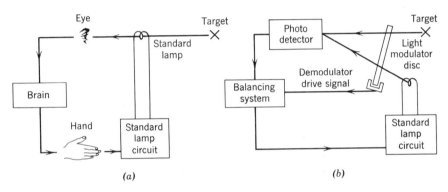

Figure 8.9 Operational diagrams of (*a*) manually-adjusted optical pyrometer, and (*b*) automatic optical pyrometer (after Leeds and Northrup).

Accuracies obtainable with commercial automatic optical pyrometers are on the order of ±7°F for 1500–2250°F, and of ±12°F from 2250–3200°F.

The operation of an automatic optical pyrometer can best be understood by referring to Figure 8.10. A revolving mirror alternately scans the target and the standard lamp filament at high speed. The optical system thus projects alternately an image of the target and of the lamp filament onto the photo multiplier tube. One technique for determining brightness temperature is to adjust the reference lamp current until a null intensity is sensed in the photo multiplier output current. Another is to maintain the constant lamp current and determine brightness temperature by sensing an off-balance meter reading. In either case, a direct reading of brightness temperature is forthcoming, and hence the designation *automatic optical pyrometer* [8].

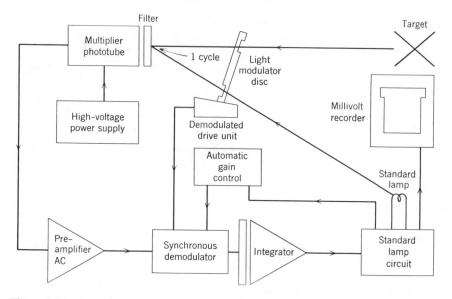

Figure 8.10 Operational diagram of an automatic optical pyrometer (after Leeds and Northrup).

8.6   References

[1] M. Jakob, *Heat Transfer,* Vol. 1, Wiley, New York, 1949, p. 23.

[2] H. J. Kostkowski and R. D. Lee, "Theory and Methods of Optical Pyrometry," *NBS Monograph 41,* March 1962.

[3] P. H. Dike, W. T. Gray, and F. K. Schroyer, "Optical Pyrometry," *L and N Tech. Pub. A1.4000,* 1966.

[4] L. J. Neuringer, "Infrared Fundamentals and Techniques," *Electro-Technical.,* March 1960, p. 101.

[5] J. King, T. Limperis, J. Morgan, F. Polcyn, and W. Wolfe, "Infrared," *Int. Science Tech.,* April 1963, p. 26.

[6] W. T. Gray, "Precision and Accuracy in Radiation Pyrometry," Leeds and Northrup paper presented at Symposium on Precision Electrical Measurements, June 1963.

[7] "Optical Pyrometers," Chapter 7 of *Temperature Measurement,* Instruments and Apparatus Supplement to ASME Performance Test Codes, PTC 19.3, 1974.

[8] J. R. Branstetter, "Some Practical Aspects of Surface Temperature Measurement By Optical and Ratio Pyrameters, "NASA TN D-3604, September 1966.

[9] G. D. Nutter, "Radiation Thermometry, Part 1, "*Mechanical Engineering,* June 1972, p. 16.

[10] G. D. Nutter, "Radiation Thermometry, Part 2," *Mechanical Engineering,* July 1972, p. 12.

[11] R. H. Tourin and W. S. Tandler, "Monochromatic Radiation Pyrometry for Gas Temperature Measurement," Paper presented at 11th Annual Symposium, ISA, April 1959.

[12] "Photoelectric Pyrometer," Staff article, *Instr. Control Systems,* **35,** July 1962, p. 178.

[13] H. J. Kostkowski, "The Accuracy and Precision of Measuring Temperatures Above 100 K," *Proc. Int. Symposium on High Temperature Technology,* McGraw-Hill, October 1959, p. 33.

[14] H. J. Hoge, "Temperature Measurement in Engineering," in *Temperature, Its Measurement and Control in Science and Industry,* Reinhold, New York, Vol. II, 1955, p. 287.

[15] W. T. Gray, "Beyond the Gold Point," *Leeds and Northup Technical Journal,* Spring 1969.

Nomenclature

Roman

c constant
exp base of natural logarithm e, raised to exponent in ( )
J intensity of radiation
T absolute temperature

Greek

ϵ emissivity
λ wavelength
μ micron

Subscripts

b black body
λ particular wavelength

Problems

1. Find the true temperature of liquid gold when the temperature as indicated by an optical pyrometer with a standard red filter is 2000°F.
   Ans. $T = 2258$°F.

2. Estimate by numerical means the error in the platinum point if the brightness temperature were assumed to be correct.
   Ans. $\Delta T \sim 200$°C.

3. The temperature indicated by an optical pyrometer when sighted on an ingot of steel was 1000°C. Estimate its true temperature.
   Ans. $T = 1083$°C.

4. An optical pyrometer, when sighted on a cavity in a furnace wall,

reads 1200°C. When sighted on the furnace wall adjacent to the cavity, it reads 1100°C. What is the effective emissivity of the furnace wall?

Ans. $\epsilon' = 0.3375$.

5. What are the temperature errors introduced by using Wien's law in place of Planck's law at $\lambda = 0.655 \ \mu$ and $T = 2000, \ 3000, \ 4000$ K?

    Ans. $\Delta T = 0.0031, \ 0.271, \ 3$ kelvins.

Chapter 9

CALIBRATION OF TEMPERATURE SENSORS

"... I recommended also the calibration of the couples in terms of the fixed points of boiling or fusion of certain pure substances..."

Henri LeChatelier (1885)

9.1 General Remarks

As seen in Chapter 4, temperature is defined by certain standards. We want to transfer this information to some more rugged, faster, cheaper, smaller device that is also temperature sensitive, and that may be carried from the standards laboratory to a job site.

Calibration consists of determining the indication or output of a temperature sensor with respect to that of a standard at a sufficient number of known temperatures so that, with acceptable means of interpolation, the indication or output of the sensor will be known over the entire temperature range of use [1].

Calibration problem areas are immediately apparent. There must be available: (1) means for measuring output of the temperature sensor; (2) a satisfactory temperature standard; (3) controlled temperature environments; and (4) a scheme for interpolating between calibration points.

As a result of proper attention to application details, the means for measuring indications or outputs of all common temperature sensors within acceptable uncertainties are available (see, for example, Chapters 5, 6, 7).

A practical, realizable temperature standard has been described in the text of the IPTS (see Chapter 4) wherein temperature is defined by certain fixed points, certain standard instruments, and certain standard interpolating equations.

The availability and use of controlled temperature environments wherein sensor outputs are determined, and the bases of several schemes that are useful for interpolating between the calibration points, are discussed in the next sections.

9.2 Controlled Temperature Environments

Calibration environments can be divided into two classes depending on the method of determining the temperature of the test sensor. In one case, the sensor is exposed to a fixed-point environment that, under certain prescribed conditions, naturally exhibits a state of quasi-thermal equilibrium whose temperature is established numerically by the IPTS without recourse to a temperature standard. In the other case, the test sensor and a temperature standard are exposed simultaneously to a controlled temperature environment whose temperature is established by the standard instrument.

Fixed-point Baths

Three general types of fixed-point baths are commercially available. These are the liquid-solid (freezing-point), the liquid-vapor (boiling-point), and the solid-liquid-vapor (triple-point) baths. In these fixed-point baths, use is made of the reproducible temperature-time plateau that exists and signifies the coexistence of two or more phases of a substance. Provided the sensor is not contaminated by the fixed-point material, that sufficient immersion exists, that the fixed-point sample is pure and at a uniform temperature, the temperature assigned to the bath (and hence to the sensor) is reliable.

Triple Point of Water

Of the triple-point baths, only that of water is of general interest. This bath provides the primary defining point on IPTS-68, and has been described in great detail in the various texts defining the IPTS (see Chapter 4). A pressure-temperature diagram is given in Figure 9.1a, and indicates that the three phases of water can coexist only at a pressure of 4.58 mm Hg. A cross-sectional diagram of the cell is given in Figure 9.1b. Essentially, it consists of a glass cylinder that has been evacuated at the required low pressure, charged with high-purity, gas-free water, and hermetically sealed.

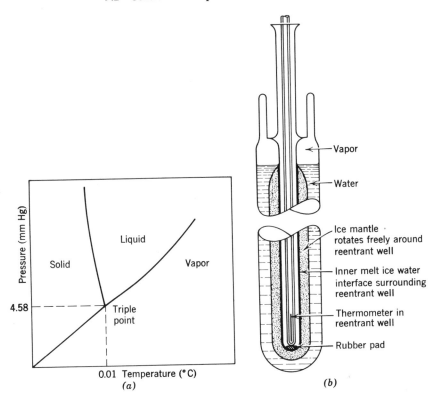

Figure 9.1 Triple-point realization. (*a*) Phase diagram for H_2O. (*b*) Inner container showing triple-point condition. (From Trans-Sonics, Bulletin 130.)

The construction and preparation of the triple-point cell are described in the literature [2]–[7]. The following is a brief review. The cell is first chilled in an ice bath. The reentrant tube of the cell serves several purposes. It is used as a thermometer well, and it is used to set up the triple point. Freezing of a portion of the sealed-in water is accomplished by pouring powdered dry ice (i.e., CO_2) into the reentrant tube. When the mantle of ice, frozen around the reentrant thermometer well, reaches a thickness of about 5 mm, the dry ice is removed. The cell is again placed in an ice bath. A rod, initially at room temperature, is inserted in the reentrant well, and this melts some of the ice mantle so that a thin layer of very pure water forms adjacent to the well wall. When the ice mantle is free from the reentrant tube, the triple point is established. To assure good thermal contact between the three-phase water and the thermometer, the well is filled with alcohol or mineral oil. It has been reported [6] that a properly aged ice mantle may be kept for several months with no appreciable drift in the triple point temperature.

Several triple-point baths are commercially available, but such baths are not normally used as routine calibration environments because of the time and care required in their preparation and use. For standards work they are, of course, a necessity.

Freezing Point of Water

The importance of the freezing point of water (ice bath) cannot be overemphasized. The ice bath is used extensively as the reference junction environment for most thermocouple systems, and as a calibration reference temperature for all other temperature sensing systems. The thermocouple reference junction must be maintained at a constant and known temperature, and this temperature must be stated as a necessary part of the calibration results. Caldwell of the NBS [8] presents useful curves on the effects of wire size, thermal conduction, and immersion depths on the reference junction temperatures of various chromel-alumel thermocouples in ice baths. He indicates that the most potent variable is the size of the copper wire used to connect the reference junctions to the measuring instrument. He concludes that uncertainties in the reference junction temperature of less than 0.1°F are assured by establishing at least 6 in. immersion in the ice-water mixture when using copper wires no larger than 20 gauge.

All freezing-point baths are virtually independent of barometric pressure variations. For example, the ice-point temperature is affected less than 0.001°F by atmospheric pressure variations from 28.5 to 31 in. Hg. The ice bath has already been shown in Figure 7.12.

Use of tap water in place of distilled water should not introduce errors greater than 0.01°C. It is important, however, that excess water be removed periodically and more ice added so that the reference junctions are never below the ice-water mixture. Such water, at the bottom of the ice bath, may be as much as 4°C above the ice point.

Another source of error when using an ice bath with thermocouples is the galvanic action that may be set up when water comes in contact with the thermocouple wires [8], [9]. Such galvanic cells may introduce voltages large enough to affect the thermocouple output. The use of insulated wires should minimize this error.

Freezing Point of Metals

Other freezing-point baths that are relatively simple in construction detail are commercially available (see Figure 9.2), and may be used by

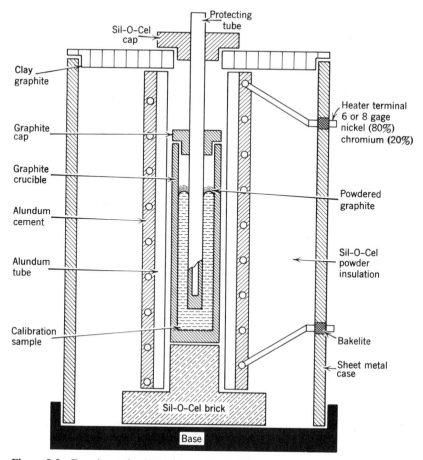

Figure 9.2   Freezing-point bath. (According to NBS.)

uninitiated personnel without great difficulty. Cooling curves for all freezing-point baths are characterized by a plateau of approximately zero slope in the temperature-time plot [10], [11]. Professor Joseph Black, in 1790, gave the first description [12] of this phenomenon, "... when we freeze a liquid, a very large quantity of heat comes out of it, while it is assuming a solid form, the loss of which heat is not to be perceived by the common manner of using the thermometer ..."

The tin and zinc baths are common examples of freezing-point baths used in thermometry. They provide additional fixed points on IPTS-68. Each provides an equilibrium state that exists between the liquid and solid phases of a metal. The freezing temperature is essentially independent of atmospheric pressure variations, is highly reproducible, and the useful temperature-time plateau can be made to persist for an extended period of time. A cross-sectional view of a commercially available freezing-point bath is shown in Figure 9.3. A graphite crucible is charged with a high-purity metal sample and, in use, dry nitrogen is introduced to retard oxidation of the metal sample.

Detailed operating procedures are given in the literature [7], [10]–[13]; however, a brief review of the tin and zinc baths is included here for completeness.

Zinc. The metal sample is first completely melted to about 10°C above the liquidus point. The sample is then cooled until nucleation begins. This is indicated by a gradual rise in temperature. At this point, heat is extracted from the sample by use of a cooling rod. The quick transfer of heat causes a thin mantle of solid metal to form on the graphite crucible and releases sufficient latent heat to raise the metal to its liquidus temperature. The temperature now remains essentially constant for an extended period of time. A typical temperature-time plot of a zinc plateau is given in Figure 9.4.

Tin. The operation of the tin bath is similar to that of the zinc bath except that tin tends to supercool. The bath must be raised out of the furnace when the metal approaches the freezing point and allowed to nucleate outside. Nucleation is indicated by a gradual rise in temperature. At this point, the cell is lowered back into the furnace where it will reach its liquidus point. Typical temperature-time plots of the tin plateau are given in Figure 9.5. These curves further indicate that freezing does not occur at a fixed temperature under all conditions, and that the plateau varies in its time extent. Such variables as purity of material, initial temperature (above melting) from which cooling begins, first heat versus reheat, immersion of test sensor, induced versus normal cooling, and so on, all influence the observed freezing-point temperature, and the duration of the plateau. For calibration work, employing freezing-point baths, accuracies ranging from 0.1 to 0.5°C are obtainable. To achieve such accuracies, the following conditions must be met. The thermometer must be protected from contamination by the freezing-point sample. The thermometer must be immersed in the sample sufficiently far so as to minimize conduction heat transfer errors. The freezing-point sample must be pure, and the sample temperature must be essentially uniform throughout the freezing [14], [15].

Note, as already implied, that the time for taking observations is limited by the period of freezing after which the material must be melted again before further observations are taken [1].

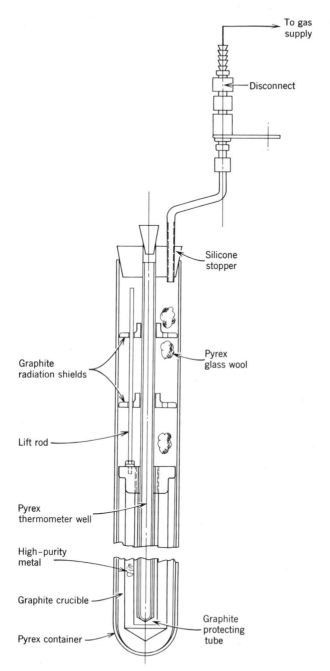

To gas
supply

Disconnect

Silicone
stopper

Pyrex
glass wool

Graphite
radiation shields

Lift rod

Pyrex
thermometer well

High-purity
metal

Graphite crucible

Pyrex container

Graphite
protecting
tube

Figure 9.3 Cross-sectional view of a typical freezing point bath (according to Leeds and Northrup)

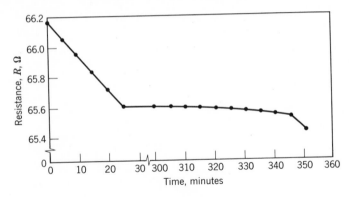

Figure 9.4 Nominal temperature-time plateau, zinc.

Boiling Point of Water

The boiling point of water (steam bath or hypsometer) also provides one of the primary fixed points on the IPTS. However, as is true for all boiling-point baths, the steam-point temperature is greatly affected by barometric pressure variations (see Table 4.1). Compared to the freezing-point baths, these baths are more elaborate in construction details (see, for example, the Leeds and Northrup Hypsometer in Figure 9.6) and, for inexperienced personnel, more difficult to use. The routine determination of the steam point is one thing, whereas the precise determination of the temperature of equilibrium between liquid sulfur and its vapor presents an even greater challenge; for example, Evans at the NBS [2] notes that it is now usual procedure to boil the sulfur actively for a period of ten days before making measurements. Note that, unlike the freezing-point baths, the boiling-point baths offer a continuous plateau with no limit in the time available for observations, since the material can be boiled continuously.

Controlled Temperature Baths

Up to 1000°F calibrations usually are made in electrically heated, temperature-controlled, stirred liquid baths wherein test sensors and standard thermometers are placed and their resulting outputs compared. In this method of calibration, by comparison [16], the validity of the calibration points depends on how closely the test sensor and the standard thermometer are brought to the same temperature. In liquid baths this is not usually a problem, since gradients in such environments are extremely small (under 0.1°F) at given levels of immersion [17]. Many such baths

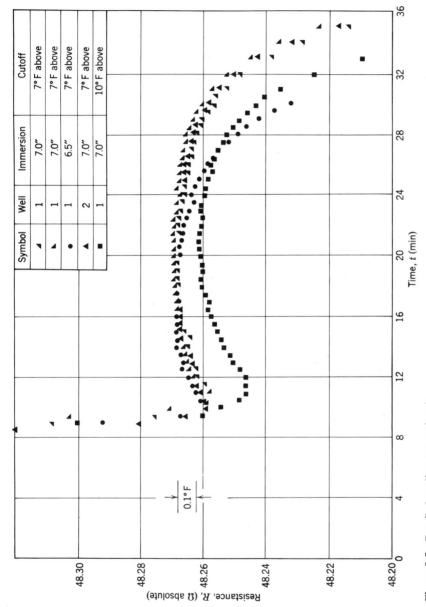

The following data table appears within the figure:

Symbol	Well	Immersion	Cutoff
▲	1	7.0″	7° F above
▲	1	7.0″	7° F above
●	1	6.5″	7° F above
▲	2	7.0″	7° F above
■	1	7.0″	10° F above

Figure 9.5 Detailed cooling curves for tin.

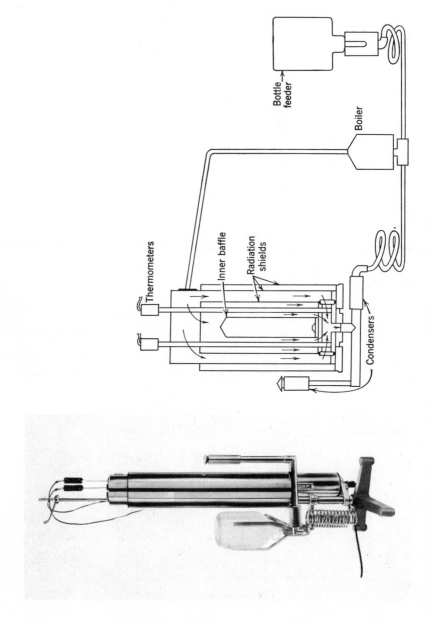

Figure 9.6 A hypsometer designed by Leeds and Northrup Co., Philadelphia, Pa.

are on the order of 18 in. deep so that immersion depths can be quite adequate to minimize conduction effects. Water can be used up to its boiling point as a satisfactory calibration liquid. Certain nonconducting oils are useful below their flash points. (Dow-Corning 550 Synthetic Oil with a flash point of about 580°F is a good example). Molten salts are frequently used between 500 and 1000°F as calibration environments. (Lavite, a heat treating and tempering salt that melts at about 400°F, is one example.) Of the liquid metals, tin, which melts at 450°F, is the one most commonly used as the calibration medium in controlled temperature baths. All of these baths are available commercially with proportional controllers that anticipate temperature changes and activate electrical heaters accordingly to provide uniform temperatures within ±0.1°F over long periods of time.

Electrically heated, temperature-controlled, gas-environment furnaces are used almost exclusively for calibration work in the temperature range of 1000 to 2000°F. For less precise calibrations they are also used below these temperatures in the realm of the stirred-liquid baths. In such gas furnaces, success in bringing test sensors and standard to the same temperature is not so easily achieved as in the liquid baths. The NBS recommends forming a common measuring junction between one or more test thermocouples and the standard thermocouple when calibrating these sensors in a gas furnace. In such cases, all wires must be insulated from each other except at the measuring junction. If this procedure is not applicable, feasible, or convenient, the various test sensors can be arranged around the standard thermometer and secured in place by wire wrapped around the assembly. Sometimes as an alternative, the sensors and standard are inserted into holes drilled in a large metal block that is placed in the furnace and is intended to serve as a temperature equalizer [18]. However, gases have such low specific heat capacities compared with liquids, and circulation of gases is usually by natural convection currents only; thus it is not uncommon to find large temperature gradients in the commercial gas furnaces. Great care must be exercised in obtaining calibration points in such furnaces.

A special class of controlled temperature baths has recently become commercially available. Such baths employ a mass of aluminum oxide particles that are confined in a container and can be "fluidized" by blowing air up through the fine solid particles (see Figure 9.7). These baths offer the advantages of a "liquid" over a gas. In particular, the fluidized solid is characterized by a high specific heat capacity and by low temperature gradients. Such baths are often safer to use than oil baths which can flash or salt baths which corrode and can explode upon overheating [19].

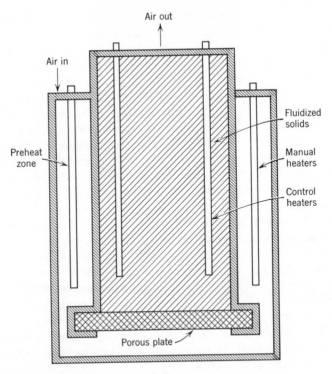

Figure 9.7 Schematic of a fluidized-solid bath (after Callahan [19]).

A specially designed air furnace is used at the NBS for the routine calibration of thermocouples up to 2282°F. The heating element of the furnace consists of an 80% nickel–20% chromium tube clamped between two water-cooled terminals in a horizontal position. The tube, with a $\frac{13}{16}$ in. inside diameter, $1\frac{5}{16}$ in. outside diameter, is 24 in. long, and is heated electrically; the tube itself serves as the heating element or resistor. The large current necessary to heat the tube is obtained from and controlled by transformers. A large cylindrical shield of sheet metal is mounted around the heating tube to reduce the radiation loss. In order to minimize time lag (see Chapter 13), no thermal insulation is used between the heating tube and the radiation shield. The middle part of this furnace is said to be at a practically uniform temperature for about 18 in. This furnace can be heated to 2192°F in about 10 min with 12 kW of power and, with power off, will cool from this temperature to 572°F in about the same time.

Many of these methods of calibration are treated in detail in the various professional society references (see for examples, [15] and [20]).

9.3 Interpolation Methods

It is impossible in a practical sense to obtain enough calibration points to define a continuous relationship between temperature and the output of a test sensor. Thus a suitable interpolation method must be used between the calibration points. Since the two independent measurements upon which each calibration point is based are not without their experimental uncertainties, a weighting scheme also must be employed to minimize effects of poor experimental points. In this section we consider several interpolation methods, as applied to the calibration data of a thermocouple. Although these examples apply to a specific temperature sensor, they serve to illustrate general methods.

An experimental thermocouple calibration consists of a series of voltage measurements determined at a finite number of known temperatures. If a test thermocouple were compared with a standard temperature instrument at 100 temperatures within a 10°F range, there would be little need for interpolation between the calibration points. However, if four to ten calibration points are all that can be afforded in a given range of interest, what is needed to characterize an individual thermocouple is a continuous relation between voltage and temperature. Efforts to obtain such a continuous relation appear thwarted from the start because of the small number of discrete calibration points available. However, interpolation between the calibration points is possible, since the emf changes only slowly and smoothly with temperature.

One can present raw calibration data directly in terms of temperature T and voltage E_{couple} on a scale so chosen that the information appears well represented by a single curve (see, for example, Figure 9.8) or by a simple mathematical equation. For example, for the highest accuracy in the range 630 to 1064°C with the Type S thermocouple, this method is that prescribed in the IPTS. An equation is used of the form $E = a + bT + cT^2$, where a, b, and c are constants determined by calibration at the freezing points of gold, silver, and antimony (see Chapter 4). By calibrating the thermocouple also at the freezing point of zinc and using an equation of the form $E = a' + b'T + c'T^2 + d'T^3$, the temperature range can be extended down to 400°C without introducing an uncertainty of more than 0.1°C in the range 630 to 1064°C. However, in general, this practice of directly representing thermocouple characteristics does not always yield results within the required limits of uncertainty.

Often, differences between observed values and values obtained from standard reference tables are used in calibration work [21]. The reference tables and the mathematical means for generating them are presented in Chapter 7, Section 8. The data of Figure 9.8 are replotted in Figure 9.9 in terms of differences from the proper reference table.

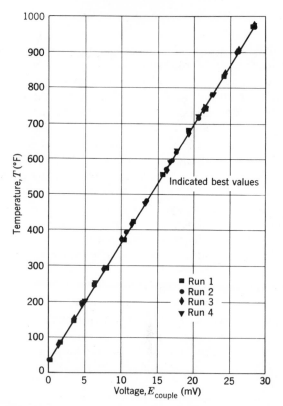

Figure 9.8 Calibration data presented on a scale chosen so that the information appears on a single curve.

Difference plots are in order whenever the range of temperature is so great that the desired readability is precluded by the use of a T−emf plot or whenever it is impossible to find a single satisfactory empirical interpolating equation for the whole T−emf range. Such difference plots allow magnification of the calibration data and can be fit by single empirical equations of low degree.

The maximum spread between points taken at the same level (replication) but obtained in random order with respect to time and level (randomization) is taken as the uncertainty envelope. This information, taken from Figure 9.9, is plotted in Figure 9.10, and constitutes a vital bit of information about the particular thermocouple and the calibration system. In lieu of an experimental determination of the uncertainty, one must rely on judgment or on the current literature for this information.

Usually, only a single set of calibration points is available. Typical

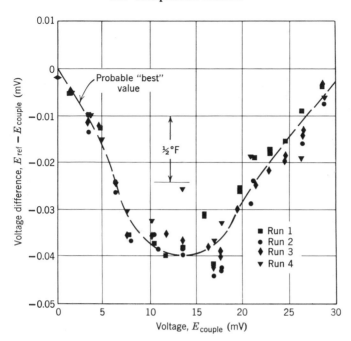

Figure 9.9 The data in Figure 9.8 replotted in terms of differences from the proper reference table.

points would be those taken from one run shown in Figures 9.8 or 9.9. These are shown in Figure 9.11 along with five of the many possible methods for representing the thermocouple difference characteristic. Although at first it appears that the most probable relation characterizing a given thermocouple is sensibly indeterminate from a single set of calibration points, it is an important fact that all experimental points must be contained within the uncertainty envelope when the uncertainty envelope is centered on the most probable interpolation equation. (If the total uncertainty were determined, and this could only result from a very large sample experiment, it would follow from an assumed normal error distribution that no experimental point could depart from the thermocouple characteristic (best value) by more than $\pm\frac{1}{2}$ the uncertainty envelope. For small-sample experiments, however, such as are usually the case, one should expect that a few poor calibration points might fail to conform with the above assertion.)

Another important fact is that overall experimental uncertainties will be minimized by using the least squares technique. Here the most probable values for a given set of data are obtained by weighing the

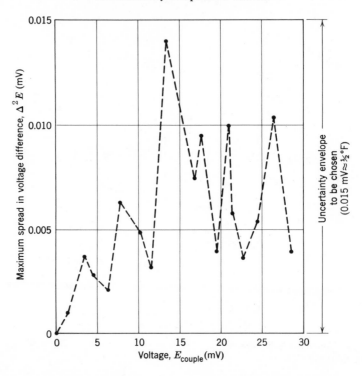

Figure 9.10   Typical determination of uncertainty envelope. (From data in Figure 9.9.)

relative value of each experimental point according to the laws of probability and then passing a precisely determined equation of prescribed degree through the data (see example in Section 9.4).

Making use of the least squares technique along with the principle that all experimental points must be contained within the uncertainty envelope, we can begin a systematic search for the most probable interpolation equation. First a least squares equation of the first degree is passed through the experimental data. A check is then made to ascertain whether all experimental points are contained within the uncertainty envelope which is centered on the linear interpolation equation (see Figure 9.12). We proceed, according to the results of the foregoing check, to the next highest degree equation, stopping at the lowest degree least squares equation that satisfies the uncertainty requirements. For the example given here, a third-degree interpolation equation was required (see Figure 9.13). By obtaining voltage differences from the least squares fit of any set of calibration points, the uncertainty in the thermocouple difference characteristic will be within one-half the uncertainty envelope.

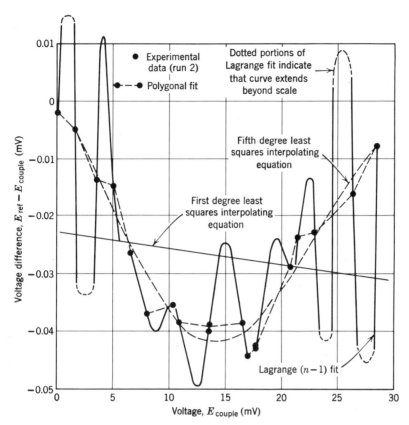

Figure 9.11 Various possible empirical representations of the thermocouple characteristic (run 2).

Generally, the form of the uncertainty envelope and the degree of the most probable least squares interpolation equation are strongly dependent on the amount of calibration data available and on the temperature range under consideration. It is recommended that the number of distinct calibration points available should be at least 2 (degree + 1). The factor 2 is arrived at from numerical analysis reasoning. A distinct calibration point is arbitrarily defined as one that is separated temperaturewise from all other points in the set by as much as $\frac{1}{10}$ the difference in temperature between the maximum and minimum temperatures of the particular run. The choice of $\frac{1}{10}$ presupposes a maximum practical degree of four for the least squares interpolation equation, in keeping with the low degree requirement of numerical analysis. Indeed, if the data cannot be fit by a fourth-degree interpolation equation, one should increase the uncertainty

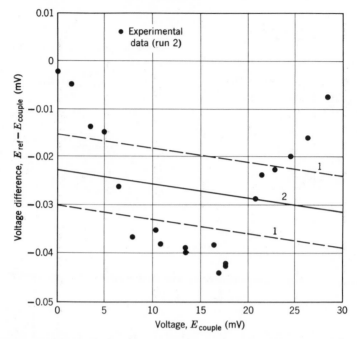

Figure 9.12 Checking to ascertain whether all experimental points are contained within the uncertainty envelope. (1) Uncertainty envelope does not include all experimental data; therefore most probable interpolating equation has not been obtained. (2) First-degree least squares interpolating equation.

interval and start the fitting procedure again. Difference curves often offer greater precision in thermocouple calibration work, for a given number of calibration points, than can be obtained from the use of the raw calibration data alone.

Although this last scheme has been particularized for thermocouple calibration data, the same four premises concerning (a) difference plot with respect to reference table; (b) determination and application of uncertainty envelope; (c) least squares technique; and (d) number of distinct points at least 2 (degree + 1) may be applied successfully to many forms of experimental data, over and above the calibration of temperature sensors, as, for example, the calibration of pressure and flow sensors.

9.4 Obtaining Temperature from Calibration Data

It is one thing to obtain a calibration of a temperature sensor, as just described, and quite another to obtain temperatures from thermometer

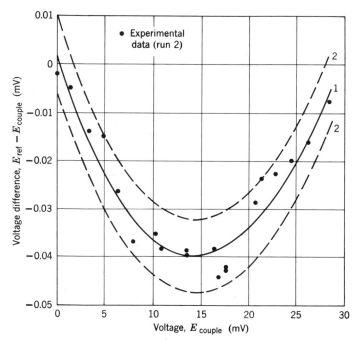

Figure 9.13 A third-degree interpolation equation fits the data of the example. (1) Third-degree least squares interpolating equation. (2) Uncertainty envelope includes all experimental data; therefore the third-degree interpolating equation best represents this calibration run.

outputs. This important distinction between calibration and interpretation will be amplified in terms of a thermocouple. In this case, the question becomes: what is the temperature corresponding to an emf generated by a thermocouple? Several methods are available [22].

Method 1. In Chapter 7 (Tables 7.4 and 7.5), NBS equations are given to represent the seven standard thermocouple types. Hence one method that immediately suggests itself is based on the use of an equation of the form $E = f(T)$. The NBS equations represent the nominal emf temperature characteristics of thermocouples.

A difficulty is encountered, since T is desired rather than E, but this problem can be solved by *iteration*. That is, a guess of T is made and E is forthcoming. This process is repeated until a T is guessed that yields E (as close as is desired to the measured E). With digital computers, this is a rapid, routine method of solution. However, according to the thermocouple standards themselves [23], [24], the emf–temperature relationship of an individual thermocouple may differ from that given by the nominal reference table values by as much as 7.2°F (4°C) at the 1000°F (538°C) level. This recognized difference, which depends on the type of materials involved and on temperature level, is tabulated for various material combinations in a limit of error Table 9.1. Thus we see that use of the NBS

Table 9.1   Limits of Error for Thermocouples (after ANSI MC96.1–1975)

		Limits of error	
Thermocouple Type	Temperature Range, °C	Standard (whichever is greater)	Special (whichever is greater)
T	0–350	±1°C or ±0.75%	±0.5°C or ±0.4%
J	0–750	±2.2°C or ±0.75%	±1.1°C or 0.4%
E	0–900	±1.7°C or ±0.5%	±1°C or ±0.4%
K	0–1250	±2.2°C or ±0.75%	±1.1°C or ±0.4%
R	0–1450	±1.5°C or ±0.25%	±0.6°C or ±0.1%
S	0–1450	±1.5°C or ±0.25%	±0.6°C or ±0.1%
B	800–1700	±0.5%	—

thermocouple reference equation alone is not satisfactory for obtaining a precise temperature from thermocouple emf because an iteration scheme is involved, and the method does not account for individual thermocouple characteristics, which can differ markedly from the nominal reference table values.

Method 2.   A second method removes one of the difficulties encountered in Method 1, namely, the iteration problem. This is accomplished by making use of the direct approximation functions developed by NBS of the form $T = f(E)$, also given in Table 7.4 for the base metal thermocouples.

It should be emphasized that such equations provide only approximations to the functional relationship between E and T. The difference between the solutions of Methods 1 and 2 is small, however, when compared to the obvious, and potentially larger, error caused by overlooking the individual thermocouple characteristics.

Thus the use of the inverse NBS relation of $T = f(E)$ is not really satisfactory for obtaining temperature from thermocouple emf, because the method does not account for individual thermocouple characteristics that can differ markedly from the inverse solution values.

Method 3.   Another general method for obtaining temperature from thermocouple emf combines the individual thermocouple characteristic (i.e., the calibration) with an acceptable thermocouple reference table to obtain a mathematical relation between ΔE $(= E_{Ref} - E_{Couple})$ and E_{Couple} (see Section 9.3).

To obtain temperature, one adds E_{couple} to ΔE (from the calibration curve) to obtain E_{Ref}, and then solves the pertinent reference function, $(E = f(T))$, for T, interatively, as described in Method 1.

Use of a reference table together with individual thermocouple data yields satisfactory temperatures from an accuracy viewpoint, but does involve the complication of an iterative scheme.

Method 4.   One could overlook the apparent advantages of the difference method of presenting calibration data and go directly to a mathematical fit of the raw calibration data. This amounts to fitting $T = f(E_{couple})$.

As with Method 2, Method 4 avoids any iteration; however, unlike Method 2, the individual thermocouple characteristics are accounted for in Method 4.

Thus we see that application of the direct fit of $T = f(E_{\text{couple}})$ is convenient in that the use of thermocouple reference tables is avoided. However, such direct fits, over any extended range, invariably are of higher-degree equations than the $\Delta E = f(E_{\text{couple}})$ equations. In this day of computers, this is not always a problem.

Summary

Methods 1 and 2 should not be used for accurate temperature measurements, since individual thermocouple characteristics are neglected. Method 3 is very useful, especially if a graphical presentation of thermocouple characteristics is required. Where computers are available, the direct method (i.e., Method 4) of temperature–emf conversion is often preferred.

Example.   The following calibration data is available for a type J thermocouple.

$T(°F)$	32	100	200	300	450
$E_{\text{couple}}(\text{mv})$	0	1.954	4.915	7.969	12.589

Apply Methods 3 and 4 to this data to obtain the thermocouple characteristics, and then use the characteristics to obtain temperatures at the given points, and compare.

As discussed in Section 9.3, a least squares equation is required to best fit the calibration data (see Figure 9.14).

Method 3.   According to Figure 9.14, the least squares *first*-degree equation of the form $\Delta E = f(E)$ can be obtained from

$$\Sigma \Delta E = 5a + b\Sigma E_c$$
$$\Sigma E_c \Delta E = a\Sigma E_c + b\Sigma E_c^2.$$

In terms of the numerical values available from Table 9.2, these equations become

$$-0.066 = 5a + 27.427b$$
$$-0.532548 = 27.427a + 249.963223b.$$

Simultaneous solution leads to the best *straight* line representing this thermocouple characteristic, namely,

$$\Delta E = -0.0038012 - 0.0017134E_c. \tag{9.2}$$

To obtain temperatures by (9.2), the given values of E_c will be used.

1. E_c yields ΔE via (9.2)
2. $E_{\text{ref}} = E_c + \Delta E$
3. T_c is obtained by iteration via the NBS defining relation $E = f(T)$ or, more simply, by NBS reference table interpolation (see Table 7.6).
4. The temperature error (ϵ_T) of Method 3 is determined by comparing the calculated temperature with the given temperature.

Least Squares Principle

Given: $y = a + bx + cx^2$.
For an exact fit: $y - a - bx - cx^2 = 0$.
For a misfit: $y - a - bx - cx^2 = r$, a residual.
Best fit is obtained when sum (R) of squares of residuals (r^2) is minimum. This occurs when

$$\frac{\partial R}{\partial a} = 0, \qquad \frac{\partial R}{\partial b} = 0, \qquad \frac{\partial R}{\partial c} = 0.$$

The resulting series of equations is:

$$\left.\begin{array}{l} \Sigma y = aN + b\Sigma x + c\Sigma x^2 \\ \Sigma xy = a\Sigma x + b\Sigma x^2 + c\Sigma x^3 \\ \Sigma x^2 y = a\Sigma x^2 + b\Sigma x^3 + c\Sigma x^4 \end{array}\right\} \qquad (9.1)$$

where N is the number of calibration points available. Equations 9.1 are solved simultaneously to yield a, b, c.

Gauss Reduction Scheme

To solve the least squares equations, a mechanical scheme of reduction can sometimes be used to advantage.

1. Set up an array of the numerical values available in the three equations of (9.1) according to the columns y, a, b, c.
2. Get coefficients in the a column equal to unity by division.
3. Subtract first row of coefficients from rows 2 and 3 to obtain 2 zeros in the a column.
4. Divide last 2 rows by first real number in that row to get 2 ones in the b column.
5. Subtract second row from row 3 to get 1 zero in the b column.
6. Then $c = \dfrac{\text{No. in } y}{\text{No. in } c}$; $b = \text{No. in } y - (\text{No. in } c)(c)$; and $a = \text{No. in } y - (\text{No. in } b)(b) - (\text{No. in } c)(c)$.

Figure 9.14   Least squares principle and the Gauss reduction scheme.

Method 4a.   According to Figure 9.14, the least squares *first*-degree equation of the form $T = f(E)$ can be obtained from

$$\Sigma T = 5a + b\Sigma E_c$$

$$\Sigma E_c T = a\Sigma E_c + b\Sigma E_c^2.$$

In terms of the numerical values available from Table 9.2, these equations become

$$1082 = 5a + 27.427b$$

$$9234.15 = 27.427a + 249.963223b.$$

Simultaneous solution leads to the best *straight* line representing this thermocouple characteristic

$$T = 34.5579 + 33.1502E_c. \qquad (9.3)$$

Table 9.2

T	E_c	E_{ref}	ΔE	E_c^2	E_c^3	E_c^4	$E_c \Delta E$	$E_c T$	$E_c^2 T$
32	0	0	0	0	0	0	0	0	0
100	1.954	1.942	−0.012	3.818116	7.4605987	14.5780099	−0.023448	195.4	381.8116
200	4.915	4.906	−0.009	24.157225	118.7327609	583.57152	−0.044235	983.0	4831.445
300	7.969	7.947	−0.022	63.504961	506.071034	4032.88007	−0.175318	2390.7	19051.4883
450	12.589	12.566	−0.023	158.482921	1995.14149	25116.83625	−0.289547	5665.05	71317.31445
Sum 1082	27.427	—	−0.066	249.963223	2627.405884	29747.86585	−0.532548	9234.15	95582.059

This work is summarized in Table 9.3.

177

The calculated temperatures and the resulting temperature errors of Method 4a are summarized in Table 9.3.

Table 9.3

Given			Method 3			Method 4a		Method 4b	
$T(°F)$	$E_c(mv)$	ΔE (9.2)	E_{ref}	T_{calc}	ϵ_T	T_{calc} (9.3)	ϵ_T	T_{calc} (9.4)	ϵ_T
32	0	−0.0038	−0.0038	31.86	0.14	34.5579	−2.56	32.4095	−0.41
100	1.954	−0.0071	1.946851	100.17	−0.17	99.3334	0.67	99.5112	0.49
200	4.915	−0.0122	4.90278	199.89	0.11	197.4911	2.51	199.5702	0.43
300	7.969	−0.0174	7.951545	300.15	−0.15	298.7318	1.27	300.7224	−0.72
450	12.589	−0.0254	12.56363	449.92	0.08	451.8858	−1.88	449.7867	0.21

Method 4b. The temperature errors resulting from (9.3) indicate that a higher-degree least squares equation should be used. According to Figure 9.14, the least squares *second*-degree equation of the form $T = f(E)$ can be obtained from

$$\Sigma T = 5a + b\Sigma E_c + c\Sigma E_c^2$$
$$\Sigma E_c T = a\Sigma E_c + b\Sigma E_c^2 + c\Sigma E_c^3$$
$$\Sigma E_c^2 T = a\Sigma E_c^2 + b\Sigma E_c^3 + c\Sigma E_c^4.$$

This series of equations is solved, in terms of the numerical values available from Table 9.2, according to the Gauss reduction scheme given in Figure 9.14. The steps and results are given in Table 9.4. Thus the best *second*-degree equation representing this thermocouple characteristic is

$$T = 32.4095 + 34.5587\, E_c - 0.111572\, E_c^2. \tag{9.4}$$

The calculated temperatures from (9.4) and the resulting temperature errors of Method 4b are summarized in Table 9.3.

The temperature errors resulting from (9.4) indicate that the use of a second-degree least squares equation is very beneficial compared with the first-degree equation, in the direct method of interpretation. Even higher degrees would be required (like a fifth degree) to get accuracies comparable to Method 3.

9.5   References

[1] R. P. Benedict, "The Calibration of Thermocouples by Freezing-Point Baths and Empirical Equations," *Trans. ASME. J. Eng. Power*, April 1959, p. 177.

[2] J. P. Evans, "The Calibration of Temperature Standards on the IPTS of 1948," Pre-print 21.1.62, 17th Annual Instrument-Automation Conference, ISA, October 1962.

[3] S. B. Williams, "Triple-Point-of-Water Temperature Reference," *Instruments and Control Systems*, **33**, 1960.

[4] H. F. Stimson, "Precision Resistance Thermometry and Fixed Points," *Temperature, Its Measurement and Control in Science and Industry*, Reinhold, New York, Vol. 2, 1955, p. 141.

Table 9.4

Operation	T	a	b	c	Equations
Original array	1082	5	27.427	249.963	
	9234.15	27.427	249.963	2627.406	$T = a + bE_c + cE_c^2$
	95582.059	249.963	2627.406	29747.866	
Dividing all coefficients in each row by the number in a	216.4	1	5.4854	49.9926	$c = \dfrac{-0.1237}{1.1087} = -0.111572$
	336.681	1	9.11376	95.796	
	382.3848	1	10.5112	119.0091	
Subtracting first row from rows 2 and 3	216.4	1	5.4854	49.9926	$b = 33.15024 - 12.6237(-0.111572)$
	120.281	0	3.62836	45.8034	
	165.9848	0	5.0258	69.0165	$b = 34.5587$
Divide last two rows by the first real number in row	216.4	1	5.4854	49.9926	$a = 216.4 - 5.4854(34.5587)$
	33.15024	0	1	12.6237	$-49.9926(-0.111572)$
	33.02654	0	1	13.7327	
Subtracting second row from the third	216.4	1	5.4854	49.9926	$a = 32.4095$
	33.15024	0	1	12.6237	
	-0.1237	0	0	1.1087	

[5] R. H. Muller, "New Precise Temperature Standard," *Analytical Chemistry*, **32**, November 1960, p. 103A.

[6] L. L. Sparks and R. L. Powell, "Calibration of Capsule Platinum Resistance Thermometers at the Triple Point of Water," *Temperature*, ISA, Vol. 4, Part 2, 1972, p. 1415.

[7] R. P. Benedict and R. J. Russo, "Calibration and Application Techniques for Platinum Resistance Thermometers," *Trans. ASME, J. Basic Eng.*, June 1972, p. 381.

[8] F. R. Caldwell, "Temperatures of Thermocouple Reference Junctions in an Ice Bath," *J. Res. Nat. Bur. Std.* 69C, April–June 1965, p. 95.

[9] D. L. McElroy, "Thermocouple Research Report for the Period Nov. 1, 1956 to Oct. 31, 1957," Progress Rept., ORNL 2467, Oak Ridge National Laboratory.

[10] H. Preston-Thomas, "The Zinc Point as a Thermometric Fixed Point," in *Temperature*, Reinhold, New York, Vol. 2, 1955, p. 171.

[11] H. M. Terwilliger, "A New Metal Freezing-Point Temperature Standard," ISA Preprint 12.11-1-66, October 1966.

[12] R. P. Benedict, "Temperature and Its Measurement," *Electro-Technol.*, July 1963, p. 71.

[13] E. H. McLaren, "The Freezing Points of High Purity Metals as Precision Temperature Standards," *Temperature*, Reinhold, New York, Vol. 3, Part 1, 1962, p. 185.

[14] ASTM E-563-76, "Recommended Practice for Preparation and Use of Freezing Point Reference Baths," 1976.

[15] ASTM STP 470A, *Manual on the Use of Thermocouples in Temperature Measurement*, 1974, pp. 115–118.

[16] "Calibration of Thermocouples by Comparison Techniques," E 220-72, from 1975 Book of ASTM Standards, Part 44, p. 523.

[17] L. G. A. Sims, "Measuring Fine Temperature Changes," *Engineering*, January 1954, p. 15.

[18] W. F. Roeser and S. T. Lonberger, "Methods of Testing Thermocouple Materials," Nat. Bur. Std. Circular 590, February 1958.

[19] J. T. Callahan, "Heat Transfer Characteristics in Air-Fluidized Solids up to 900°F," *Trans. ASME, J. Basic Eng.*, June 1971, p. 165.

[20] "Temperature Measurement," ASME PTC 19.3, Performance Test Code, Instrument and Apparatus Supplement, 1974.

[21] R. P. Benedict and H. F. Ashby, "Empirical Determination of Thermocouple Characteristics," *Trans. ASME, J. Eng. Power*, January 1963, p. 9.

[22] R. P. Benedict and T. M. Godett, "A Note on Obtaining Temperatures from Thermocouple EMF Measurements," *Trans. ASME, J. Eng. Power*, October 1975, p. 516.

[23] "Temperature Measurement Thermocouples," American National Standards Institute, ANSI MC 96.1, 1975.

[24] "Standard Temperature-Electromotive Force (EMF) Tables for Thermocouples," ASTM E-230-73, 1973.

Nomenclature

Roman

a, b, c coefficients

E electromotive force

emf   electromotive force
T empirical temperature

Greek

σ standard deviation

Problems

For the following test data for a type J thermocouple:

$T(°F)$	32	85	145	194	244	295
$E_c(mV)$	0	1.521	3.270	4.743	6.292	7.865

1. Find temperature errors at the calibration points if Method 2 (NBS direct equations, $T = f(E_c)$) is used.

   Ans. $\Delta T = 0, -0.35, -0.18, -0.60, -1.86, -2.38°F$.

2. Find temperature errors at the calibration points if Method 3 ($\Delta E = f(E_c)$) is used with a least squares first-degree fit.

   Ans. $\Delta E = +4.59558 \times 10^{-3} - 8.812855 \times 10^{-3} E_c$,
   $\Delta T = -0.16, -0.18, +0.57, +0.64, -0.14, -0.20°F$.

3. Find temperature errors at the calibration points if Method 4 ($T = f(E_c)$) is used with a least squares first-degrees fit.

   Ans. $T = 33.9095 + 33.41113 E_c$,
   $\Delta T = -1.91, +0.27, +1.84, +1.62, -0.13, -1.69°F$.

Chapter 10

UNCERTAINTIES AND STATISTICS

"...an intrinsic variability exists which prevents us from getting exactly the same observation or measurement on repeated trials..."

J. Stuart Hunter (1968)

10.1 General Remarks

If there is one premise basic to instrumentation engineering it is this: *No measurement is without error.* Hence neither the exact value of the quantity being measured nor the exact error associated with the measurement can be ascertained. In engineering, as in physics, the uncomfortable principle of indeterminacy exists. Yet, as we have seen in our discussion of interpolation methods (Section 9.3), uncertainties can be useful and, like friction, are often a blessing in disguise. It is toward a methodical use of measurement uncertainties as a guide to approaching true values that this section is addressed.

The output in most experiments is a measurement. The reliability of the measurement depends not only on variations in controlled inputs but also, in general, on variations in factors that are uncontrolled and perhaps unrecognized. Some of these factors that might unwittingly affect a measurement are: the experimenter, the supporting equipment, and conditions of the environment. Thus in addition to errors caused by the device under test, and in addition to errors caused by variations in the quantity being measured, extraneous factors might introduce errors in the experiment that would cloud the results.

Errors caused by the experimenter and the supporting equipment can be *minimized* by securing independent readings by different observers and by the use of different measuring equipment. Effects of those variables that are not part of the study can be further minimized by taking observations in a random order. This is called *randomization.*

The important task of measuring the remaining significant errors is approached by taking a number of independent observations of the output at fixed values of the controlled input. This is called *replication* Replication is shown to be at the heart of statistics.

Stating the above ideas in mathematical terms, each measurement (X) can be visualized as being accompanied by an error (ϵ) such that

$$\bar{X}' = X \pm \epsilon \qquad (10.1)$$

where $\bar{X}'$ represents the true value of the quantity being measured. The measurement error, in turn, is usually expressed in terms of two components: a *random* error, and a *systematic* error.

Random Errors

When repeated measurements are taken, random errors will show up as scatter about the *average* of these measurements. The scatter is caused by characteristics of the measuring system and/or changes in the quantity being measured. Random errors will always be observed as long as the readout equipment has adequate discrimination. The term *precision* is used to characterize random errors. Precision is quantified by the true standard deviation (σ) of the whole population of measurements or, more often, by its estimator, S, the standard deviation of the data available. These statistical terms will be defined shortly, by working equations. Now it is sufficient to understand that a large standard deviation means a lot of scatter in the data, and conversely, a small standard deviation means high precision.

Systematic Errors

Over and above the random errors involved in all measurements, there are also errors that are consistently either too high or too low with respect to the accepted true value. Such errors, which are termed fixed errors or systematic errors, are characterized by *bias.* When bias can be quantified, it can be used as a correction factor to be applied to all measurements. A zero bias implies that there is no difference between the true value and the mean of many observations. Systematic errors can often be minimized by various methods, as by calibration, for example

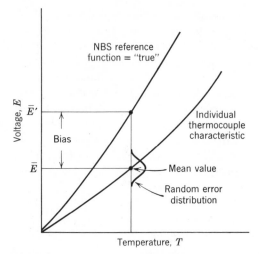

Figure 10.1 Systematic and random errors illustrated for the case of a thermocouple calibration.

(see Figure 10.1). Calibrations are usually accomplished by comparing a test instrument to a standard instrument. Since such comparisons are not always direct or perfect, we may not succeed in totally reducing the bias to zero. Yet there are many cases in engineering practice when we can presume that the bias is removed, that all errors are of the random type, and that hence the errors can be treated statistically [1], [2].

10.2 Statistical Relations

It is clear that we are to be denied by the nature of things the ability to measure directly the true value of a variable. Thus it becomes our job to extract from the experimental data at least two vital bits of information. First, we must form an estimate of the *best value* of the variable. This will be denoted by $\bar{X}$. And closely coupled with this requirement, we must give an estimate of the *intervals*, centered on $\bar{X}$, within which the true value is expected to lie. This will be denoted by the uncertainty margin that we tack on to $\bar{X}$ [3].

Best Value at a Given Input

When an output X is measured many times at a given input, the mean value of X is simply

$$\bar{X} = \frac{1}{N} \sum_{i=1}^{N} X_i \tag{10.2}$$

where X_i is the value of the ith observation, and N is the number of observations in the sample. It is a mathematical fact that the arithmetic average defined by (10.2) is the *best* representation of the given set of X_is. Note that, when the estimated best value of X is taken as $\bar{X}$, the sum of the squares of the deviations of the data from their estimate is a minimum (this is essentially the least squares principle). However, whereas $\bar{X}$ represents an unbiased estimate of the true arithmetic mean $\bar{X}'$ of all possible values of X, there is no assurance that $\bar{X}$ is the true value, $\bar{X}'$. (Incidentally, good agreement, that is, high precision, in small sample replication does not imply that $\bar{X}$ is close to $\bar{X}'$, that is, high accuracy). Nevertheless, from any viewpoint, the best estimate of the value of the population mean at a given input is the average of the available measurements (see Figure 10.2).

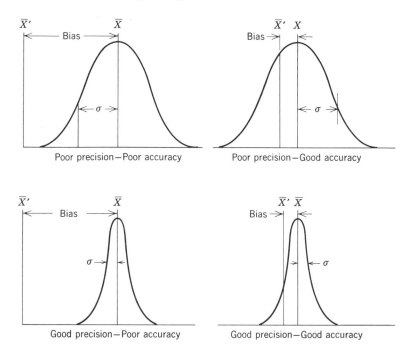

Figure 10.2 Distinction between precision and accuracy.

Confidence Intervals

Having decided on the best available value of X, we inquire next as to its worth as an estimate of the true value of X. C. G. Darwin [4] has noted the importance of confidence intervals as follows: "It seemed to me that there was a defect in the habit of thought of many in the engineering profession, and that some sort of campaign was needed to inculcate in people's minds the idea that every number has a fringe, that it is not to be regarded as exact but as so much plus or minus a bit, and that the size of this bit is one of its really important quantities."

To get at these confidence intervals, we of course need replicate data, and we note in this regard that these intervals will differ according to the size and number of samples available.

1. SAMPLES OF VERY LARGE *N.* Many times in engineering, a tabulation of how often the various values of *X* occur in replication is well approximated by the Gauss-Laplace *normal distribution* relation [5]

$$f(X) = \frac{1}{\sqrt{2\pi}\,\sigma} \exp\left[-\frac{1}{2}\left(\frac{X - \bar{X}'}{\sigma}\right)^2\right] \qquad (10.3)$$

where the factor $1/\sqrt{2\pi}\,\sigma$ has the normalizing effect of making the integral of $f(X)$ over all values of *X* equal unity, and where σ represents the true standard deviation of *X* which is well approximated in turn by:

$$\sigma = \left[\frac{1}{N}\sum_{i=1}^{N}(X_i - \bar{X}')^2\right]^{1/2}. \qquad (10.4)$$

The standard deviation of a normal distribution of *X*:

1. measures the scatter of *X* at a given input,
2. usefully has the same units as the data,
3. is the square root of the average of the sum of the squares of the deviations of all possible observations from the true arithmetic mean.

For the normal distribution, σ has the direct interpretation that 68.26% of all possible values of *X* will fall within the $\pm 1\sigma$ interval centered on $\bar{X}'$. For many engineering applications this is not good enough, and wider intervals must be employed to express greater confidence. For example, 95.46% of the data can be expected to fall within the $\pm 2\sigma$ interval, and 99.73% within $\pm 3\sigma$ (see Figure 10.3).

We are assured that $\bar{X}$ is a very good estimate of $\bar{X}'$ by the large size of the sample. We may ask, however: How typical is a single observation of *X*? As we have just seen, one answer is

$$\bar{X}' = X \pm 3\sigma (99.73\%) \qquad (10.5)$$

Equation 10.5 brings up the important point that, to be most meaningful, a measurement should be given in *three* parts [6], [7].

1. A magnitude (the indicated value of *X*).

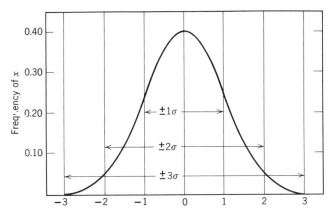

Figure 10.3 Normal distribution of *X*.

2. An uncertainty margin [i.e., a confidence interval, which is your estimate of what the error might be; $\pm 3\sigma$ in the example of (10.5)].

3. A probability statement [an indication of your confidence that the actual error will be within the uncertainty margin chosen; 99.73% in the example of (10.5)].

2. SINGLE SAMPLE OF SMALL *N*.   We now face the unpleasant fact that, in a practical experiment, we usually obtain only a relatively small sample from all possible values of *X*. This means that we cannot obtain the true arithmetic mean, $\bar{X}'$, and hence we cannot form the true standard deviation, σ. Instead of the inaccessible *deviations* $(X_i - \bar{X}')$ of (10.3) and (10.4), we can determine only the *residuals* $(X_i - \bar{X})$. We note in this regard that the sum of the residuals, being always a minimum according to the least squares principle previously mentioned, is always less than the sum of the deviations. The standard deviation of the single sample is defined in terms of the residuals and is patterned after (10.4) as

$$S = \left[\frac{1}{N-1} \sum_{i=1}^{N} (X_i - \bar{X})^2 \right]^{1/2} \tag{10.6}$$

where the factor $(N-1)$ is used in place of the usual *N* to compensate for the negative bias that results from using $\bar{X}$ in place of $\bar{X}'$ in forming the differences. However, a negative bias unfortunately still remains in the small sample standard deviation, and *S*, the obtainable, does not equal σ, the desired.

STUDENT t DISTRIBUTION.   Recognizing this deficiency, a method was developed by the English Chemist W. S. Gosset (writing in 1907 under the pseudonym, "*Student*") by which confidence intervals could be based on the single small sample S. He introduced the "Student" t distribution whose values have been tabulated as a constant depending only on degree of freedom, ν, and on the desired degree of confidence or probability, p (see Table 10.1 and Figure 10.4).

Careful perusal of these values will show that the t statistic inflates the confidence interval (i.e., the uncertainty margin) so as to reduce the

Table 10.1 *Student t Distribution*

ν	$t_{90\%}$	$t_{95\%}$	$t_{99\%}$
1	6.314	12.706	63.657
2	2.920	4.303	9.925
3	2.353	3.182	5.841
4	2.132	2.770	4.604
5	2.015	2.571	4.032
6	1.943	2.447	3.707
7	1.895	2.365	3.499
8	1.860	2.306	3.355
9	1.833	2.262	3.250
10	1.812	2.228	3.169
11	1.796	2.201	3.106
12	1.782	2.179	3.055
13	1.771	2.160	3.012
14	1.761	2.145	2.977
15	1.753	2.131	2.947
16	1.746	2.120	2.921
17	1.740	2.110	2.898
18	1.734	2.101	2.878
19	1.729	2.093	2.861
20	1.725	2.086	2.845
21	1.721	2.080	2.831
22	1.717	2.074	2.819
23	1.714	2.069	2.807
24	1.711	2.064	2.797
25	1.708	2.060	2.787
26	1.706	2.056	2.779
27	1.703	2.052	2.771
28	1.701	2.048	2.763
29	1.699	2.045	2.756
30	1.697	2.042	2.750
40	1.684	2.021	2.704
60	1.671	2.000	2.660
120	1.658	1.980	2.617
∞	1.645	1.960	2.576

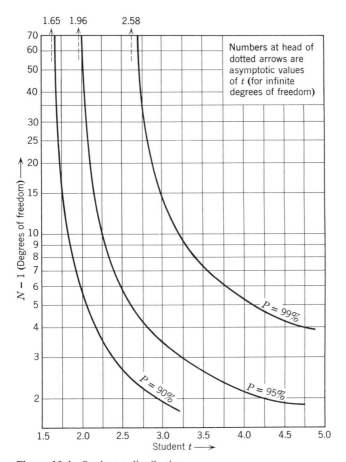

Figure 10.4   Student t distribution.

effect of underestimating the standard deviation when a *small* sample is used to calculate S (see Figure 10.5).

Degrees of freedom can be defined in general as the number of observations minus the number of constants calculated from the data. According to (10.2), $\bar{X}$ has N degrees of freedom, whereas by (10.6), S has $N-1$ degrees of freedom because one constant, $\bar{X}$, is used to calculate S.

The answer to the question: How typical is a single observation of X? is, in terms of S and t

$$\bar{X}' = X \pm t_\nu S \text{ (to given probability)} \tag{10.7}$$

where ν signifies the degrees of freedom in S. The counterpart of (10.7)

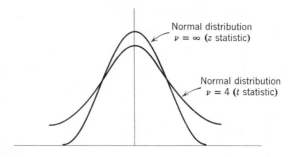

Figure 10.5 Comparison of z and t statistics (there is a normal distribution for each degree of freedom, ν).

was given in (10.5) in terms of σ and can be generalized in terms of the Z statistic as

$$\bar{X}' = X \pm Z\sigma \text{ (to given probability).} \tag{10.8}$$

The plus and minus quantities in (10.7) and (10.8) should be recognized as the confidence intervals on the individual measurements of X. Appropriate values of Z are given in Table 10.2. The Z and t distributions are compared in Figure 10.5.

Table 10.2   Values of the Normal Deviate

Probability (%)	80	90	95	98	99	99.8
Normal deviate Z	1.282	1.645	1.960	2.326	2.576	3.090

3. SEVERAL SAMPLES OF SMALL N. A useful measure of scatter in multiple sample experiments is called the *standard deviation of the mean,* patterned after (10.4) as

$$S_{\bar{X}} = \left[\frac{1}{M} \sum_{i=1}^{M} (\bar{X}_i - \bar{\bar{X}})^2 \right]^{1/2} \tag{10.9}$$

where M is the number of sets involved, $\bar{X}_i$ is the mean of the ith sample, and $\bar{\bar{X}}$ is the average of the set formed by the average values, that is,

$$\bar{\bar{X}} = \frac{1}{M} \sum_{i=1}^{M} \bar{X}_i = \frac{1}{MN} \sum_{i=1}^{MN} X_i. \tag{10.10}$$

The factor $\bar{\bar{X}}$, being actually the average of all available measurements, is naturally to be used in all multiple sample experiments as the *best estimate* of the value at a point.

It is an observable fact that the means of different sets of measurements are alway much closer to each other than are the individual values of a single set. Equation 10.9 defines the precision of a set of M values of $\bar{X}$, whereas (10.6) defines the precision of a set of N values of X. Statistical theory gives us an important relation between these two *sample* standard deviations, namely,

$$S_{\bar{X}} = \frac{S_X}{\sqrt{N}} \qquad (10.11)$$

where N is consistently the number of observations common to the M samples.

Equation 10.11 says in effect: The standard deviation of the average value of a sample (i.e., of $\bar{X}$) is always less than that of a typical measurement in the sample (i.e., of X) by the factor $1/\sqrt{N}$.

Patterned directly after (10.11) is the standard deviation of the average value of M sets of samples (i.e., of $\bar{\bar{X}}$) which can be given as

$$S_{\bar{\bar{X}}} = \frac{S_{\bar{X}}}{\sqrt{M}} = \frac{S_X}{\sqrt{MN}}. \qquad (10.12)$$

Equation 10.12 presents one apparent problem: Which S_X should be used in a multiple set experiment? One answer is to chose any one S_{Xi} at random, but this leads to wide variation in $S_{\bar{\bar{X}}}$. A more satisfying answer, and one recommended here, is to define a weighted average of the S_is, the weights being the appropriate degrees of freedom, that is,

$$\bar{\bar{S}} = \left[\frac{\nu_1 S_1{}^2 + \nu_2 S_2{}^2 + \cdots + \nu_M S_M{}^2}{\nu_1 + \nu_2 + \cdots + \nu_M} \right]^{1/2}. \qquad (10.13)$$

Since it is often common practice to keep all samples of the same size, that is, to keep N the same for each sample, and since this practice assures us that $\bar{\bar{S}}$ is a minimum, (10.13) can be rewritten in the useful and simplified form as

$$\bar{\bar{S}} = \left[\frac{1}{M} \sum_{i=1}^{M} S_i{}^2 \right]^{1/2} \qquad (10.14)$$

On the basis of the t statistic, and in terms of (10.7, 10.11, 10.12, and 10.14), best estimates of the true value of X can be given in terms of the *mean* of a single set of measurements as

$$\bar{X}' = \bar{X} \pm \frac{t_{N-1,p} S_X}{\sqrt{N}} = \bar{X} \pm t_{N-1,p} S_{\bar{X}} \qquad (10.15)$$

and in terms of the *mean* of M sets of measurements as

$$\bar{X}' = \bar{\bar{X}} \pm \frac{t_{MN-M,p}\bar{\bar{S}}}{\sqrt{MN}}. \tag{10.16}$$

The subscripts of t indicate degrees of freedom (ν) and probability (p), and the $\pm$ quantity indicates the confidence interval (CI) of the best value estimate.

Several examples are given here to illustrate the ideas embodied thus far in these statistical relations.

Example 1. For the observations, X_is = 7, 8, 7, 6, 5, 6, 7, 8, 6, 9, 8, find the best estimate of $\bar{X}'$, the standard deviation of the observations, the standard deviation of the mean, and the 95% confidence interval statement for $\bar{X}$ [8].

By (10.2)

$$\bar{X} = \frac{1}{11} \sum_{i=1}^{11} X_i = \frac{77}{11} = 7$$

which is the best estimate of $\bar{X}'$.

By (10.6)

$$s_X = \left[\frac{1}{N-1} \sum_{i=1}^{11} (X_i - 7)^2 \right]^{1/2} = \sqrt{\frac{14}{10}} = 1.18$$

which is the standard deviation of the observations.

By (10.11)

$$S_{\bar{X}} = \frac{S_X}{\sqrt{N}} = \frac{1.18}{\sqrt{11}} = 0.357$$

which is the standard deviation of the mean.

By Table 10.1

$$t_{\nu,p} = t_{10,0.95} = 2.228$$

which is the t statistic for the 95% confidence interval statement. Hence according to (10.15)

$$\bar{X}' = \bar{X} \pm t_{N-1,p}S_{\bar{X}} = 7 \pm 2.228(0.357)$$

$$= 7 \pm 0.795(95\%)$$

or $6.2 \le \bar{X}' \le 7.8$, with 95% confidence.

Example 2. Express the best value and its 95% confidence interval for a single sample experiment of eight observations. According to (10.15) and Table 10.1, at $N = 8$: $\nu = 7$, and $t_{7,0.95} = 2.365$.

Hence $$\bar{X}' = \bar{X} \pm \frac{2.365 S_X}{\sqrt{8}} (95\%)$$

Example 3. If five sets of data of the type of Example 2 were taken, how much more confidence could be placed in the best value? According to (10.16) and Table 10.1, at $M \times N = 40$: $\nu = MN - M = 35$, and $t_{\nu,p} = 2.031$.

Hence $$\bar{X}' = \bar{\bar{X}} \pm \frac{2.031 S_X}{\sqrt{40}} (95\%).$$

[This formulation presumes that S_X remains constant. In reality, it would vary as previously discussed, and one would use the weighted standard deviation, $\bar{\bar{S}}$ of (10.14).]

When the results of Examples 2 and 3 are ratioed, there results

$$\frac{2.365/\sqrt{8}}{2.031/\sqrt{40}} = \frac{0.836}{0.321} = 2.6$$

which indicates that the confidence interval could be tightened by a factor of about 3 for the multiple sample case.

Example 4. From three sets of five measurements each, the following table is derived:

M	$\bar{X}$	S
1	−0.2052	0.6719
2	−0.2232	0.6395
3	0.1030	1.2168

Find the best estimate of $\bar{X}'$ and its 95% confidence interval.

By (10.10) $\bar{\bar{X}} = \frac{1}{3}(-0.2052 - 0.2232 + 0.1030) = -0.1085.$

By (10.14) $\bar{S} = [\frac{1}{3}(0.6719^2 + 0.6395^2 + 1.2168^2)]^{1/2} = 0.8834.$

By (10.16), for $\nu = 12$,

$$CI_{\bar{X}} = \frac{2.179(0.8834)}{\sqrt{15}} = 0.4970.$$

Hence $\bar{X}' = -0.1085 \pm 0.4970 \ (95\%).$

Range. A final estimate of σ in multiple sample experiments can be obtained through the average of a set of ranges, $\bar{w}$. Range is defined as the difference between the largest and smallest values in a sample. The standard deviation can be estimated from the range by the statistical relation

$$\sigma \simeq \frac{\bar{w}}{(d_2)_N} \tag{10.17}$$

where

$$\bar{w} = \frac{1}{M} \sum_{i=1}^{M} w_i \tag{10.18}$$

and where d_2 is tabulated as a function of N only in Table 10.3.

A tacit assumption in (10.17) is that the number of samples available is large (i.e., on the order of 20). However, even when M is small, (10.17) still provides a useful estimate of σ, the standard deviation of all available observations. In fact, many times in engineering [9], [10], [11], the range of a *single* sample is used via (10.17) to estimate σ.

Confidence interval statements for the case when range is being used can be given as the counterparts of (10.15) and (10.16) as

$$\bar{X}' = \bar{X} \pm \tau_N w \tag{10.19}$$

for the *single* set, and

$$\bar{X}' = \bar{X} \pm \frac{\tau_N w}{\sqrt{M}} \tag{10.20}$$

Table 10.3 Ratio of Range to Standard Deviation (d_2), as a Function of Sample Size

N	d_2
2	1.128
3	1.693
4	2.059
5	2.326
10	3.078
20	3.735
25	3.931

in terms of the *mean* of M sets of measurements, where τ is called the substitute t statistic, and is given in Table 10.4.

Example 5. Estimate the number of ranges by which $\bar{X}$ can depart from $\bar{X}'$ by reference to Table 10.4 and (10.19), for $N = 2$, 3, and 4. For two observations in the sample:

$$CI_{\bar{X}} = \tau_2 w = 6.353 \text{ ranges (95\%)}$$
$$= 31.828 \text{ ranges (99\%).}$$

Table 10.4 Substitute t distribution (the τ Statistic)

N	$\tau_{90\%}$	$\tau_{95\%}$	$\tau_{99\%}$
2	3.175	6.353	31.828
3	0.885	1.304	3.008
4	0.529	0.717	1.316
5	0.388	0.507	0.843
6	0.312	0.399	0.628
7	0.263	0.333	0.507
8	0.230	0.288	0.429
9	0.205	0.255	0.374
10	0.186	0.230	0.333
11	0.170	0.210	0.302
12	0.158	0.194	0.277
13	0.147	0.181	0.256
14	0.138	0.170	0.239
15	0.131	0.160	0.224
16	0.124	0.151	0.212
17	0.118	0.144	0.201
18	0.113	0.137	0.191
19	0.108	0.131	0.182
20	0.104	0.126	0.175

For three observations in the sample:

$$CI_{\bar{X}} = 1.304 \text{ ranges } (95\%)$$
$$= 3.008 \text{ ranges } (99\%).$$

Here we should note how dramatically the confidence interval tightens with one additional measurement.

For four observations in the sample:

$$CI_{\bar{X}} = \tau_4 w = 0.717 \text{ ranges } (95\%)$$
$$= 1.316 \text{ ranges } (99\%).$$

Thus we conclude that the use of three readings over two greatly improves our understanding of the dispersion of X, and the worth of additional measurements becomes primarily an economic question.

Example 6. For two sets of five measurements each, the following table is derived.

M	w	$\bar{X}$
1	1.516	0.4506
2	0.815	0.0478.

Find the best estimate of $\bar{X}'$ and its confidence interval.

By (10.10) $\bar{\bar{X}} = \frac{1}{2}(0.4506 + 0.0478) = 0.2492.$

By (10.18) $\bar{w} = \frac{1}{2}(1.516 + 0.815) = 1.1655.$

By (10.20) $CI_{\bar{X}} = \dfrac{\tau_N \bar{w}}{\sqrt{M}} = 0.507\,\dfrac{(1.1655)}{\sqrt{2}} = 0.4178.$

Hence $\bar{X}' = 0.2492 \pm 0.4178 \ (95\%).$

4. SAMPLES OF SMALL N WITH σ KNOWN. Often σ can be considered known in the sense of being established by experience. In such favorable situations, even a single small sample can yield reliable estimates of $\bar{X}'$. For example, in terms of σ, the counterparts of (10.15), (10.16), and of (10.19) and (10.20) are

$$\bar{X}' = \bar{X} \pm \frac{Z\sigma}{\sqrt{N}} \tag{10.21}$$

for the *single* set, and

$$\bar{X}' = \bar{\bar{X}} \pm \frac{Z\sigma}{\sqrt{MN}} \tag{10.22}$$

for multiple sets.

All of the confidence intervals developed thus far are summarized in Table 10.5.

Table 10.5   Summary of Various Confidence Interval Statements

Available Statistic	Multiplying Statistic	Confidence Interval on Single Set Mean, $\bar{X}$	Confidence Interval on Multiple Set Mean, $\bar{\bar{X}}$
σ	Z	$Z_p \dfrac{\sigma}{\sqrt{N}}$	$Z_p \dfrac{\sigma}{\sqrt{MN}}$
S	t	$t_{N-1,p} \dfrac{S}{\sqrt{N}}$	$t_{MN-M,p} \dfrac{\bar{S}}{\sqrt{MN}}$
w	τ	$\tau_{N,p} w$	$\tau_{N,p} \dfrac{\bar{w}}{\sqrt{M}}$

Example 7. A certain temperature measurement yields an average value of 150.75°F, based on four readings. By experience, it is felt that no temperature measurement is off more than ±0.5°F, with 95% assurance. What confidence interval statement can be made concerning the true temperature?

Based on a *single* measurement, according to (10.8) and Table 10.2

$$\bar{T}' = T \pm 1.96\sigma \ (95\%).$$

But, $1.96\sigma = 0.5°F$.

Based on the *mean* of four measurements, according to (10.21),

$$\bar{T}' = 150.75 \pm \frac{1.96\,\sigma}{\sqrt{4}}$$

$$= 1\,50.75 \pm \frac{0.5}{2}.$$

Thus the following confidence interval statement can be made:

$$150.5°F \le \bar{T}' \le 151.0°F \ (95\%).$$

In words, the most believable value of T is 150.75°F. Furthermore, 95% of the time, the true temperature is believed to lie between 150.5 and 151.0°F.

10.3   Parametric Variations

We have been speaking of uncertainties in *individual* measurements taken at a point in space by a single measuring system when the variable was constant over the time period of the measurement.

In this section, we consider uncertainties in ascertaining the best value of a *parameter*, such as temperature, pressure, or flow rate, when the measurements can vary with time, with spacial location, and with the installation or instrument system being used.

To say it another way, whereas *individual* measurements always show random variations (e.g., the measurements of temperature show a random scatter about a mean value), *parametric* values also can show differences because of:

1. random *time* variations (e.g., the temperature being measured is not a constant, but its value varies randomly with time).

2. *spacial* variations (e.g., the temperature at one location differs randomly from that at another).

3. variations among different *instruments* or among different *installations* (e.g., a thermocouple and a resistance temperature detector respond differently to their environment).

It is difficult to distinguish between the random variations of a parameter with time and the random variations associated with the measuring system. However, such random variations can be lumped together and accounted for by the rules of statistics, and there should be no problem in characterizing the precision of such measurements.

Spacial variations signify that the average values, measured at various locations, differ from each other. This variability is over and above the random variations associated with time and the instrument system at a given location.

Different instruments, of the same or different types, or different installations, may yield different samples from the total population of measurements. Such samples may show random time-dependent variations or random installation variations that differ from each other as well as from the time and spacial variations.

Such instrument and spacial variations should not be taken as biases, but can be considered to be randomly distributed variations that must be factored in to the overall precision statement. These possible variations can be expressed mathematically, according to [9], as

$$P = f(t, s, i) \qquad (10.23)$$

meaning that the determination of the value of a parameter, P, is a function of *when* the measurement was taken (signified by t), *where* it is taken (signified by s), and by *what* instrument or installation it was taken (signified by i).

For simplicity, and with much reality, it is again presumed that all the systematic errors (i.e., biases) of all the measurements initially have been reduced to zero. Then, it is a recognized concept of statistics that the precision indicator of a result, R, is given by

$$\sigma_R = \left[\left(\frac{\partial R}{\partial V_1} \sigma_1 \right)^2 + \left(\frac{\partial R}{\partial V_2} \sigma_2 \right)^2 + \cdots + \left(\frac{\partial R}{\partial V_N} \sigma_N \right)^2 \right]^{1/2} \qquad (10.24)$$

where $V_1, V_2, \ldots, V_N$ are independent variables.

Equation 10.24 has been expressed [9] in terms of the best value of a parametric measurement, $\bar{\bar{P}}$, as

$$\sigma_{\bar{P}} = [\sigma_{\bar{P},t}^2 + \sigma_{\bar{P},s}^2 + \sigma_{\bar{P},i}^2]^{1/2}. \qquad (10.25)$$

This is valid whenever the variations with time, space, and instrument can be considered independent, and as long as the changes in $\bar{\bar{P}}$ caused separately by changes in each of the independent variables can be considered unity.

However, the standard deviations, σs, called for in (10.25) may not always be available, and furthermore, we have already seen that the standard deviation is not necessarily the best measure of the uncertainty in a result. A more useful expression is the *uncertainty interval* of the result, UI_R, given to the same odds (i.e., with the same degree of confidence) as the uncertainty intervals of each of the variables. According to [7]

$$UI_R = \pm \left[\left(\frac{\partial R}{\partial V_1} UI_1 \right)^2 + \left(\frac{\partial R}{\partial V_2} UI_2 \right)^2 + \cdots + \left(\frac{\partial R}{\partial V_N} UI_N \right)^2 \right]^{1/2}$$

$$(10.26)$$

can be used in place of (10.24) to provide more meaningful information about a result.

Under the restrictions mentioned in regards to (10.25), and in terms of the *confidence intervals* already developed, (10.26) becomes

$$CI_{\bar{P}} = \pm [CI_{\bar{P},t}^2 + CI_{\bar{P},s}^2 + CI_{\bar{P},i}^2]^{1/2}. \qquad (10.27)$$

It is interesting to note that (10.27) reduces to a form of (10.25) whenever σs are available, for then all CIs are of the form $Z\sigma$, and (10.27) becomes

$$CI_{\bar{P}} = \pm Z\sigma_{\bar{P}} = \pm Z[\sigma_{\bar{P},t}^2 + \sigma_{\bar{P},s}^2 + \sigma_{\bar{P},i}^2]^{1/2}. \qquad (10.28)$$

To particularize (10.27), we can consider that estimates of the standard deviation (i.e., Ss) are available. Then, with N measurements common to each instrument at each location, and M instruments (or installations) available at each location, and with L spacial locations involved, according to (10.11) and (10.12), there results

$$CI_{\bar{P},s} = \frac{t_{\nu_s,p}S_s}{\sqrt{L}}$$ (10.29)

$$CI_{\bar{P},i} = \frac{t_{\nu_i,p}\bar{S}_i}{\sqrt{LM}}$$ (10.30)

$$CI_{\bar{P},t} = \frac{t_{\nu_t,p}\bar{\bar{S}}_t}{\sqrt{LMN}}.$$ (10.31)

It must be realized that Ss are not always available. Sometimes ranges (i.e., ws) and sometimes σs will be available. These change the confidence interval expressions according to Table 10.5.

Examples may perhaps illustrate these concepts more clearly than further discussion.

Example 8.   A certain test is planned, based on the use of one instrument (i.e., $M = 1$) at each of four spacial locations (i.e., $L = 4$). The spacial precision is predetermined to be $\sigma_{p,s} = 4$, and time variations (i.e., $\sigma_{p,t}/\sqrt{N}$) are believed to be negligible. Two instrument types are available: one of $(\sigma_{p,i})_1 = 2$, and the other of $(\sigma_{p,i})_2 = 0.4$. Is use of the more precise instrument warranted?

Since σs are available, (10.25) is applicable. This, together with (10.29) and (10.30), yields

$$\sigma_{\bar{P}} = [\sigma_{\bar{P},s}^2 + \sigma_{\bar{P},i}^2]^{1/2}$$

or

$$\sigma_{\bar{P}} = \left[\frac{\sigma_{p,s}^2}{L} + \frac{\sigma_{p,i}^2}{LM}\right]^{1/2}.$$

In terms of the information available:

$$(\sigma_{\bar{P}})_{\text{instrument}_1} = [\tfrac{1}{4}(16+4)]^{1/2} = 2.236$$

and

$$(\sigma_{\bar{P}})_{\text{instrument}_2} = [\tfrac{1}{4}(16+0.16)]^{1/2} = 2.010.$$

Since the precision of the parameter $\bar{P}$ is improved by about 11% by the use of the more precise instrument, its use would seem warranted.

As a general rule, improvement in an overall measurement can be expected until the ratio $K = \sigma_{P,s}/\sigma_{P,i}$ exceeds 10, according to [9]. For example, if an instrument of $\sigma_{P,i} = 0.2$ were available: K would equal 20; the resulting $\sigma_{\bar{P}}$ would equal 2.002; and the improvement over instrument 2 would be only 0.4%. This confirms the claim that K need not be greater than 10.

Example 9.   Would the use of *two* of the more precise instruments of Example 8, at each of the spacial locations, be warranted, other things remaining the same?

From the previous example, with one instrument/location, $\sigma_{\bar{P}} = 2.010$.

With two instruments/location, the resulting expression is

$$\sigma_{\bar{P}} = [\tfrac{1}{4}(16+\tfrac{1}{2}(0.4)^2)]^{1/2} = 2.005.$$

Since the precision of the parameter $\bar{P}$ is improved by only 0.25%, the installation of four additional instruments does not appear warranted.

Example 10. Using the single instrument/location, and the same set-up of Example 9, ten measurements were taken with each instrument. The precision indicator of these measurements with time was found to be $\sigma_{P,t} = 4$. Given that one could improve $\sigma_{\bar{P}}$ by increasing L, M, N, what one single change would be most effective?

According to (10.25), it is clear that we need only examine the relative magnitudes of the three terms under the square root, namely,

$$\sigma_{\bar{P},s}^2 = \frac{\sigma_{P,s}^2}{L} = \frac{16}{4} = 4.00$$

$$\sigma_{\bar{P},i}^2 = \frac{\sigma_{P,i}^2}{LM} = \frac{0.16}{4 \times 1} = 0.04$$

$$\sigma_{\bar{P},t}^2 = \frac{\sigma_{P,t}^2}{LMN} = \frac{16}{4 \times 1 \times 10} = 0.40.$$

It can be seen that the random spacial variations are ten times more influential in setting the precision of $\bar{P}$ than are the random variations in the parameter with time. Thus one should increase the number of spacial samplings if the precision of $\bar{P}$ is to be improved.

Example 11. If an acceptable parameter precision were defined as $\sigma_{\bar{P}} = 1$, how many more spacial samplings must be taken for the conditions of Example 10?

According to (10.25), squaring both sides

$$1 = \frac{16}{L} + \frac{0.16}{L} + \frac{16}{10L}$$

or, $L = 16 + 0.16 + 1.6 \approx 18$. This means that about 14 more spacial samplings are required.

It may or may not be practical in this case to achieve the acceptable precision of $\sigma_{\bar{P}} = 1$, since 18 spacial samplings may be prohibitively large.

Example 12. Using the same table given in Example 4, we interpret it as based on five measurements obtained at each of three spacial locations. Find the best value of X and its confidence interval.

By (10.10) $\qquad\qquad\qquad\qquad \bar{X} = -0.1085.$

By (10.27) $\qquad\qquad\qquad\qquad CI_{\bar{X}} = \pm[CI_{\bar{X},t}^2 + CI_{\bar{X},s}^2]^{1/2}.$

By (10.14) $\qquad\qquad\qquad\qquad \bar{S} = 0.8834.$

By (10.16) $\qquad\qquad\qquad\qquad CI_{\bar{X},t} = 0.4970.$

Without forethought, one might be tempted to write, according to (10.19)

$$CI_{\bar{X},s} = \tau_M w = \tau_3(\bar{X}_3 - \bar{X}_2) = 1.304(0.3262) = 0.4254.$$

Hence according to (10.27) one would write

$$CI_{\bar{X}} = \pm[0.4970^2 + 0.4254^2]^{1/2} = \pm0.6542.$$

But an important step in logic was bypassed, namely, to question whether the differences between the resulting $\bar{X}$s were significant. That is, do the data support any spacial variation at all?

Difference Between Two Means. The confidence interval between two means has been given [12] in terms of

$$(\bar{X}_1' - \bar{X}_2') = (\bar{X}_1 - \bar{X}_2) \pm t_{\nu_1+\nu_2,p}\frac{\bar{S}}{\sqrt{N/2}}. \tag{10.32}$$

The hypothesis that $\bar{X}_1' = \bar{X}_2'$ can be tested by comparing the observed $t_{\nu_1+\nu_2}$, that is,

$$t_{\nu_1+\nu_2} = \left| \frac{(\bar{X}_1 - \bar{X}_2)\sqrt{N/2} - 0}{\bar{S}} \right| \tag{10.33}$$

with the t statistic (see Table 10.1), for the same degrees of freedom and the desired confidence level.

In terms of the information available in Example 12, (10.33) yields

$$t_{8 \text{ observed}} = \frac{0.3262\sqrt{5/2}}{0.8834} = 0.5838$$

whereas, $t_{8,0.95}$ from Table 10.1 equals 2.306. Since $t_{\text{observed}} \ll t_{\text{table}}$, we can conclude that this is not at all a rare event, that these means could very well be from the same population, that the hypothesis that there is but one $\bar{X}'$ is supported by the data, and specifically that there are no significant spacial variations.

Hence, we must revise our estimate of $CI_{\bar{X}}$ from 0.6542 down to 0.4970 which means we are really more sure of $\bar{X}$ when there are no spacial variations.

Example 13. For the same information given in Example 12, find the best value of X and its confidence interval if these values were based on 100 measurements at each spacial location.

By (10.16)
$$CI_{\bar{X},t} = \frac{t_{MN-M}\bar{\bar{S}}}{\sqrt{MN}} = \frac{1.97(0.8834)}{\sqrt{300}} = 0.1005.$$

By (10.33)
$$t_{198 \text{ observed}} = \frac{0.3262\sqrt{100/2}}{0.8834} = 2.611.$$

whereas, $t_{198,0.95}$ from Table 10.1 equals 1.97. Since $t_{\text{observed}} > t_{\text{table}}$, we conclude that the difference between the observed means is too rare for the means to be accepted as from the same population. Hence spacial variations must be considered in this example. By (10.19)

$$CI_{\bar{X},s} = \tau_M w = 1.304(0.3262) = 0.4254.$$

We should realize that some of the effects being charged to spacial variations are really caused by time variations. To consider them as independent effects is to err on the conservative side.

By (10.27)
$$CI_{\bar{X}} = \pm(0.1005^2 + 0.4254^2)^{1/2} = \pm0.4371.$$

Example 14. In the course of determining the best value of a certain parameter, two separate instruments were installed at each of three spacial locations. By experience, it is known that the random variations in P with time, for a given installation at a given location, can be described by $\sigma_{P,t} = 0.1\%$ of full scale. For eight measurements per instrument, what is the 95% confidence interval about the best value, $\bar{P}$, if the average range between the means of the instruments is 0.05% of full scale, and the standard deviation estimate of the spacial precision is 0.02% of full scale? By (10.20) (Note: when the substitute t statistic is used, we do not divide by M.)

$$CI_{\bar{P},i} = \frac{\tau_2 \bar{w}_i}{\sqrt{L}} = \frac{6.353(0.05\%)}{\sqrt{3}} = 0.1834\%.$$

By (10.22)
$$CI_{\bar{P},t} = \frac{z\sigma_{P,t}}{\sqrt{LMN}} = \frac{1.96(0.1\%)}{\sqrt{48}} = 0.0280\%.$$

By (10.15)
$$CI_{\bar{P},s} = \frac{t_{L-1} S_{P,s}}{\sqrt{L}} = \frac{4.303(0.02\%)}{\sqrt{3}} = 0.1921\%.$$

By (10.27)
$$CI_{\bar{P}} = \pm[0.1834^2 + 0.0280^2 + 0.0497^2]^{1/2} = \pm 0.1921\%.$$

This information can best be stated as

$$\bar{P}' = \bar{\bar{P}} \pm 0.2\% \ (95\%).$$

Example 15. As a *special case*, eight different installations were tested in a standards laboratory, where the spacial variations were found to be negligible. The range of the installation variations was found to be 0.5%. At the same time, a large number of measurements were taken with one installation. The random variations with time, for a certain instrumentation package, were determined to be characterized by $\sigma_{P,t} = 0.2\%$. At another time, as the *usual case*, only a single installation was tested. Three measurements were taken with the same instrumentation package, and the range with time was found to be 0.1%. What is the best estimate of the confidence interval about $\bar{P}$ at the 95% confidence level?

The only information available for a single installation is that derived from the special case data, namely, by (10.19)

$$CI_{\bar{P},i} = \sqrt{M}\tau_M w_i = \sqrt{8}(0.288)(0.5\%) = 0.4073\%.$$

where $\bar{P}$ is the average value for a single installation. From the single installation data, in terms of (10.17), it appears that

$$\sigma_{P,t} = \frac{w_t}{(d_2)_N} = \frac{0.1\%}{1.693} = 0.059\%.$$

But this is less than $\sigma_{P,t}$ derived from a large number of measurements. Hence we conclude that the larger value must be used as being more representative and more conservative. Thus according to (10.21)

$$CI_{\bar{P},t} = \frac{z\sigma_{P,t}}{\sqrt{N}} = \frac{1.96(0.2\%)}{\sqrt{3}} = 0.2263\%.$$

In view of the foregoing, the best estimate of the *CI* about $\bar{P}$ at the 95% confidence level is, according to (10.27)

$$CI_{\bar{P}} = \pm[CI_{\bar{P},i}^2 + CI_{\bar{P},t}^2]^{1/2} = \pm[0.4073^2 + 0.2263^2]^{1/2} = \pm 0.466\%.$$

Example 16.   If two installations were used in the usual case of Example 15, such that $w_i = 0.3\%$, and again for $N = 3$ measurements/installation, find the 95% CI for $\bar{\bar{P}}$. From the two-installation data

$$CI_{\bar{P},i} = \tau_2 w_i = 6.353(0.3\%) = 1.906\%.$$

This disagrees violently with the more extensive eight-installation case applied here as

$$CI_{\bar{P},i} = \frac{CI_{\bar{P},i}}{\sqrt{M}} = \frac{0.4073\%}{\sqrt{0.2}} = 0.288\%.$$

We conclude that the two-installation case does not make use of all the information available on installation uncertainties, presents a larger-than-to-be-expected range, and we choose instead the 0.288% value as being more realistic. The $CI_{\bar{P},t}$ for two installations improves, via $z\sigma_{P,t}/\sqrt{MN}$, to 0.16%. Hence

$$CI_{\bar{P}} = \pm[0.288^2 + 0.16^2]^{1/2} = \pm 0.3295\%,$$

as compared with the single installation value of 0.466%. On the basis of these figures, the worth of two installations over one is judged questionable. Furthermore, the worth of even three times the number of readings per installation is also judged questionable since $CI_{\bar{P}} = \pm 0.303\%$.
$\scriptstyle N=9$

10.4   Uncertainty Considerations

When the samples of measurements are extremely small, or when no statistical information at all is available, as when an instrument is used in an uncalibrated state, one must estimate an *uncertainty interval* in lieu of the confidence interval. The uncertainty interval represents the experimenter's best estimate of the maximum error to be reasonably associated with the measurement.

Thus one could say, for example, that the uncertainty interval of a temperature measurement is $\pm 1°F$ (95% of the time), without reference to any particular measurements, without computation of S, and without application of the t statistic. Or, one could say that all flow measurements made with uncalibrated nozzles, can be counted on to $\pm 1.5\%$ (95% of the time), and use this as the uncertainty interval.

Although it is true that the uncertainty interval, thus conceived, includes both systematic and random errors, the idea of separating these errors and dealing with them separately is too arbitrary to be practical, and we will consider uncertainty intervals to describe systematic errors alone.

Uncertainty intervals can be given on an absolute or relative basis. An *absolute* uncertainty interval is one that is expressed in the same units as the mean value of the measurement (as $150 \pm 0.5°F$). A *relative* uncertainty interval is one that is expressed on a dimensionless basis, and is formed by dividing the absolute uncertainty interval by the mean value of the measurement (as $150°F \pm 0.5/150 = 150°F \pm 1/3\%$).

10.5 Propagation of Uncertainties Into a Result

When a number of measurements are to be combined, an important question must be considered. How are the uncertainty intervals or the confidence intervals of independent measurements propagated into the result?

Much of the basis for the answer to this question has already been given in Section 10.3 beginning with (10.26) and extending to (10.31). In particular, when uncertainty intervals only are involved, (10.26) completely details the method. The so-called sensitivity factors, $\partial R/\partial V_i$, must be evaluated and used as multipliers of the uncertainty intervals of each of the independent variables [8], [13]. These products are squared, the results summed, and the overall square root taken according to (10.26). This procedure yields the uncertainty interval of the result to the same percent confidence as that of each of the variables. This is the method used to handle systematic errors propagated into a result.

When working with confidence intervals only, an equation patterned after (10.26) and (10.27) is used, namely,

$$CI_R = \pm\left[\left(\frac{\partial R}{\partial V_1} CI_1\right)^2 + \left(\frac{\partial R}{\partial V_2} CI_2\right)^2 + \cdots + \left(\frac{\partial R}{\partial V_N} CI_N\right)^2\right]^{1/2}.$$

(10.34)

This is the method used to handle random errors propagated into a result.

When one must deal with *both* random and systematic errors, as when a combination of calibrated and uncalibrated instruments are used together, an arbitrary decision must be made about the formation of an *overall* uncertainty interval. Two formulations are in common use. The more conservative estimate, according to [14], is

$$UI_{\text{overall} \atop \text{maximum}} = \pm(UI_{R \atop \text{systematic}} + CI_{R \atop \text{random}})$$

(10.35)

A more realistic estimate, presuming there will be some beneficial canceling of the errors or, to say it another way, assuming that all errors will not be in the same direction, is based on the familiar root mean square equation [8], [13], namely,

$$UI_{\text{overall} \atop \text{probable}} = \pm(UI_R^2 {\atop \text{systematic}} + CI_R^2 {\atop \text{random}})^{1/2}$$

(10.36)

Equation 10.36 is the approach recommended here for combining systematic and random errors.

Example 17.   A certain result (R) is related to its variables (V_i) by the equation

$$R = V_1 V_2^2 / V_3^{1/2} \tag{10.37}$$

The mean values of each variable, and the corresponding sensitivity factors ($\partial R / \partial V_i$) are found to be

Mean Value of Variable	Sensitivity
$V_1 = 1$	$V_2^2 / V_3^{1/2} = 5$
$V_2 = 5$	$2 V_1 V_2 / V_3^{1/2} = 2$
$V_3 = 25$	$-V_1 V_2^2 / 2 V_3^{3/2} = -0.1$

The uncertainty intervals for each variable, based on the experimenter's judgement, are best estimated as: $UI_1 = \pm 0.5\%$, $UI_2 = \pm 1\%$, $UI_3 = \pm 2\%$. Determine the *uncertainty interval* of the result. By (10.26)

$$UI_R = \pm[(5 \times 0.5)^2 + (2 \times 1)^2 + (1/5 \times 2)^2]^{1/2} = \pm 3.208\% .$$

Example 18.   Using relation (10.37) and the mean values and sensitivities of Example 17, estimate the *confidence interval* of the result if the following information applies.

Variable	Number of Measurements Taken, N	Estimate of Standard Deviation, S
1	20	0.5%
2	10	0.2%
3	5	0.1%

By (10.15)

$$CI_1 = \frac{t_{19} S_1}{\sqrt{N_1}} = \frac{2.093(0.5)}{\sqrt{20}} = 0.234\%$$

$$CI_2 = \frac{t_9 S_2}{\sqrt{N_2}} = \frac{2.262(0.2)}{\sqrt{10}} = 0.143\%$$

$$CI_3 = \frac{t_4 S_3}{\sqrt{N_3}} = \frac{2.776(0.1)}{\sqrt{5}} = 0.124\% .$$

By (10.34)

$$CI_R = \pm[(5 \times 0.234)^2 + (2 \times 0.143)^2 + (0.1 \times 0.124)^2]^{1/2} = \pm 1.204\% .$$

Example 19.   If the instrumentation used in determining the measurements of the three variables in (10.37) were such that *both* the systematic and the random errors of Examples 17 and 18 applied, what would be the maximum and probable values of the overall uncertainty interval?
By (10.35)

$$UI_{\text{overall}\atop\text{maximum}} = \pm(3.208 + 1.204) = \pm 4.412\% .$$

By (10.36)

$$UI_{\text{overall}\atop\text{probable}} = \pm(3.208^2 + 1.204^2)^{1/2} = \pm 3.426\% .$$

General Rules

Some general rules can be given for various combinations of variables [6].

RULE 1.   If a result is of the form

$$R = \pm rV_1{}^a \pm sV_2{}^b \pm tV_3{}^c + \cdots \qquad (10.38a)$$

the probable absolute uncertainty of R is given by

$$(UI_R)_{\text{probable}} = \pm\left[\left(rV_1{}^a \frac{UI_1}{V_1} a\right)^2 + \left(sV_2{}^b \frac{UI_2}{V_2} b\right)^2 + \left(tV_3{}^c \frac{UI_3}{V_3} c\right)^2 + \cdots\right]^{1/2}$$

$$(10.38b)$$

where UI_i/V_i are the relative uncertainties of the variables.

RULE 2.   If a result is of the form

$$R = rV_1{}^a V_2{}^b / V_3{}^c \qquad (10.39a)$$

the probable relative uncertainty of R is given by

$$\left(\frac{UI_R}{R}\right)_{\text{probable}} = \pm\left[\left(\frac{UI_1}{V_1} a\right)^2 + \left(\frac{UI_2}{V_2} b\right)^2 + \left(\frac{UI_3}{V_3} c\right)^2\right]^{1/2}. \qquad (10.39b)$$

RULE 3.   If a result is of the form

$$R = k \pm rV^a \qquad (10.40a)$$

where k is a constant, the absolute uncertainty of R is given by

$$UI_R = \pm rV_1{}^a \frac{UI_1}{V_1} a. \qquad (10.40b)$$

RULE 4.   If a result is of the form

$$R = rV_1{}^a - sV_1{}^b \qquad (10.41a)$$

the absolute uncertainty of R is given by

$$(UI_R)_{\text{maximum}} = \pm\left|\left(rV_1{}^a \frac{UI_1}{V_1} a\right) - \left(sV_1{}^b \frac{UI_1}{V_1} b\right)\right|. \qquad (10.41b)$$

All of these rules can be obtained by differentiating the given function provided that the uncertainties of the independent terms are added.

10.6   The Number of Measurements Required

To determine whether the number of measurements available is adequate to satisfy the test objectives, two distinct approaches are possible.

One is intended for use *before* a test has begun, and the other is for use *during* the early stages of a test.

Before a Test

One can propose a likely number of measurements per instruments per locations and, on the basis of confidence intervals established by experience and/or manufacturer's literature, using equations (10.27–10.34), determine if the proposed number of measurements and distributions among instruments and spacial locations appears sufficient.

Briefly, the procedure is as follows:

1. Form the *expected* confidence interval of the mean.
2. Form the *acceptable* confidence interval of the mean.
3. If acceptable *equals or exceeds* expected, conclude that the number of measurements proposed is adequate.

Specifically, let δ_{exp} be the expected confidence interval of the *mean* of a set of measurements (see Figure 10.6). From (10.15), one estimate of δ_{exp}, in normalized form, can be given in terms of the t statistic as

$$\left(\frac{\delta}{\bar{X}}\right)_{exp} = \frac{t_{N-1,P}(S/\bar{X})}{\sqrt{N}}. \tag{10.42}$$

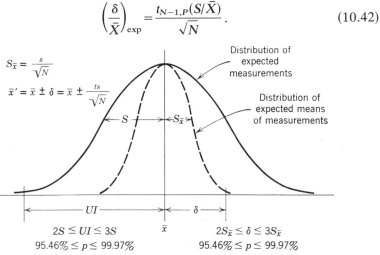

Figure 10.6 Distribution of measurements and means of measurements, with expected confidence intervals of each shown.

To arrive at δ_{exp}, we can estimate S conservatively as $CI/2$, where CI is the estimated confidence interval of the measurements [see (10.7 and 10.8)]. Thus

$$\left(\frac{\delta}{\bar{X}}\right)_{exp} = \frac{t_{N-1,p}(CI/\bar{X})}{2\sqrt{N}}. \tag{10.43}$$

Another normalized estimate of δ_{exp} can be given in terms of the substitute t statistic, according to (10.19), as

$$\left(\frac{\delta}{\bar{X}}\right)_{exp} = \tau_N(w/\bar{X}). \tag{10.44}$$

To arrive at an acceptable confidence interval of the mean, δ_{acc}, we must first distinguish between single and multiple variable experiments.

1. For single variable experiments, δ_{acc} must be specified.

2. For multiple variable experiments, it is usually an acceptable change in the result, ΔR_{acc}, that is specified. Then, via the partial differential relation, $\Delta R = (\partial R/\partial X)\delta$, it follows that

$$\left(\frac{\delta}{\bar{X}}\right)_{acc} = \frac{\Delta R/R}{\left(\frac{\partial R/R}{\partial X/\bar{X}}\right)} \tag{10.45}$$

in normalized percentage form.

The use of these ideas and these equations is illustrated by Examples 20 and 21.

Example 20.   Before a test, find if 100 measurements of X lead to an acceptable precision in a result of 0.1%. The sensitivity factor is 0.5, the mean value of X is 50, and the 95% confidence interval of measurements is estimated to be 1.

1. $\left(\dfrac{\delta}{\bar{X}}\right)_{exp} = \dfrac{t_{99,\,95\%}(S/\bar{X})}{\sqrt{N}} = \dfrac{1.99(1\%)}{\sqrt{100}} = 0.199\%,$

where $\dfrac{S}{\bar{X}} \approx \dfrac{CI}{2\bar{X}} = \dfrac{1}{2(50)} = 0.01 = 1\%.$

2. $\left(\dfrac{\delta}{\bar{X}}\right)_{acc} = \dfrac{\Delta R/R}{\text{sensitivity}} = \dfrac{0.1\%}{0.5} = 0.2\%.$

3. Since $\delta_{acc} \geq \delta_{exp}$, we conclude that 100 measurements are adequate for the required precision in the result.

Example 21.   Before a test, find if 20 measurements of X lead to an acceptable precision in a result of 0.1%. The sensitivity factor is 0.5, the mean value of X is 1000, and the 95% range of measurements is estimated at 20.

1. $\left(\dfrac{\delta}{\bar{X}}\right)_{exp} = \tau_N\left(\dfrac{w}{\bar{X}}\right) = 0.126\left(\dfrac{20 \times 100}{1000}\right) = 0.25\%.$

2. $\left(\dfrac{\delta}{\bar{X}}\right)_{acc} = \dfrac{\Delta R/R}{\text{sensitivity}} = 0.2\%.$

3. Since $\delta_{acc} < \delta_{exp}$, we conclude that 20 measurements are not adequate to yield the required precision in the result.

During a Test

In the recent literature [10], [11], [15], the suggestion has been made to take a small number of measurements and perform a quick analysis, to determine whether the number of measurements taken was adequate or how many more measurements are still required to satisfy test objectives. This type of analysis is based on (10.15), solved for N as

$$N_{\text{total}} \simeq \left(\frac{t_{N_1-1,p} S_1}{\delta_{\text{acc}}} \right)^2. \tag{10.46}$$

Briefly, the procedure is as follows:

1. From a small sample of size N_1, obtain $S_1 (\simeq w/(d_2)_N)$ according to (10.17).
2. Establish an acceptable confidence interval of the mean.
3. Estimate N_T via (10.46).
4. Then the size of the remaining number of measurements required is simply $N_2 = N_T - N_1$.

This approach is sometimes called the Stein method.

Example 22. From 50 measurements it is determined that $S_1 = 160$. How many more measurements should be taken to ensure an acceptable δ of 30?

1. $S_1 = 160$ based on $N_1 = 50$.
2. $\delta_{\text{acc}} = 30$.
3. By (10.46), $N_T = \left(\dfrac{2.01 \times 160}{30} \right)^2 \simeq 115$.
4. $N_2 = 115 - 50 = 65$ additional measurements required.

Example 23. For ten initial measurements, the range is determined to be 4. If the mean value is 30 and the sensitivity factor is 0.5, find the number of additional measurements required to ensure a $\Delta R/R$ of 0.1% at the 95% confidence level.

1. $\dfrac{\delta_{\text{acc}}}{\bar{X}} = \dfrac{\Delta R/R}{\text{sensitivity}} = \dfrac{0.1\%}{0.5} = 0.2\%$,

and $\delta_{\text{acc}} = \dfrac{\bar{X} \times 0.2\%}{100} = 0.06$.

2. $S_1 \simeq \dfrac{w}{(d_2)_{10}} = \dfrac{4}{3.078} = 1.3$.

3. By (10.46), $N_T = \dfrac{2.262(1.3)}{0.06} = 49$.

4. $N_2 = 49 - 10 = 39$ additional measurements required.

Critique

The main fault of these approaches (i.e., Before Test and During Test) is that they are essentially limited to time variations only. But variations with space and installation may far outweigh random variations with time. In such cases, the numbers predicted by the Before and During approaches may be of academic interest only. For example, we may be predicting the need for 100 measurements to reduce σ_t to 0.1% when at the same time, all unexamined, σ_s and/or σ_i are on the order of 2%.

After a Test

One should, of course, use (10.26 and 10.34) with the experimentally determined confidence intervals for the random errors, and/or the estimated uncertainty intervals for the systematic errors, to decide if the number of measurements taken were adequate. Such procedures are illustrated by Examples 9–19.

10.7 References

[1] W. J. Youden, "Uncertainties in Calibration," IRE Trans. on *Instrumentation*, December 1962, p. 133.

[2] D. G. Sanders, "Accuracy of Type K Thermocouple Wire Below 500°: A Statistical Analysis," ISA Trans. Vol. 13, No. 3, 1974, p. 202.

[3] R. P. Benedict, "Engineering Analysis of Experimental Data," *Trans. ASME, J. Eng. Power*, January 1969, p. 21.

[4] E. L. Grant, *Statistical Quality Control*, 3rd ed., McGraw-Hill, New York, 1964.

[5] ASTM STP-15-C-1967, "Presentation of Data and ± Limits of Accuracy of an Observed Average."

[6] R. P. Benedict, "Uncertainties in Measurement," *Electro-Technol.*, October 1964, p. 51.

[7] S. J. Kline and F. A. McClintock, "Describing Uncertainties in Single-Sample Experiments," *Mech. Eng.*, Vol. 75, January 1953.

[8] G. D. Johnson et al., "Evaluation of Measurement Uncertainties in Performance Testing of Hydraulic Turbines and Pump/Turbines," *Trans. ASME, J. Eng. Power*, April 1975, p. 145.

[9] J. S. Wyler, "Estimating the Uncertainty of Spacial and Time Average Measurements," *Trans. ASME, J. Eng. Power*, October 1975, p. 473.

[10] W. H. Rousseau and E. L. Milgram, "Estimating Precision in Heat Rate Testing," *Trans. ASME, J. Eng. Power*, July 1974, p. 223.

[11] S. Sigurdson and D. E. Kimball, "Practical Method for Estimating Number of Test Readings Required," ASME Paper 75-WA/PTC-1, December 1975.

[12] J. Stuart Hunter, *Design of Experiments*, Vol. 2, Elements of Experimental Design and Analysis, Westinghouse Electric Corporation, 1968, pp. 1–22.

[13] "Calculation of the Uncertainty of a Measurement of Flow Rate," ISO/TC-30, January 1975.

[14] D. L. Colbert, B. D. Powell, and R. B. Abernethy, "Measurement Uncertainty Handbook for Liquid Rocket Engines," PWA FR-2974B, 30, April 1969 (Obtained from Defense Documentation Center AD 85127).

[15] M. G. Natrella, *Experimental Statistics*, NBS Handbook 91, August 1963.

Nomenclature

Roman

a, b, c	exponents
r, s, t	coefficients
CI	confidence interval
d_2	statistic used with range
i	instrument or installation
L	number of spacial locations
M	number of sets of observations, number of instruments or installations
N	number of observations in sample
p	probability
P	value of parameter
$\bar{\bar{P}}$	best value of parameter
R	result
s	space
S	estimate of standard deviation
$\bar{\bar{S}}$	weighted average of M sets of S
t	student distribution, time
UI	uncertainty interval
V	variable
w	range of observations
$\bar{w}$	average of M sets of w
X	value of measurement
$\bar{X}$	mean of single set of measurements
$\bar{\bar{X}}$	mean of M sets of measurements
$\bar{X}'$	true value of quantity
Z	normal deviate

Greek

δ_{acc}	acceptable confidence interval
δ_{exp}	expected confidence interval
ϵ	error
ν	degrees of freedom
σ	true standard deviation
τ	substitute t statistic used with range

Problems

1. A thermocouple in a fixed point bath yields the following replicate data in terms of deviates (Xi) from the standard NBS Reference Table.

N	1	2	3	4	5	6	7	8	9
Xi (mV)	0.061	0.061	0.061	0.062	0.062	0.050	0.059	0.056	0.058

Find the mean value, the range, the estimated standard deviation, and the 95% confidence interval by both range and estimated standard deviation for the *individual* readings.

Ans. $\bar{X} = 0.059$ mV, $w = 0.012$ mV, $S = 0.0039$ mV, $CI_w = \pm0.0092$ mV, $CI_S = \pm0.0090$ mV.

2. If the thermocouple of problem 1 was a type J with a slope of 0.03085 mV/°F, express the confidence intervals of Problem 1 in terms of temperature.

Ans. $CI_w = \pm0.298°F$, $CI_S = \pm0.292°F$.

3. Given three sets of ten readings each, find the 95% confidence interval of the mean if: (a) $\sigma = 0.5$, (b) $\bar{\bar{S}} = 0.5$, (c) $\bar{w} = 0.5$.

Ans. (a) $CI_\sigma = 0.179$, (b) $CI_{\bar{S}} = 0.187$, (c) $CI_w = 0.066$.

4. Two instruments are used at each of four spacial locations, with ten readings taken on each. The average range between instruments is 0.1%. The spacial standard deviation is estimated at 0.1%. The actual standard deviation with respect to time is 0.2%. Find the 95% confidence interval about the best value of the parameter being measured.

Ans. $CI_{\bar{p}} = 0.35796\%$.

5. The mass flow rate is given in terms of $Ap/\sqrt{T}$. If $A = 10$, $p = 20$, and $T = 484$, in compatible units, and the uncertainties are respectively: $\pm0.1\%$, $\pm0.2\%$, and $\pm1\%$, find the probable uncertainty interval in $\dot{m}$.

Ans. $UI_{\dot{m}} = \pm0.129\%$.

Chapter 11

TEMPERATURE
MEASUREMENT
IN MOVING FLUIDS

". . . it must be obvious that a thermometer placed in the wind registers the temperature of the air, plus the greater portion, but not the whole, of the temperature due to the vis viva of its motion . . ."

James Prescott Joule and William Thomson (1860)

Thermodynamic temperatures are defined for thermal states of statistical equilibrium only. These are seldom encountered in practice; thus the usual temperature concept must be modified. A common nonstatic condition for which temperature measurement requires special definitions is that in which a directed kinetic energy of flow exists in a fluid. Methods for practical measurement of temperature in real moving fluids using real temperature-sensing probes can be developed by making certain necessary modifications of the concepts of idealized fluids, temperature relations, and probes.

11.1 Idealized Gases

Because of its simplicity, an expression of the form

$$pv = RT \tag{11.1}$$

has been used since 1834 as an idealized equation of state of a gas, and may be accepted as a first approximation to the equation of state of a real gas (see 2.10, 2.13, 2.15, 3.4) [1]. In this connection we note a most important thermodynamic relationship that holds for all real gases and has never been contradicted by experiment, namely

$$(pv)^0 = RT, \tag{11.2}$$

where the superscript ^{0} refers to the zero pressure intercept (see 3.11). By comparing (11.1) and (11.2), we may observe some of the implications of using the ideal equation of state to represent a real gas [2], [3]:

1. The expression $pv = RT$ serves as an increasingly exact relation for all real gases as the pressure approaches zero along the respective isotherm.

2. Since at these low pressures (large specific volumes) both the size of the molecules and the intermolecular forces are negligible, the transport properties such as viscosity and thermal conductivity which depend on molecular size and interaction must likewise be considered negligible.

11.2 Idealized Gas-Temperature-Sensing Probes

An idealized gas-temperature-sensing probe is defined arbitrarily at this point as one that will completely stagnate a moving gas continuum locally (i.e., ideal geometry), and is isolated, in terms of heat transfer, from all its surroundings (i.e., adiabatic).

11.3 Idealized Gas-Temperature Relations

Consider several cases (Figure 11.1a), each involving the following: a system with fixed boundaries across which no mechanical work or heat is transferred, an idealized gas continuum ($pv/T = $ a constant, $c_p = $ a constant), and an idealized gas-temperature-sensing probe (ideal geometry, adiabatic).

1. For the case in which both the probe and the gas are at rest with respect to the system boundaries, the probe will indicate the gas temperature. This temperature may be visualized as a measure of the average random translational kinetic energy of the continuum molecules (but see, for example, [4]).

2. For the case in which both the probe and the gas are in identical motion with respect to the system boundaries, the probe again will indicate the gas temperature. However, this local temperature must be

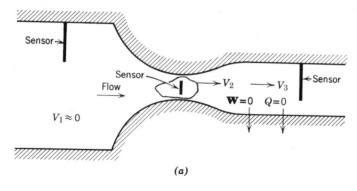

(a)

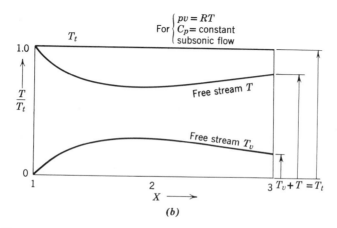

(b)

Figure 11.1 Three cases illustrating idealized gas temperature relations: (*a*) the physical arrangement of the elements and (*b*) the temperatures involved.

distinguished carefully from the temperature of case 1. The gas temperature of case 2 is lower than that of case 1 by an amount equivalent to that part of the random thermal energy which now appears in the form of directed kinetic energy of the gas continuum.

3. For the case in which the gas is in direct motion with respect to both the probe and the system boundaries, the probe will indicate not only the gas temperature (as in case 2) but, in addition, will indicate a temperature equivalent to the directed kinetic energy of motion of the gas continuum. This latter part of the net indicated temperature is obtained by stagnating the gas continuum locally. The probe thus reconverts the directed kinetic energy back to random thermal effects. Note that this net indicated temperature is identical to the gas temperature of case 1.

Three separate temperatures must be distinguished in the aforementioned three cases (Figure 11.1*b*):

1. Static Temperature (T). This is the actual temperature of the gas at all times (in motion or at rest). It has been considered as a measure of the average random translational kinetic energy of the molecules. The static temperature will be sensed by an adiabatic probe in thermal equilibrium and at rest with respect to the gas.

2. Dynamic Temperature (T_v). The thermal equivalent of the directed kinetic energy of the gas continuum is known as the dynamic temperature.

3. Total Temperature (T_t). This temperature is made up of the static temperature plus the dynamic temperature of the gas. The total temperature will be sensed by an *idealized* probe, at rest with respect to the system boundaries, when it stagnates an *idealized* gas.

That these three temperatures are related in the manner stated may be seen by applying the steady-flow general energy equation to a situation in which the idealized gas continuum is in direct motion with respect to the system boundaries [5]:

$$\delta Q + \delta \mathbf{W} = du + d(pv) + \frac{V\,dV}{g_c}. \tag{11.3}$$

Or, introducing the enthalpy definition, $h = u + pv$, (11.3) becomes

$$\delta Q + \delta \mathbf{W} = dh + \frac{V\,dV}{g_c}, \tag{11.4}$$

where δQ = heat transferred across system boundaries;

$\quad\;\; \delta \mathbf{W}$ = mechanical work transferred across system boundaries;

$\quad\;\; dh$ = enthalpy change between two thermodynamic states within the system;

In general:

$$dh = c_p\,dT + \left[v - \left(\frac{\partial v}{\partial T} \right)_p T \right] dp; \tag{11.5}$$

and for the ideal gas, $dh = c_p\,dT$ (since $(\partial v/\partial T)_p T = v$).

$\dfrac{V\,dV}{g_c}$ = net change in directed kinetic energy between two thermodynamic states within the system.

All are on a per pound mass basis. In the absence of heat transfer and mechanical work across the system boundaries, (11.4) may be integrated as

$$c_p(T_2 - T_1) = (V_1{}^2 - V_2{}^2)/2Jg_c, \tag{11.6}$$

where J is the mechanical equivalent of heat (778 ft lb/Btu). Thus the general energy relation indicates that a change in the directed kinetic energy of the gas continuum is always accompanied by a change in the static temperature of the gas [6]. Furthermore, if the subscript 2 in (11.5) refers to the stagnant condition, we note that the temperature in a stagnant gas ($T_2 = T_t$) is always greater than the temperature in a moving gas ($T_1 = T$) by an amount equivalent to the directed kinetic energy of the gas continuum, that is,

$$T_t = T + \frac{V^2}{2Jg_c c_p} = T + T_v, \tag{11.7}$$

and thus the dynamic temperature is particularized as $V^2/2Jg_c c_p$.

11.4 Idealized Liquids

In the absence of heat transfer across the boundaries of a system in which a frictionless (inviscid), incompressible fluid flows, the internal energy remains constant throughout any process. This may be seen by applying the first law of thermodynamics to such a situation

$$\delta Q + \delta F = du + p \, dv. \tag{11.8}$$

Consequently, the temperature of an idealized liquid must also remain constant throughout any process. This follows from the basic assertion that the internal energy, generally $u = f(T, v)$, is a function of temperature only for an incompressible fluid [7]. Thus it is incorrect to speak of the various temperatures (static, dynamic, total) of an inviscid, incompressible fluid; the liquid temperature (T_l) is the only one of any significance. No change in the directed kinetic energy of a liquid continuum will effect a change in this liquid temperature, and any adiabatic probe will indicate this liquid temperature.

In Summary: (a) If the idealized equation of state ($pv = RT$) well represents the thermodynamic quantities (p, v, T) of a particular *gas*, and, if the specific heat capacity c_p of the gas may be considered constant over the temperature range involved, the total temperature will be constant in any adiabatic, workless change in the thermodynamic state of the gas. Furthermore, any adiabatic probe that completely stagnates this gas locally will indicate the total temperature, that is, $T_{pi} = T_t = T + T_v$

(idealized gas and idealized probe), where T_{pi} is the equilibrium temperature sensed by a stationary, ideal-geometry, adiabatic probe.

(b) If a *liquid* can be considered inviscid and incompressible, its temperature will remain constant in any adiabatic, workless change in the thermodynamic state of the liquid. Furthermore, any adiabatic probe will indicate this liquid temperature.

11.5 Real Gas Effects

Idealized relations are useful in that they allow us to draw certain broad conclusions with a minimum of effort, but in the measurement of temperature in moving fluids we must also consider many perturbation effects, especially those arising because of a deviation of the fluid characteristics from assumed idealized relations.

Departures from ideal conditions in gases are immediately encountered, for we do not always test at near-zero pressures. Thus, in general, both the size of the molecules and the intermolecular forces become important. Along with these realities, the associated transport properties (i.e., the viscosity and the thermal conductivity) must also be considered [8], [9], [10].

A second approximation to the true equation of state of a real gas was given in 1873 by van der Waals as

$$\left(p + \frac{a}{v^2}\right)(v + b) = RT, \tag{11.9}$$

where the term a/v^2 was introduced to account for intermolecular forces, and the term b for the finite size of the molecules [11].

The actual equation of state of a real gas is naturally more complex than any of its approximations; yet even with the van der Waals gas, and even with a constant specific heat capacity c_p, the static temperature no longer remains constant in an isenthalpic (i.e., constant enthalpy), workless change of state. This phenomenon, wherein the static temperature changes in a constant enthalpy (i.e., throttling) expansion of any real fluid, is known as the Joule-Thomson effect [12] (see Figure 11.2).

Furthermore, at a point we cannot, in general, realize the total temperature, although the real gas (or even the van der Waals gas) is completely stagnated locally by an isolated temperature-sensing probe. This is because of the combined effects of aerodynamic stagnation, viscosity, and thermal conductivity, which set up temperature and velocity gradients in the fluid boundary layers surrounding the probe [13].

The consequent rise in temperature of the inner fluid layers that results

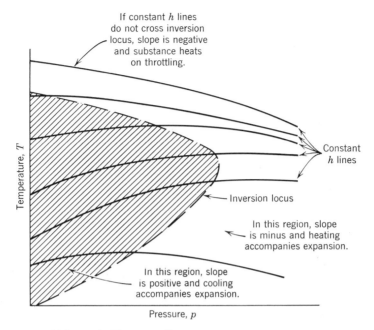

Figure 11.2 Joule-Thomson effect.

from a combination of viscous shear work on the fluid particles and the impact-conversion of directed kinetic energy to thermal effects is necessarily accompanied by a heat transfer through the gas, away from the adiabatic probe. These opposing effects tend to upset the simple picture of total temperature recovery previously given.

The usual thermodynamic simplification introduced to allow the continued definition of total temperature in this situation is the assumption of an isentropic process of deceleration, signifying a reversible, adiabatic stagnation. By whatever designation, however, the implications are unrealistic in practical thermometry. The term "adiabatic" is meant to indicate absence of heat transfer, both to and from the probe and to and from the fluid, but we have seen that a temperature gradient must exist in the gas. The term "reversible" is reversed for quasi-static (slow) or "frictionless" processes, but the stagnation of a moving gas is not a quasi-static process. Furthermore the viscous shear forces that are present are synonomous with friction forces in a fluid. Thus boundary-layer effects, associated with viscosity and thermal conductivity, conflict with the isentropic assumption, and the total temperature simply cannot be realized by a measurement in a real gas.

Note that viscosity and conductivity are simply perturbation effects on the main flow, and their relative importance is indicated by the Prandtl number; that is, the Prandtl number is a ratio of the fluid properties governing transport of momentum by viscous effects (because of a velocity gradient) to the fluid properties governing transport of heat by thermal diffusion due to a temperature gradient, that is,

$$\text{Pr} = \frac{\text{kinematic viscosity}}{\text{thermal diffusivity}} = \frac{c_p \mu}{k}. \tag{11.10}$$

By replacing the isentropic assumption with the assumption of a Prandtl number of one, we need not discount the effects of conductivity or viscosity; both may be actively present, but they will be counterbalancing effects. This is the same requirement as for the Reynolds analogy, where heat and momentum are transported in the same manner in a fluid. Thus we have avoided the mental stumbling block that an isentropic assumption creates when we work with real gases. But the total temperature *still* will not be sensed by an idealized probe, at rest with respect to system boundaries, when it stagnates a *real* gas even if the Prandtl number is one.

11.6 Real Liquid Effects

With liquids, as with gases, we never encounter a real fluid of zero viscosity or of zero thermal conductivity. We must therefore modify our assertion that was based on an idealized liquid to the effect that there is only one significant temperature T_l in a liquid.

The first consideration is the Joule-Thomson effect. For liquids, the Joule-Thomson coefficient is generally negative (true for water below 450°F). As pressure drops isenthalpically, therefore, temperature rises (see Figure 11.2), in contrast to the case for most gases. This means that the liquid temperature will not remain constant in any throttling change of state of a liquid. Furthermore, for reasons previously given concerning the interplay between viscous shear work and heat transfer in the liquid boundary layers surrounding any probe immersed in any real liquid, we cannot in general realize the liquid temperature. Even when the Prandtl number of the real liquid is one, the adiabatic probe will not sense the idealized liquid temperature T_l.

In Summary: (a) The static temperature will not be constant in an isenthalpic, workless change in the thermodynamic state of any real fluid. This is explained by the Joule-Thomson effect. Furthermore, even if the Prandtl number of a *gas* is one, the adiabatic probe that completely stagnates such a gas locally will still fail to indicate the local total temperature.

(b) An adiabatic probe immersed in a real *liquid*, even if its Prandtl number is one, also will fail to indicate the local liquid temperature.

11.7 The Recovery Factor

The fluids we test are not always characterized by a Prandtl number of one (e.g., the Prandtl number of air varies between 0.65 and 0.70; the Prandtl number of steam varies between 1 and 2; and the Prandtl number of water varies between 1 and 13. See Table 11.1). Therefore we must again alter the simplified picture of temperature recovery. Total temperature in a gas and liquid temperature in a liquid are seen more and more in their true light, as idealized concepts or devices rather than as physically measurable quantities.

Joule and Thomson [12] as early as 1860 noted ". . . it must be obvious that a thermometer placed in the wind registers the temperature of the air, plus the greater portion, but not the whole, of the temperature due to the vis viva of its motion. . ." Of course they were close to the idea of a recovery factor, which we now introduce to account for deviations from previously stated ideal conditions in real fluids.

Table 11.1 Typical Values of Prandtl Number for Various Fluids

		Prandtl Number					
$T(°F)$	Air (Low Pressure)	Water (Saturated)	15	100	Steam (psia) 200	500	1000
0	0.720	—	—	—	—	—	—
100	0.705	4.35	—	—	—	—	—
200	0.694	1.89	—	—	—	—	—
300	0.686	1.14	0.98	—	—	—	—
400	0.681	0.90	0.95	1.04	1.16	—	—
500	0.680	0.85	0.94	0.99	1.04	1.24	—
600	0.680	1.02	0.93	0.95	0.98	1.08	1.30
700	0.681	—	0.92	0.92	0.94	1.00	1.12
800	0.684	—	0.91	0.91	0.92	0.95	1.01
900	0.687	—	0.90	0.90	0.91	0.92	0.95
1000	0.690	—	0.89	0.89	0.90	0.91	0.92

At a Stagnation Point

Even when we limit our attention to the fluid stagnation point, the total temperature of a gas and the liquid temperature of a liquid can never be measured directly. This assertion can be substantiated by the following development. For an isentropic workless process, (11.4) yields

$$dh_s = -\frac{V \, dV}{g_c} \tag{11.11}$$

and (11.8) yields

$$0 = (du + p \, dv)_s \tag{11.12}$$

where the subscript s signifies an isentropic process. By (11.12), the enthalpy change (generally, $dh = du + p \, dv + v \, dp$) also can be expressed as

$$dh_s = v \, dp_s. \tag{11.13}$$

When the general definition of enthalpy change, (11.5), is combined with (11.13), these results

$$dT_s = T\left(\frac{\partial v}{\partial T}\right)_p \frac{dp_s}{c_p} \tag{11.14}$$

which, when combined with (11.11) and (11.13) yields

$$dT_s = -\left[\frac{T}{v}\left(\frac{\partial v}{\partial T}\right)_p\right]\frac{V \, dV}{Jg_c c_p}. \tag{11.15}$$

On integrating (11.15) between the stagnation and free stream conditions, we obtain

$$T_{\text{stagnation}} = T_{\text{free stream}} + S \times T_{\text{dynamic}} \tag{11.16}$$

where the stagnation factor

$$S = \frac{T}{v}\left(\frac{\partial v}{\partial T}\right)_p \tag{11.17}$$

is often considered a constant, or its mean value may be used [14]. The factor S is shown graphically in Figure 11.3 for air, water, and steam.

It is clear from a comparison of (11.7) and (11.16) that $T_{\text{stagnation}}$ does not equal T_{total} for real gases. However, for the idealized gas, since $(\partial v/\partial T)_p = R/p = v/T$, the stagnation factor does equal unity, and hence $(T_{\text{stagnation}})_{\text{ideal gas}}$ does reduce to T_t as already indicated.

Since (11.16) applies equally well to liquids, gases, and vapors, it is likewise clear that $T_{\text{stagnation}}$ does not equal the liquid temperature T_l which, of course, is the same as the free stream temperature. However, for the idealized liquid, since $(\partial v/\partial T)_p = 0$, the stagnation factor equals zero, and hence $T_{\text{stagnation}}$ does reduce to T_l, as already indicated.

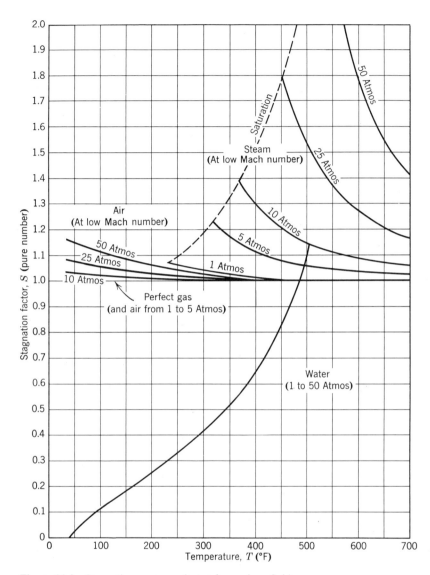

Figure 11.3 Stagnation recovery factor for various fluids.

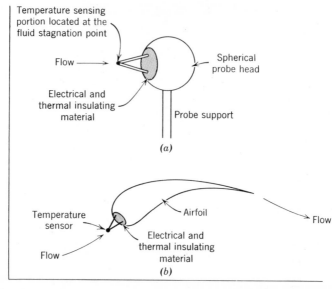

Figure 11.4 (a)

Figure 11.4 (b)

Figure 11.4 Several schematic arrangements of a stagnation temperature probe.

The stagnation factor S, as given by (11.17), and as used in (11.16), we now define as the *recovery factor* at a fluid stagnation point. Any probe that is designed so that the temperature-sensing portion is located at an isentropic stagnation point will be characterized by a recovery factor of S and will yield the stagnation temperature (see Figure 11.4).

Example 1. A temperature stagnation probe, as shown in Figure 11.4*a*, is exposed to atmospheric air flow of 500 ft/sec, where $c_p = 0.24$ and the static temperature is 200°F. Find the deviation between the probe stagnation temperature and the thermodynamic total temperature.

By (11.7) $$T_t = T + V^2/2Jg_c c_p$$

where $T =$ static temperature = free stream temperature = $200 + 460 = 660$°R, and $T_v =$ dynamic temperature = $500^2/2 \times 778 \times 32.174 \times 0.24 = 250{,}000/12{,}015 = 20.8$°R.

Hence, $T_t =$ total temperature = $660 + 20.8 = 680.8$°R

By (11.16) $$T_{stag} = T + ST_v$$

where $S =$ stagnation recovery factor = 1 from Figure 11.3.

Hence, $$T_{stag} = 660 + 1 \times 20.8 = 680.8°\text{R}.$$

In this example, the difference between the measured $T_{stagnation}$ and the idealized T_{total} is 0°F.

Example 2. A turbine blade, arranged as in Figure 11.4*b*, is exposed to a steam flow of 500 ft/sec where $c_p = 0.58$, the static temperature is 400°F, and the static pressure is 150 psia. Find the difference between the airfoil stagnation temperature and the thermodynamic total temperature.

By (11.7) $$T_t = T + T_v$$

where $$T = 400 + 460 = 860°R,$$

and $$T_v = 500^2/2 \times 778 \times 32.174 \times 0.58 = 250,000/29,106.5 = 8.59°R.$$

Hence, $$T_t = 860 + 8.6 = 868.6°R.$$

By (11.16) $$T_{stag} = T + ST_v$$

where $S = 1.28$ from Figure 11.3.

Hence, $$T_{stag} = 860 + 1.28 \times 8.59 = 871.0°R.$$

In this example, the difference between the measured $T_{stagnation}$ and the idealized T_{total} is 2.4°F.

Example 3. For the stagnation probe shown in Figure 11.4*a*, what is the expected difference between the stagnation temperature and the liquid temperature when water flows at a velocity of 20 ft/sec at a temperature of 200°F?

By (11.16) $$T_{stag} = T + ST_v$$

where $$T = 200 + 460 = 660°R,$$

and $$S = 0.26 \text{ from Figure 11.3,}$$

and $$T_v = 20^2/2 \times 778 \times 32.174 \times 1 = 400/50,062.7 = 0.008°R.$$

Hence $T_{stag} = 660 + 0.26 \times 0.008 = 660.002°R$, and the difference between the measured $T_{stagnation}$ and the idealized liquid temperature is 0.002°F.

Over a Flat Plate

Much work has been done on the viscous frictional recovery factor that characterizes the flow of gas over a flat plate. Patterned directly after (11.16) we have

$$T_{\substack{\text{adiabatic} \\ \text{flat plate}}} = T_{\text{free stream}} + r \times T_{\text{dynamic}}. \tag{11.18}$$

However, although the recovery factor is expressed as some percentage of the dynamic temperature, there is no implication that the recovery factor indicates only the degree of conversion of directed kinetic energy to thermal effects, or that a recovery of one is the maximum attainable. For example, for gases having Prandtl numbers below one, thermal-conduction effects overshadow viscous effects, and the adiabatic flat plate will sense a temperature less than the total temperature (i.e., $r < 1$). Conversely, for Prandtl numbers above one, the flat plate will come to equilibrium at a temperature greater than the total temperature (i.e., $r > 1$). The effects of various recovery factors are shown in Figure 11.5.

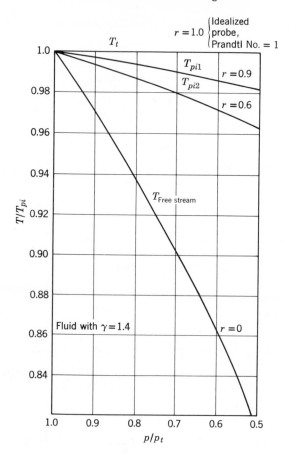

Figure 11.5   The effects of various viscous recovery factors.

For liquids, although not commonly discussed, the recovery factor r might still be applied, where r accounts not at all for the degree of conversion of directed kinetic energy to thermal effects, but for the net thermal effect resulting from the viscous shear work and heat transfer in the boundary layers.

Several particular recovery factors have been distinguished for gases [15]. Very little information on recovery factors for liquids flowing over flat plates appears in the literature.

The fractional recovery factor has been found to be very nearly independent of the Mach and Reynolds numbers. Pohlhausen, in 1921, found that $r = f(\text{Pr}) = 0.844$ for flat plates in a laminar flow of air.

Emmons and Brainerd [16], in 1941, and Squire [17], in 1942, found that

$$r_{\text{laminar}} = \text{Pr}^{1/2} \tag{11.19}$$

for flat plates in a laminar flow of fluid having a Prandtl number between 0.5 and 2, and for Mach numbers ranging between 0 and 10. For air, $\text{Pr}^{1/2}$ yields 0.846 for the recovery factor.

In a turbulent boundary layer we know that, effectively, both thermal diffusivity (d) and kinematic viscosity (ν) increase because of the mechanism of eddy diffusion (ϵ) in addition to molecular diffusion. If we assume, as is usually done, that the increase in apparent viscous effects is identical with the increase in apparent heat-transfer effects, an effective Prandtl number may be defined as

$$\text{Pr}_{\text{eff}} = \frac{(\nu + \epsilon)}{(d + \epsilon)}. \tag{11.20}$$

Thus, if $\text{Pr} > 1$, d will increase by the greater percentage, and the effect will be a lower plate temperature in turbulent flow than in laminar flow. The converse is true if $\text{Pr} < 1$. Consequently, for any Prandtl number, the plate in turbulent flow at adiabatic equilibrium will reach a temperature that is always closer to the total temperature than that attained by the plate in laminar flow. Emmons and Brainerd and others found that the local recovery factor could be represented very nearly as

$$r_{\text{turbulent}} = \text{Pr}^{1/3} \tag{11.21}$$

for flat plates at the higher Reynolds numbers (i.e., in turbulent boundary layers).

Over a Cylinder

Eckert and Weise [18] indicate that r is also very nearly constant for the laminar flow of air over right cylinders (the usual geometry for temperature probes and wells). That is, the frictional recovery factor for a cylinder is similar to the recovery factor for a flat plate up to the point of flow separation, or until the transition to a turbulent boundary layer. On this basis, we would use (11.19) and (11.21) to estimate the recovery factor for cylindrical temperature-sensing probes.

In opposition to this simplified viewpoint, however, El Agib, et al. [19] suggest the use of an overall recovery factor R, which is operationally defined by

$$T_{\substack{\text{adiabatic}\\ \text{probe}}} = T_{\text{free stream}} + RT_v. \tag{11.22}$$

Reasoning that the *heat-conducting* cylindrical temperature probe will sense some mean temperature as a result of stagnation heating at a point and viscous heating around the surface of the probe, they combine the stagnation recovery factor S and the frictional recovery factor r to arrive at an overall recovery factor

$$R = \bar{P}S + (1 - \bar{P})r \qquad (11.23)$$

where $\bar{P}$ is the average value of the pressure coefficient around the circular cylindrical surface, defined in turn as

$$\bar{P} = \frac{1}{\pi} \int_0^\pi P_\theta \, d\theta. \qquad (11.24)$$

P_θ is the local value of the pressure coefficient (see Figure 11.6).

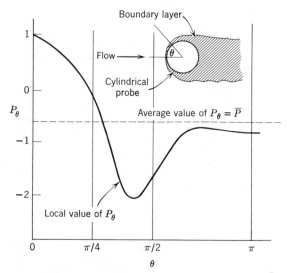

Figure 11.6 Pressure coefficient for a cylinder. $P_\theta = \dfrac{p_\theta \text{ gauge}}{p \text{ dynamic}}$ $\bar{P} = \dfrac{1}{\pi} \displaystyle\int_0^\pi P_\theta \, d\theta.$

Actually, both the average and the local pressure coefficients vary widely with Reynolds number, free stream turbulence, and the nature of the boundary layer surrounding the probe. However, in lieu of more specific information, we can often use to advantage the representative value, $\bar{P} = -0.7$, for the cylindrical probe.

The $\bar{P}S$ term in (11.23) accounts for the stagnation effects at the probe leading edge, and the $(1 - \bar{P})r$ term accounts for the frictional effects around the surface of the heat conducting probe.

Confidence in the (11.23) formulation of R is gained when it is noted that:

(a) for *flat plates*, the average value of the pressure coefficient is 0, and the overall recovery is seen to reduce to the frictional recovery factor r, as expected.

(b) at the *stagnation point*, as at the leading edge of the probe, the pressure coefficient is 1, and the overall recovery is seen to reduce to the stagnation recovery factor S, as expected.

Example 4. An adiabatic cylindrical temperature probe is installed perpendicular to a flowing stream of water. If the temperature is 100°F, and the Prandtl number is 4.44, and the probe boundary layer is laminar, find the overall recovery factor. By Figure 11.3, $S = 0.12$.

By (11.19), $$r_{laminar} = Pr^{1/2} = 4.44^{1/2} = 2.11.$$

Using $\bar{P} = -0.7$, as discussed, there results by (11.23)

$$R = -0.7(0.12) + 1.7(2.11) = 3.5.$$

Example 5. The same probe of Example 4 is installed in a moving air stream. If the temperature is 300°F, and the Prandtl number is 0.686, and the probe boundary layer is turbulent, find the overall recovery factor.

By Figure 11.3, $$S = 1.$$

By (11.21), $$r_{turbulent} = Pr^{1/3} = 0.686^{1/3} = 0.88.$$

Using $\bar{P} = -0.7$, as discussed, there results by (11.23)

$$R = -0.7(1) + 1.7(0.88) = 0.8.$$

11.8 The Dynamic Correction Factor for Gases

A real probe, immersed in a real gas, tends to radiate to its surroundings. Also, there is a tendency for a conductive heat transfer along the probe stem (and wires if they are part of the thermometer). These two effects are, of course, just balanced by a "convective" heat transfer between the probe and the gas, which must be considered whenever the actual probe temperature differs from the adiabatic probe temperature (see Chapter 12). In addition, real probes do not always stagnate a moving gas effectively. This is important when the fluid is compressible; that is, although there will be stagnation at a point, most temperature-sensing probes indicate a mean temperature, and the effect of the entire probe geometry must be considered.

Thus it is not more realistic to consider an idealized probe than it is to consider a fluid in which effects of viscosity and thermal conductivity are nil. Once more we must revise our ideas concerning temperature recovery, this time to account for real-probe heat transfer effects.

A dynamic correction factor K, correcting for the performance of a diabatic probe that attempts to stagnate a real flowing gas, may be defined so that

$$T_p = T + K T_v, \qquad (11.25)$$

where T_p is the equilibrium temperature sensed by the stationary, real probe. Here it is appropriate to remark that all reported experimental measurements of temperature-recovery factors are, of necessity, determinations of the dynamic correction factor, where differences between K and R indicate deviations from the idealized probe assumptions. Unlike the recovery factor, the dynamic correction factor is especially affected by variations in the Mach and Reynolds numbers, and by the surrounding, fixed-boundary temperatures. At low directed velocities, the probe temperature will be influenced more by stem conduction and radiation than by convective heat transfer, and the dynamic correction factor may assume any value. However, as the directed velocity increases, the probe temperature is dominated by convective heat transfer, and the dynamic correction factor approaches the overall recovery factor (see Figure 11.7). The temperature indicated by a real probe immersed in a real gas may be expressed in terms of a deviation from the ideal total temperature as

$$T_t - T_p = (1 - K) T_v. \qquad (11.26)$$

In summary, in the practical measurement of temperature in real moving gases, a real probe will indicate an equilibrium temperature

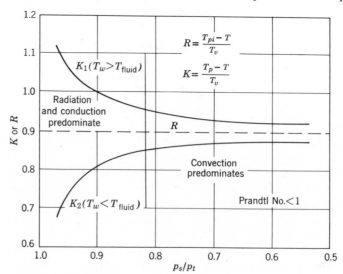

Figure 11.7 General dynamic correction factors.

which, in general, differs from the total temperature (real gas, real probe); that is,

$$T_p = T + KT_v,$$

where K is the dynamic correction factor that accounts for impact effects, viscosity and thermal conductivity effects, and diabatic probe effects, and may take on *any* value depending on the relative importance of these effects.

11.9 Specific Dynamic Correction Factors

By the experimental setup illustrated schematically in Figure 11.8, dynamic correction factors can be determined in air for various probe geometries. For the three probe geometries indicated in Figure 11.9 the

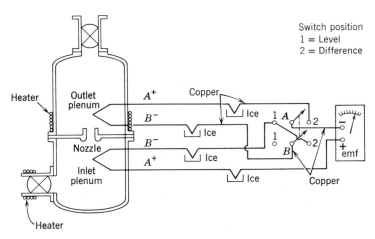

Figure 11.8 Experimental arrangement for determining dynamic correction factors.

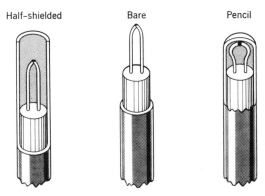

Figure 11.9 Three probe types.

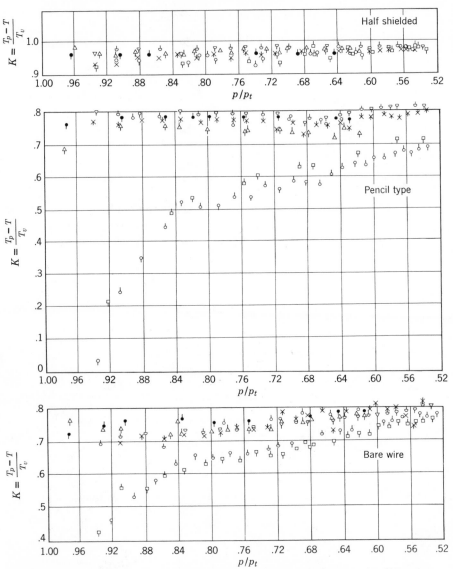

Figure 11.10 Dynamic correction factors for iron-constantan thermocouples (18 gauge—ceramic insulated).

× T/C 1 reference ⎫ air & surroundings at room temperature
○ T/C 2 reference ⎭ back pressure = atmospheric
△ T/C 1 reference ⎫ air & surroundings at room temperature
◇ T/C 2 reference ⎭ back pressure = atmospheric + 10″ Hg.
○ T/C 1 reference ⎫ air at 200°F, surroundings at room temperature
□ T/C 2 reference ⎭ back pressure = atmospheric
◔ increasing Mach number
◗ decreasing Mach number

following has been reported [20] (see Figure 11.10):

1. The half-shielded type is least sensitive to radiation, and to Mach and Reynolds effects (under test conditions it had a K of approximately 0.96).

2. The pencil type is most sensitive to radiation, and to Mach and Reynolds effects (it had a K between 0 and 0.82 under test conditions).

3. The bare-wire type is intermediately sensitive to radiation, and to Mach and Reynolds effects (it had a K between 0.42 and 0.82 under test conditions).

11.10 Applying the Dynamic Correction Factor

Example 6. Assume that a thermocouple is in thermal equilibrium with a fluid and its surrounding boundaries. The velocity, as determined by pressure measurements, is of the order of 1000 ft/sec. For the same thermocouple, immersed in the same fluid (assume $c_p = 0.6$ Btu/°R-lbm), under conditions of identical fluid-wall temperatures, identical pressures, and at a velocity of 1000 ft/sec, the dynamic correction factor was determined as $K = 0.9$. What is the difference between probe and total temperatures?

$$T_t - T_p = (1 - K)T_v$$

$$= (1 - 0.9)\frac{V^2}{2Jg_cc_p}$$

$$= \frac{0.1 \times 10^6}{2 \times 778 \times 32.174 \times 0.6} = 3.33°F.$$

If for thermodynamic reasons the total temperature is required, and if dynamic effects are overlooked, this difference (3.33°F) would constitute an error greater than any associated with static calibrations. The dynamic correction factor, determined for one set of fluid-wall conditions, simply cannot be applied (uncorrected) when the sensing probe is used under different conditions. The reason is that radiation and conduction effects may vary the factor K by large amounts. Similarly, in different fluids (different Prandtl numbers), the factor K may vary greatly. Finally, variations in fluid density (different Reynolds numbers) affect the value of the K factor.

Example 7. Let the dynamic correction factor of a bare-wire couple be determined in air, when both the air and its surroundings are at 80°F. The factor K is found to be essentially constant at 0.65. Since the probe has little tendency to radiate or conduct thermal energy to its surroundings, it behaves like an adiabatic probe and K represents, in effect the recovery factor R. Let the same probe be exposed to the varying velocities of a gas turbine at a point at which the total temperature is 1500°F and the wall temperature is 1000°F. What are the approximate differences in temperature (i.e., between total and probe temperatures) predicted by the use of this room-temperature-determined K as compared to actual test data? The comparisons are shown below, where column I is the directed velocity in ft/sec; column II is predicted ΔT in °F (using $K = R = 0.65$); column III is actual ΔT in °F from [21]; and column IV is actual K (computed).

I	II	III	IV
150	0.7	75	−36.5
800	18	50	0
1500	63	78	+0.57
2000	112	125	+0.61

Note that the unrestricted application of the conventional laboratory determined dynamic correction factor to any and all situations is especially dangerous at low velocities. Note too that at the higher velocities K approaches R.

11.11   References

[1] R. Roseman and S. Katzoff, "The Equation of State of a Perfect Gas," *J. Chem. Ed.*, **2**, June 1945, p. 350.

[2] R. P. Benedict, "Essentials of Thermodynamics," *Electro-Technol.*, July 1963, p. 107.

[3] J. A. Beattie, "The Thermodynamic Temperature of the Ice Point," in *Temperature*, Vol. 1, Reinhold, New York, 1941, p. 76.

[4] A. G. Worthing, "Is Temperature a Basic Concept?" in *Temperature*, Vol. 1, Reinhold, New York, 1941, p. 41.

[5] N. P. Bailey, "Abrupt Energy Transformations in Flowing Gases," *Trans. ASME*, October 1947, p. 750.

[6] W. J. King, "Measurement of High Temperatures in High-Velocity Gas Streams," *Trans. ASME*, July 1943, p. 421.

[7] W. G. Steltz and R. P. Benedict, "Thermodynamics of Constant-Density Fluids," *Trans. ASME, J. Eng. Power*, January 1962, p. 44.

[8] M. Fishenden and O. A. Saunders, *An Introduction to Heat Transfer*, Oxford, 1950, p. 71.

[9] A. H. Shapiro, *The Dynamics and Thermodynamics of Compressible Fluid Flow*, Ronald, New York, 1954, p. 1043.

[10] O. Reynolds, "On the Extent and Action of the Heating Surface for Steam Boilers, *Proc. Manchester Lit. Phil. Soc.*, **14**, No. 7, p. 1874.

[11] F. W. Sears, *An Introduction to Thermodynamics, The Kinetic Theory of Gases, and Statistical Mechanics*, Addison-Wesley, Reading, Mass., 1950, p. 17.

[12] J. P. Joule and W. Thomson, "On the Thermal Effects of Fluids in Motion," Pt. 2, in *Mathematical and Physical Papers of W. Thomson*, Vol. 1, Cambridge, 1882, p. 357.

[13] W. F. Hilton, *High Speed Aerodynamics*, Longmans, Green, New York, 1951, p. 525.

[14] A. A. R. El Agib, "A General Expression for the Isentropic Stagnation Temperature of Fluids, *Nature*, London, **204**, (4962), 1964, p. 989.

[15] H. A. Johnson and M. W. Rubesin, "Aerodynamic Heating and Convective Heat Transfer," *Trans. ASME*, July 1949, p. 447.

[16] H. W. Emmons and J. G. Brainerd, "Temperature Effects in a Laminar Compressible-Fluid Boundary Layer Along a Flat Plate," *Trans. ASME*, **63**, 1941, p. A-105.

[17] H. B. Squire, "Heat Transfer Calculation for Aerofoils," Aeronautical Research Council Reports and Memoranda, R and M 1986, November 1942, London.

[18] E. Eckert and W. Weise, "The Temperature of Unheated Bodies in a High-Speed Gas Stream," NACA TM 1000, December 1941.

[19] A. A. R. El Agib, A. J. Binnie, and T. R. Foord, "Effects of Recovery Factor on

Measurement of Temperature of Moving Fluids," *Proc. Instn. Mech. Engrs.* Vol. 180, Pt. 3F, 1965–1966, p. 174.

[20] R. P. Benedict, "Temperature Measurement in Moving Fluids," ASME paper 59A-257, December 1959.

[21] E. F. Fiock and A. I. Dahl, "The Measurement of Gas Temperature by Immersion-Type Instruments," *J. ARS*, May–June 1953, p. 155.

Nomenclature

Roman

a, b	constants
c_p	specific heat capacity at constant pressure
d	exact differential; also thermal diffusivity
f	function of
F	friction
g_c	gravitational constant $\left(= 32.174 \dfrac{\text{lbm ft}}{\text{lbf sec}^2} \right)$
h	specific enthalpy
J	mechanical equivalent of heat
k	thermal conductivity
K	dynamic correction factor
M	mass
p	absolute pressure
$\bar{P}$	average pressure coefficient
P_θ	local pressure coefficient
Pr	Prandtl number
Q	quantity of heat
r	functional recovery factor
R	specific gas constant, overall recovery factor
S	entropy, stagnation recovery factor
T	absolute temperature
u	specific internal energy
v	specific volume
V	directed velocity
W	work

Greek

δ	inexact differential
ϵ	eddy diffusion
μ	dynamic viscosity

θ angular displacement from stagnation
ν kinematic viscosity

subscripts

1, 2   stations in flow path
l liquid
pi ideal probe
p real probe, constant pressure
s isentropic, static
t total
v dynamic

Superscript

0   zero pressure intercept

Problems

1. Treating the atmosphere as an ideal gas, find the total temperature on the forward face of a Gemini-type space craft if it is traveling at 10,000 miles per hour at an ambient temperature of −60°F. Assume $c_p = 0.25$ Btu/lbm°F.

    Ans. $T_t = 17,587°$R.

2. Find the difference between the total temperature and the temperature at the stagnation point of a turbine blade operating in steam at an oncoming velocity of 1200 ft/sec, free stream temperature of 450°F, and a static pressure of 25 atmospheres. Assume $c_p = 0.75$ Btu/lbm°F.

    Ans. $\Delta T = 28.4°$R.

3. The adiabatic temperature of a flat plate in air is required for heat transfer calculations. The free stream temperature is 200°F, the air velocity is 500 ft/sec, the Prandtl number is 0.694, and a laminar boundary layer prevails. Assume $c_p = 0.25$ Btu/lbm°F.

    Ans. $T_{adi} = 676.6°$R.

4. Estimate the temperature attained by an adiabatic cylindrical probe in air if the static temperature is 500°F, the free stream velocity is

800 ft/sec, the Prandtl number is 0.68, and the boundary layer is essentially turbulent. Assume $c_p = 0.25$ Btu/lbm°F.

Ans. $T_{adi} = 1000.4$°R.

5. Find the difference between the total temperature and the actual probe temperature if the dynamic correction factor is determined to be -10, the velocity is 100 ft/sec, and $c_p = 0.25$ Btu/lbm°F.

Ans. $\Delta T = 8.8$°R.

Chapter 12

INSTALLATION EFFECTS ON TEMPERATURE SENSORS

"... the key to obtaining accurate gas temperature measurements from thermocouples is an understanding of the behavior of a bare wire thermocouple in response to its environment ... radiation shields and stagnation tubes do not change a junction's sensitivity to its environment, they change its environment..."

Robert J. Moffat (1961)

In this chapter we consider separately installations in fluids and installations in solids as they affect the temperature sensor. In fluid applications, the sensor faces a complex heat transfer effect [1]–[7]. Convective heat transfer between fluid and sensor is balanced against radiative heat transfer between fluid, its enclosing walls, and the sensor, and simultaneously against conductive heat transfer between sensor and its supports. In surface applications, on the other hand, radiation and convection are assumed to affect sensor and surface roughly alike; thus conduction between sensor and its supports becomes the dominant mode of heat transfer to consider. Only secondarily must the convection effects between sensor supports and the ambient fluid be considered.

12.1 Combined Heat Transfer Equations for Fluid Installations

For a physical model, consider a gas flowing in an enclosure into which is immersed a temperature sensor and its support (hereafter referred to simply as the sensor). To be specific let $T_{gas} > T_{sensor} > T_{wall}$, although in a

238

numerical case consideration of algebraic signs will allow variations in this restriction. The temperature sensor can receive heat by convection and radiation. It can lose heat by conduction and radiation. The modes and quantities of heat transfer involved are affected by boundary conditions, gas properties, the thermodynamic state of the gas, and by the nature of the directed motion of the gas. A model of the problem to be examined is represented schematically in Figure 12.1 For a differential element of the

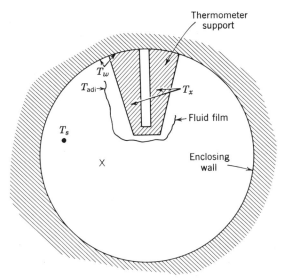

Figure 12.1 Schematic model of thermal problem (X indicates flow perpendicular to plane of the drawing).

sensor (see Figure 12.2), we can write a general heat balance expressing the conservation of thermal energy as

$$dq_c = dq_r + \frac{dq_k}{dx}\, dx, \tag{12.1}$$

where the subscripts have the following meanings: c = convection, r = radiation, and k = conduction. Thus our first job is to determine reasonable expressions for the various terms in (12.1).

Convection

Heat will be transferred to the sensor from the moving gas by forced convection. This phenomenon has been described by Newton's cooling equation as

$$dq_c = h_c\, dA_c (T_s - T_x), \tag{12.2}$$

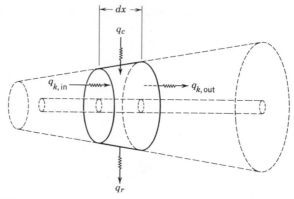

Figure 12.2 Differential element of sensor support for heat balance.

where T_s is the static temperature of the gas, and T_x is the temperature of a differential element of the sensor. If the gas moves with an appreciable velocity, however, the temperature even an adiabatic sensor attains will not be the static temperature of the gas (see Chapter 11). Instead, the thermally isolated sensor will sense the adiabatic temperature of the gas, which can be defined in terms of (11.22) as

$$T_{adi} = T_s + RV^2/2Jg_cc_p, \tag{12.3}$$

where c_p must be considered constant over the temperature range ($T_{adi} - T_s$). Thus it is clear that the modified Newton cooling equation to use in the heat balance of (12.1) is

$$dq_c = h_c \, dA_c(T_{adi} - T_x). \tag{12.4}$$

The temperature distribution through the gas surrounding the sensor can be visualized as in Figure 12.3. The two coefficients of (12.4), h_c and R, are discussed at greater length in Section 12.3.

Radiation

There will be an interchange of radiant energy between the sensor, the gas, and the enclosing walls. For a black body, this phenomenon is described by the Stefan-Boltzmann radiation equation

$$dq_r = \sigma \, dA_r T^4. \tag{12.5}$$

SENSOR EMISSION. The sensor will radiate energy according to its absolute temperature, as indicated by (12.5), modified, however, by the emissivity of the sensor, which accounts for its non-black body characteristics. Some of this energy will be absorbed by the gas and the

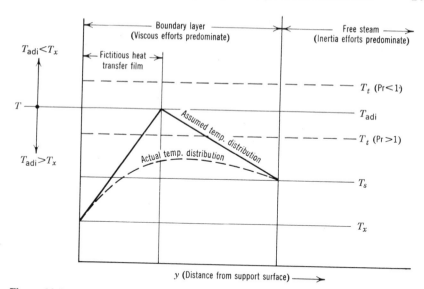

Figure 12.3 Temperature distribution in fluid surrounding well.

remainder by the enclosing walls; that is,

$$dq_{r,x} = dq_{r,(x \to g)} + dq_{r,(x \to w)} = \alpha_g \, dq_{r,x} + (1 - \alpha_g) \, dq_{r,x}. \tag{12.6}$$

But this is not the complete story. There are reflections (radiant interchanges) between the gas and the solid gray body (the sensor) that must be considered. Effects of these reflections are included by adjusting the sensor emissivity as recommended by McAdams [8] and by Grober [9]; thus

$$dq_{r,(x \to g)} = \alpha_{g,x} \sigma \left(\frac{\epsilon_x + 1}{2} \right) dA_r T_x^4, \tag{12.7}$$

where $\alpha_{g,x}$ signifies absorptivity of the gas, to be evaluated at the sensor temperature. For the sensor emission absorbed by the enclosing walls we have

$$dq_{r,(x \to w)} = (1 - \alpha_{g,x}) \sigma \epsilon_x \, dA_r T_x^4. \tag{12.8}$$

When (12.7) and (12.8) are combined according to (12.6), we have

$$dq_{r,x} = \sigma \epsilon_x \, dA_r T_x^4 [\alpha_g F + (1 - \alpha_{g,x})], \tag{12.9}$$

where $F = (\epsilon_x + 1)/2\epsilon_x$. Equation 12.9 represents the net rate of radiant heat transfer from the sensor.

GAS EMISSION. The gas will radiate energy according to its absolute temperature and its emissivity. The sensor will absorb some fraction of this energy incident on its surface.

$$dq_{r,(g \to x)} = \alpha_x \sigma \epsilon_{g,g} \, dA_r T_{adi}^4, \qquad (12.10)$$

where $\epsilon_{g,g}$ signifies emissivity of the gas, to be evaluated at the gas temperature. By Kirchhoff's law for solid bodies, α_x may be replaced by ϵ_x; and again considering the reflections between the fluid and the sensor, we replace ϵ_x with $(\epsilon_x + 1)/2$ to obtain

$$dq_{r,(g \to x)} = \sigma \left(\frac{\epsilon_x + 1}{2} \right) \epsilon_{g,g} \, dA_r T_{adi}^4. \qquad (12.11)$$

WALL EMISSION. The enclosing walls will radiate according to their absolute temperature only, considering such enclosures to be black. However, the gas intercepts and absorbs some of the radiant energy emitted by the walls so that the net radiation received by the sensor from the walls is

$$dq_{r,(w \to x)} = \sigma(1 - \alpha_{g,w})\epsilon_x \, dA_r T_w^4, \qquad (12.12)$$

where $\alpha_{g,w}$ signifies absorptivity of the gas, to be evaluated at the wall temperature.

NET SENSOR EMISSION. By combining (12.9), (12.11), and (12.12), we obtain the expression for the net emission from the sensor by radiation, as required by the heat balance of (12.1).

$$dq_r = \sigma \epsilon_x \, dA_r \{[1 + \epsilon_{g,x}(F-1)]T_x^4 - F\epsilon_{g,g}T_{adi}^4 - (1 - \epsilon_{g,w})T_w^4\}. \qquad (12.13)$$

For convenience, (12.13) also can be expressed in terms of a radiation coefficient as

$$dq_r = h_r \, dA_r(T_x - T_w) \qquad (12.14)$$

patterned after Newton's cooling law, where

$$h_r = \frac{\sigma \epsilon'(T_x^4 - T_w^4)}{T_x - T_w} \qquad (12.15)$$

and where

$$\epsilon' = \epsilon_x \left[1 + \frac{\epsilon_{g,x}(F-1)T_x^4 - \epsilon_{g,g}T_{adi}^4 + \epsilon_{g,w}T_w^4}{T_x^4 - T_w^4} \right]. \qquad (12.16)$$

The coefficients h_r and ϵ' are discussed further in Section 12.3.

Conduction

Heat will be transferred from the tip of the sensor (in the gas) to its base (at the wall) by means of conduction. This phenomenon is described by Fourier's conduction equation

$$q_k = -kA_k \frac{dT_x}{dx}. \tag{12.17}$$

For an element of the sensor in the steady state (i.e., for the case of zero heat storage), the one-dimensional expression for the net conductive heat transfer, as required by the heat balance of equation (12.1), is

$$\frac{dq_k}{dx} dx = -kA_k \frac{d^2T_x}{dx^2} dx - k \frac{dT_x}{dx} \frac{dA_k}{dx} dx. \tag{12.18}$$

The Heat Balance

When (12.4), (12.14), and (12.18) are combined according to (12.1), we obtain

$$\frac{d^2T_x}{dx^2} + a_1(x) \frac{dT_x}{dx} - a_2(x, y)T_x = -a_2 a_3(x, y), \tag{12.19}$$

where

$$a_1(x) = \frac{dA_k}{A_k \, dx}$$

$$a_2(x, y) = \frac{dA_c(h_r + h_c)}{kA_k \, dx},$$

$$a_3(x, y) = \frac{h_c T_{\text{adi}} + h_r T_w}{h_c + h_r}.$$

12.2   Solutions to Combined Heat Transfer Equation

Equation 12.19 is a second-order, first-degree, nonlinear differential equation, and as such has no known closed-form solution. It is nonlinear because the coefficients a_2 and a_3 are both functions of the dependent variable T_x. The offender in both cases is the radiation coefficient h_r. There are at least three approaches to a solution of (12.19).

Tip Solution

Here all conduction effects are neglected, and (12.19) reduces to

$$T_{\text{tip}} = a_3 = \frac{h_r T_w + h_c T_{\text{adi}}}{h_r + h_c}. \tag{12.20}$$

This solution leads to a sensor tip temperature that is usually too high, since any conduction tends to reduce T_{tip}.

Overall Linearization

Here h_r is based on an average T_x, justified by noting that $(T_{adi} - T_w) \ll T_w$ in any practical problem. Thus for specifying the radiation coefficient only, we can approximate T_x, which is bounded by T_{adi} and T_w, by T_{adi}, T_w, or its average value

$$\bar{T}_x = \frac{\int_0^L T_x \, dx}{L}. \tag{12.21}$$

If, in addition, a right circular cylinder is assumed for the geometry of the sensor and support, the area dependence on x is removed. Under these conditions, (12.19) is linearized, the coefficients a_1, a_2, and a_3 are constants, and the closed solution is

$$\frac{T_x - a_3}{T_w - a_3} = \frac{e^{mx}}{1 + e^{2mL}} + \frac{e^{-mx}}{1 + e^{-2mL}}, \tag{12.22}$$

where
$$m = \left[\frac{4D(h_r + h_c)}{k(D^2 - d^2)} \right]^{1/2}. \tag{12.23}$$

This solution leads to quick, approximate answers for the case in which the gas can be considered transparent to radiation (i.e., for $\epsilon_g \approx 0$), but, in general, overall linearization leads to unreliable results.

Stepwise Linearization

Here the solution is based on dividing the sensor and its support, lengthwise, into a number of elements, as indicated in Figure 12.4. The temperature T_x at the center of each element is taken to represent the temperature of that entire element. The number of lengthwise divisions can be made as large as desired to enhance the finite difference approximation to the nonlinear equation (12.19). Three heat balance equations will describe the heat transfer completely through the model of Figure 12.4.

TIP ELEMENT.   The heat balance for element 1 is

$$q_{c,0} + q_{c,1} = q_{r,0} + q_{r,1} + q_{k,1}$$

or

$$h_c(A_{c,0} + A_{c,1})(T_{adi} - T_1) = h_{r,1}(A_{r,0} + A_{r,1})(T_1 - T_w) + \frac{kA_{k,1}(T_1 - T_2)}{\Delta x},$$

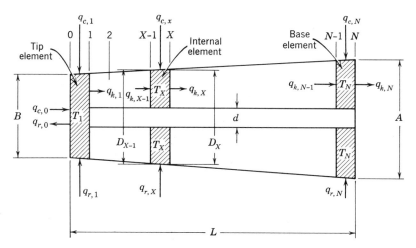

Figure 12.4 Finite-difference model for stepwise linearization.

from which we obtain

$$T_2 = T_1 - \left(\frac{A_{c,0} + A_{c,1}}{A_{k,1}}\right)\frac{\Delta x}{k}[h_c T_{\text{adi}} - (h_c + h_{r,1})T_1 + h_{r,1} T_w]. \quad (12.24)$$

INTERNAL ELEMENT. The heat balance for any internal element is

$$q_{c,x} + q_{k,x-1} = q_{r,x} + q_{k,x}$$

or $h_c A_{c,x}(T_{\text{adi}} - T_x) + \dfrac{k A_{k,x-1}(T_{x-1} - T_x)}{\Delta x}$

$$= h_{r,x} A_{r,x}(T_x - T_w) + \frac{k A_{k,x}(T_x - T_{x+1})}{\Delta x},$$

from which we obtain

$$T_{x+1} = \left(1 + \frac{A_{k,x-1}}{A_{k,x}}\right)T_x - \left(\frac{A_{k,x-1}}{A_{k,x}}\right)T_{x-1}$$

$$- \left(\frac{A_{c,x}}{A_{k,x}}\right)\left(\frac{\Delta x}{k}\right)[h_c T_{\text{adi}} - (h_c + h_{r,x})T_x + h_{r,x}T_w]. \quad (12.25)$$

BASE ELEMENT. The heat balance for element N is

$$q_{c,N} + q_{k,N-1} = q_{r,N} + q_{k,N}$$

or

$$h_c A_{c,N}(T_{\text{adi}} - T_N) + \frac{k A_{k,N-1}(T_{N-1} - T_N)}{\Delta x}$$

$$= h_{r,N} A_{r,N}(T_N - T_w) + \frac{k A_{k,N}(T_N - T_w)}{\Delta x/2}$$

from which we obtain

$$T'_w = \left(1 + \frac{A_{k,N-1}}{2A_{k,N}}\right) T_N - \left(\frac{A_{k,N-1}}{2A_{k,N}}\right) T_{N-1} - \left(\frac{A_{c,N}}{2A_{k,N}}\right)\frac{\Delta x}{k}$$
$$\times [h_c T_{adi} - (h_c + h_{r,N}) T_N + h_{r,N} T_N], \quad (12.26)$$

where T'_w is the calculated value of the enclosing wall temperature.

AREA CALCULATIONS.   For convection and radiation calculations, it is the surface area of the sensor that is required. Considering a right circular cylinder, we have

$$A_{c,x} = A_{r,x} = \pi D \Delta x. \quad (12.27)$$

For conduction calculations it is the cross-sectional area of the sensor that is required. Again for a right circular cyclinder we have

$$A_{k,x} = \frac{\pi D_x^2}{4}. \quad (12.28)$$

SOLUTION METHOD.   An initial T_1 is assumed (T_{adi} is a good first choice), and all other temperatures (T_2, T_3, . . . , T_N, and T'_w) are obtained according to (12.24), (12.25), and (12.26). The calculated wall temperature T'_w is compared to the given wall temperature T_w, and adjustments in T_1 are made for a second try. Iteration schemes (such as Newton's method) rapidly lead to a unique solution for the steady-state temperature distribution throughout the sensor-support combination [10].

Comparison of the Methods

The *tip* solution, in which conduction effects are neglected, leads to a sensor temperature that is always too high. The *overall linearized* solution is sometimes adequate (when ϵ_g is small), leads at times to sensor temperatures that are higher than the *tip* solution (when $\epsilon_g \approx 0.5$), and sometimes yields imaginary solutions (when $\epsilon_g \approx 1$), in which case the ratio

$$(T_x - T_w)/(T_{adi} - T_w)$$

exceeds 1. Of course the *stepwise linearized* solution represents the most reliable solution of the three.

Trend Curves

Because of the complex relationships between the many variables involved in the heat transfer analysis, it is easy to lose track of the physical effects of the controlling variables. In an effort to give a clearer picture of these effects, trend curves are presented in Figure 12.5 for a particular practical problem. In Figure 12.5*a*, note the highly beneficial effect of ensuring a substantial convective film coefficient over the thermometer well. In Figure 12.5*b*, the less dramatic effect of well thermal conductivity is seen. In Figure 12.5*c*, the importance of fluid and well emissivity is shown; note the insulating effect the fluid emissivity has on the thermometer well, simplifying the temperature measuring problem in

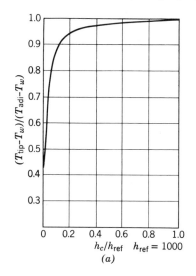

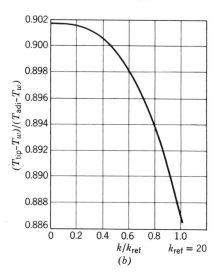

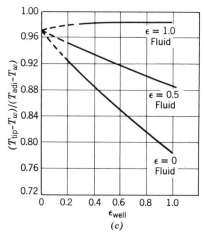

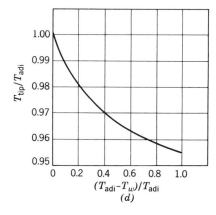

Figure 12.5 Trend curves.

steam, for example. In Figure 12.5d, the importance of insulating the enclosing walls, at least in the vicinity of the thermometer well, is seen.

Graphical Solution

In Figure 12.6, a summary curve of many problems solved by the *step linearized method* (based on 20 steps) is given. The coordinates are the parameters of (12.22) and (12.23), that is, of the *overall linearized* solution. Because the actual problem is nonlinear, no exact graphical solution is possible. Even under the extremes of all the variables, however, as noted in Figure 12.6, the limits of uncertainty that attend the summary curve are relatively narrow. Thus some useful information concerning a given sensor installation can be obtained rapidly from Figure 12.6, as illustrated in several examples to follow.

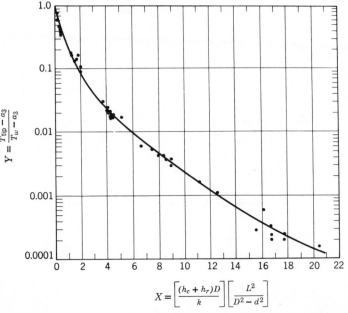

$$X = \left[\frac{(h_c + h_r)D}{k}\right]\left[\frac{L^2}{D^2 - d^2}\right]$$

Figure 12.6 Summary curve of 20 step-linearized solutions. Limits of variables $0 < \epsilon_{well} < 1$; $0 < \epsilon_{fluid} < 1$; $1 < k < 20$; $1 < h_c < 1000$; 2 in. $< L < 12$ in.: $\frac{1}{16}$ in. $< D < 2$ in.; $0 < d < \frac{1}{4}$ in.; 70°F $< T < 1200$°F.

12.3   Some Useful Heat Transfer Coefficients

Often we must estimate a heat transfer coefficient as an item of secondary importance in a particular job [11]. A case in point is the

Table 12.1 For Flow Normal to a Single Cylinder

a	b	c	Re	Source
0.26	0.6	0.3	1000 to 40,000	McAdams [8]
0.193	0.618	0.31	4000 to 40,000	Eckert [15]
0.02653	0.805	0.31	40,000 to 400,000	Grober [9]
0.478	0.5	0.3	250 to 30,000	Scadron and Warshawsky [12]

requirement of obtaining the coefficients h_c, α, and h_r in the equations of Section 12.2. There is such a confusing array of coefficients, exponents, and equations available in the various texts (see for example, Table 12.1) that it is believed best to present satisfactory values in graphical form for quick, ready estimates of these heat transfer coefficients.

Forced Convection Film Coefficients

The dimensional analysis that, for forced convection, yields an expression of the form

$$Nu = a\, Re^b\, Pr^c \tag{12.29}$$

is well known, and advantages of the use of (12.29) in correlating large amounts of data for many different fluids are obvious [12]–[15]. The convective film coefficient h_c, however, is often desired, and this requires an evaluation of the Reynolds number (GD/μ), the Prandtl number $(c_p\mu/k)$, and then an arithmetic operation to obtain h_c from the Nusselt number (h_cD/k).

To simplify the procedure, the following is suggested. From (12.29)

$$h_c = \frac{ak}{D}\left(\frac{GD}{\mu}\right)^b\left(\frac{c_p\mu}{k}\right)^c. \tag{12.30}$$

Equation 12.30 can be rearranged to give

$$h_c = \left(\frac{ak^{1-c}c_p^{\,c}}{\mu^{b-c}}\right)\frac{G^b}{D^{1-b}}. \tag{12.31}$$

Thus through the term in parentheses the film coefficient is seen to be a function of the thermodynamic state of the fluid. If a particular reference state is chosen for a particular fluid (e.g., 500 psi, 500°F steam), a unique family of curves can be plotted in terms of $h_c' = f(G, D)$. This has been done for appropriate values of a, b, and c as reported in the literature for

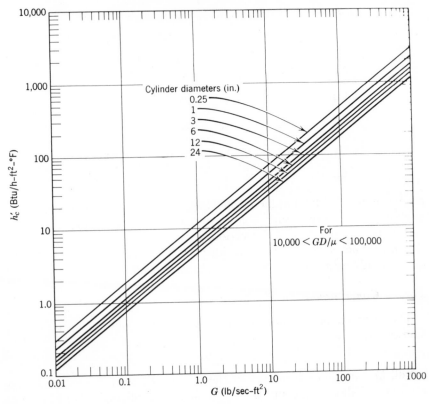

Figure 12.7 Forced convection film coefficients inside cylinders ($Nu = 0.023\,Re^{0.8}\,Pr^{0.4}$).

(a) forced convection inside cylinders (see Figure 12.7); and (b) for forced convection across single cylinders (see Figures 12.8 and 12.9).

To obtain the actual film coefficient h_c for any fluid at any state, it is necessary only to multiply the plotted reference value of h_c' by the ratio

$$\frac{h_c}{h_c'} = \frac{(ak^{1-c}c_p{}^c/\mu^{b-c})^{\text{actual fluid}}_{\text{actual state}}}{(ak^{1-c}c_p{}^c/u^{b-c})_{\text{reference value}}}. \tag{12.32}$$

Correction curves representing this ratio for steam and air are given in Figures 12.10 through 12.12. These curves are required to account for variations in the thermodynamic state properties μ, k, c_p.

Recovery Factor

In Section 11.7 the recovery factor was discussed at some length. Briefly, there are several recovery factors to choose from. The frictional

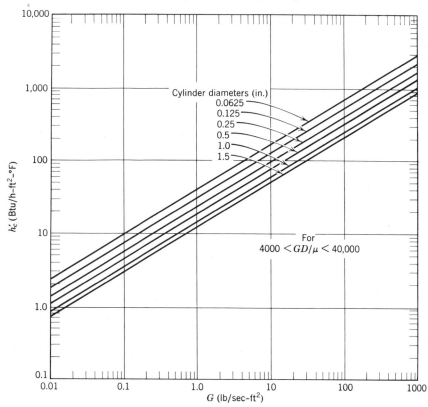

Figure 12.8 Generalized convective film coefficients, low Re, across cylinders. (Nu = $0.193 \, Re^{0.618} \, Pr^{0.31}$.)

recovery factor r, based on the local adiabatic temperature, remains constant around the periphery of a cylinder. This convenient-to-use recovery factor is identical to the flat-plate recovery factors of (11.19) and (11.21).

Another recovery factor is based on the undisturbed free stream velocity, the static temperature, and the mean adiabatic temperature. This is the correct recovery factor to use, but unfortunately it varies with the Mach number, and no systematic correlation exists. However, noting that its value for air varies from 0.55 to 0.82, which is similar to the $Pr^{1/2}$ evaluation, and considering the inherent difficulties in the use of this mean-free r, we recommend for analysis the constant valued local rs, as defined by (11.19) and (11.21). This simplification tends to introduce a slightly greater rate of convective heat transfer to the sensor than actually exists and leads in turn to uncertainties that are on the optimistic side.

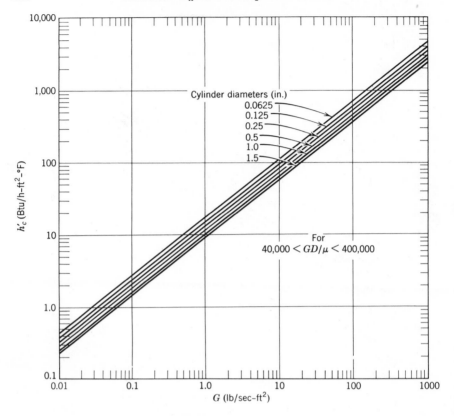

Figure 12.9 Generalized convective film coefficients, high Re, across cylinders. (Nu = $0.02653\,\mathrm{Re}^{0.805}\,\mathrm{Pr}^{0.31}$.)

Radiation Coefficients

The coefficients h_r and ϵ' of (12.15) and (12.16) also can be represented as the ratio h_r/ϵ' in convenient graphical form, as indicated in Figure 12.13. In view of the scarcity of information on gaseous emissivity, it is suggested that advantage be taken of the simplification $\epsilon_{g,x} = \epsilon_{g,g} = \epsilon_{g,w}$. This, together with the approximations that F of (12.9) equals 1, leads to a more manageable form of (12.16), namely,

$$\epsilon' \approx \epsilon_{\text{well}}\left(1 - \epsilon_{\text{fluid}}\frac{T_{\text{adi}}^4 - T_{\text{wall}}^4}{T_{\text{tip}}^4 - T_{\text{wall}}^4}\right). \qquad (12.33)$$

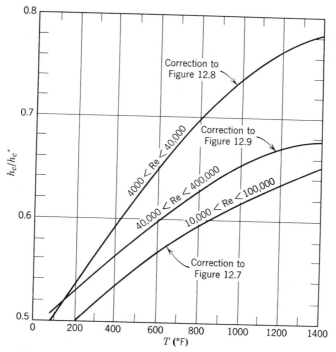

Figure 12.10 Correction curves for convective films in air.

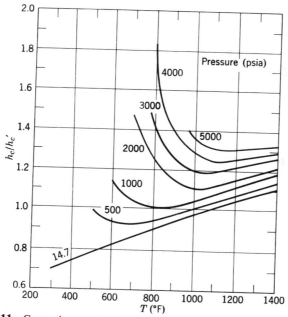

Figure 12.11 Correction curves for convective films in steam at low Re.

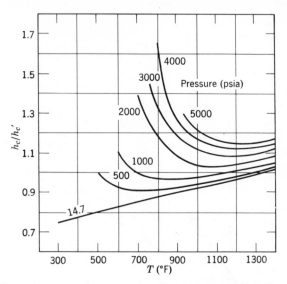

Figure 12.12 Correction curves for convective films in steam at high Re.

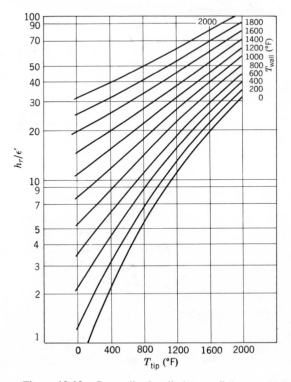

Figure 12.13 Generalized radiation coefficients.

12.4 Fluid Applications and Examples

Several examples are given to illustrate the use of the graphical heat transfer coefficients in conjunction with the various combined heat transfer analyses toward solutions to typical fluid installations.

Example 1. Tip Solution (Involving no Conduction). Given that a spherical sensor of surface area = 0.8 in.^{2}, disk area = 0.2 in.^{2}, and emissivity = 1 is installed in a duct of radius 2 in. and length 6 ft, the sensor indicates 250°F when the duct wall is at 200°F, and the convective heat transfer coefficient is 20 Btu/h-ft^2-°F. Find the gas temperature.

SOLUTION (based on use of a radiation coefficient)

$$q_c = q_r,$$

$$h_c A \underset{\substack{\text{surface} \\ \text{of sensor}}}{} (T_{gas} - T_{sensor}) = h_r A \underset{\substack{\text{surface} \\ \text{of sensor}}}{} (T_{sensor} - T_{wall}),$$

or

$$T_{gas} = T_{sensor} + (T_{sensor} - T_{wall}) \frac{h_r}{h_c},$$

in agreement with (12.20).

Now from Figure 12.13,

$$\frac{h_r}{\epsilon'} = f(T_{sensor}, T_{wall}) = f(250°F, 200°F) = 2.3.$$

From (12.16) $\epsilon' = 1$ if the walls and sensor are considered to be black bodies ($\epsilon_w = \epsilon_x = 1$), and if the gas is assumed to be transparent to radiation ($\epsilon_g = 0$). Thus

$$T_{gas} = 250 + (250 - 200) \frac{2.3}{20} = 255.7°F.$$

Example 2. Same as Example 1 except the Solution is now based on use of form and emissivity factors,

$$q_c = q_r,$$

$$h_c A \underset{\substack{\text{surface} \\ \text{of sensor}}}{} (T_{gas} - T_{sensor}) = F_\epsilon F_A \sigma A \underset{\substack{\text{surface} \\ \text{of sensor}}}{} (T^4_{sensor} - T^4_{wall}).$$

When the form factor F_A is given according to Schenck [14] as

$$F_A = \frac{4 L_{duct}}{(L^2_{duct} + R^2_{duct})^{1/2}} \left(\frac{A_{disk}}{A_{surface}} \right)_{sensor} = \frac{4 \times 6}{(36 + \frac{1}{36})^{1/2}} \left(\frac{0.2}{0.8} \right) \approx 1$$

and the emissivity factor is taken as unity, the result is

$$T_{gas} = T_{sensor} + \frac{\sigma(T^4_{sensor} - T^4_{wall})}{h_c}.$$

Note that it is the difference in fourth-power absolute temperatures that must be used in this form of the radiation heat transfer equation. Thus

$$T_{\text{gas}} = 250 + \frac{0.174(7.1^4 - 6.6^4)}{20} = 255.6°F.$$

We see that under the same conditions the two solutions agree within close limits.

Thermometer Well Solutions

Such problems resolve to the following. Given the installation details of a thermometer well, find whether the installation is satisfactory. The steps required are:

1. Remove area dependence on immersion length by approximating the given well by a right circular cylinder of the same surface area.
2. Determine the pertinent heat transfer coefficients.
3. Determine the value of

$$X' = \left(\frac{h_c D}{k}\right)\left(\frac{L^2}{D^2 - d^2}\right) \tag{12.34}$$

where L = length of immersed portion of thermometer well,
d = bore diameter of the well,
D = outside diameter of the well.

4. If $X' \leq 20$, enter Figure 12.6 with

$$X = \frac{(h_c + h_r)D}{k}\left(\frac{L^2}{D^2 - d^2}\right) \tag{12.35}$$

and obtain

$$Y = \frac{T_{\text{tip}} - a_3}{T_{\text{wall}} - a_3}$$

where a_3 is defined under (12.19): Now compute either T_{tip} or T_{adi}, whichever is unknown, accounting to the following:

$$T_{\text{tip}} = YT_{\text{wall}} + (1 - Y)\left[\frac{h_c T_{\text{adi}} + h_r T_{\text{wall}}}{h_c + h_r}\right] \tag{12.36}$$

$$T_{\text{adi}} = \left(\frac{h_c + h_r}{h_c}\right)\left(\frac{T_{\text{tip}} - YT_{\text{wall}}}{1 - Y}\right) - \left(\frac{h_r}{h_c}\right)T_{\text{wall}}. \tag{12.37}$$

On the basis of the difference between T_{tip} and T_{adi}, the worth of the installation can be judged.

5. If $X' > 20$, it means that conduction effects are negligible. Thus (12.36) or (12.37) are solved for T_{tip} or T_{adi} by setting $Y = 0$.

Example 3. Air flows at a rate of 1 lb/sec at a pressure of 2 atmospheres in an uninsulated 6-in. pipe. The indicated well tip temperature is 200°F, and the indicated enclosing wall temperature is 180°F. The cylindrical well has a 3-in. immersion, $\frac{1}{2}$ in. OD, $\frac{1}{8}$ in. ID, $k = 20$ Btu/h-ft-°F, $\epsilon_{well} = 0.9$, $\epsilon_{fluid} = 0$ (i.e., transparent). The installation is to yield fluid temperature to ± 1°F.

SOLUTION

$$G = \frac{\dot{w}}{A} = 1 \times \frac{4}{\pi} \times \frac{4}{1} = 5.093 \text{ lb/sec-ft}^2$$

$h_c' = 50$ from Figure 12.8 (low Reynolds number),

$\dfrac{h_c}{h_c'} = 0.53$ from Figure 12.10 (low Reynolds number),

$h_c = 50 \times 0.53 = 26.50$ Btu/h-ft²-°F,

$\dfrac{h_r}{\epsilon'} = 1.9$ from Figure 12.13,

$\epsilon' = \epsilon_{well} = 0.9$ by (12.16),

$h_r = 0.9 \times 1.9 = 1.71$ Btu/h-ft²-°F,

By (12.34)

$$X' = \left(\frac{26.5 \times \frac{1}{24}}{20}\right)\left(\frac{9}{0.25 - 0.01562}\right) = 2.120.$$

Since X' is less than 20, form X according to (12.35).

$$X = \left[\frac{26.5 + 1.71}{20 \times 24}\right]\left[\frac{9}{0.25 - 0.01562}\right] = 2.257,$$

and from Figure 12.6, $Y = 0.070$.
By (12.37)

$$T_{adi} = \left(\frac{28.21}{26.5}\right)\left[\frac{660 - (640 \times 0.07)}{0.93}\right] - \left(\frac{1.71}{26.5}\right)640$$

$$= 662.89°\text{R},$$

or
$$T_{adi} = 202.89°\text{F}.$$

This fluid temperature is not within requirements, and the installation must be changed. Insulating the wall in the well vicinity seems the simplest change. One inch of 85% magnesia ($k = 0.036$ Btu/h-ft-°F) raises T_{wall} to 196°F, although neither X nor Y changes appreciably. Now, (12.37) yields

$$T_{adi} = \left(\frac{28.21}{26.5}\right)\left[\frac{660 - (656 \times 0.085)}{0.93}\right] - \left(\frac{1.77}{26.5}\right)656$$

$$= 660.58°\text{R},$$

or
$$T_{adi} = 200.58°\text{F}.$$

This is within specified bounds so the installation is now satisfactory. The generalized curve of Figure 12.6 does not always lead to reliable graphical solutions, since the ratio $(T_x - A_3)/(T_w - a_3)$ does not reflect the radiation effect adequately, and the parameter $(h_r + h_c)DL^2[k(D^2 - d^2)]$ may mask the conduction effects when radiation is significant. Whenever conduction effects can be neglected, however (i.e., whenever

$$(h_c D/k)(L^2/(D^2 - d^2)) > 20),$$

a very reliable approximation is easily obtained. This approach, based on the tip solution, as given by (12.20), is illustrated by the following example.

Example 4. Steam flows at a rate of 1.7 lb/sec-ft^2 at a pressure of 14.7 psia in an uninsulated pipe. The indicated well tip temperature is 506.6°F, and the enclosing wall temperature is 350°F. The well has an 8-in. immersion, OD $= \frac{1}{4}$ in., ID ≈ 0, $k = 10$ Btu/h-ft-°F, $\epsilon_{well} = 0.97$, $\epsilon_{fluid} = 0.35$. Would this installation be acceptable if fluid temperature to ± 5°F were required?

SOLUTION

$h_c' = 31$ from Figure 12.8 (low Reynolds number),

$\dfrac{h_c}{h_c'} = 0.78$ from Figure 12.11,

$h_c = 24$ Btu/h-ft^2-°F,

$\dfrac{h_r}{\epsilon'} = 4.9$ from Figure 12.13,

$\epsilon_1' = \epsilon_{well}(1 - \epsilon_{fluid})$ for first try by assuming $F = 1$, and $\dfrac{T_{adi}^4 - T_w^4}{T_{tip}^4 - T_w^4} \approx$

$\dfrac{T_{adi} - T_w}{T_{tip} - T_w} = 1$ in (12.16).

Thus $\epsilon_1' = 0.97(1 - 0.35) = 0.631$

and $h_{r,1} = 0.631 \times 4.9 = 3.09$ Btu/h-ft^2-°F.

By (12.34) $X' = \left(\dfrac{24 \times 0.0208}{10}\right)\left(\dfrac{0.667}{0.0208}\right)^2 = 51,$

which is much greater than the suggested 20; therefore, set $Y = 0$ and solve (12.37). There results

$$T_{adi} = \left(\frac{27.09}{24}\right)966.6 - \left(\frac{3.09}{24}\right)810 = 986.8°R.$$

The estimate ϵ_1' is now corrected by using $T_{adi,1}$ to give

$$\begin{cases} \epsilon_2 = 0.97\left[1 - 0.35\left(\dfrac{987 - 810}{966.6 - 810}\right)\right] = 0.587, \\[2mm] h_{r,2} = 0.587 \times 4.9 = 2.87 \text{ Btu/h-ft}^2\text{-°F}, \\[2mm] T_{adi,2} = \left(\dfrac{26.87}{24}\right)966.6 - \left(\dfrac{2.87}{24}\right)810 = 985.3°R = 525.3°F. \end{cases}$$

Further iteration will not change T_{adi} significantly. Thus we conclude that this installation is unsatisfactory, since $T_{adi} - T_{tip} = 18.7$°F which is greater than the five-degree requirement.

The information presented here should be used to correct the installation (if so required) until acceptable limits of uncertainty in the temperature measurement are obtained. These solutions should never be used as corrections to a sensor indication.

12.5 Solid Applications

The main difficulty in sensing surface temperatures usually concerns the method of attachment of the sensor to the surface. That is, the sensor must attain and yet not upset the surface temperature. In Chapter 8, *Optical Pyrometry*, a method was discussed that circumvents the above difficulty. A cavity is drilled into the surface whose temperature is required in an attempt to approach black body conditions. By sighting on the cavity, one avoids the attachment problem and, hopefully, avoids the temperature perturbation problem also (recall that surface sightings require a knowledge of surface emissivity as a function of wavelength and temperature, and that such information is rarely available, or at best is highly uncertain).

Usually more mundane methods are employed in surface temperature measurement. In particular, the thermocouple method is discussed briefly as most representative of the possible approaches, although RTDs are used in increasing numbers to perform this same function. The following information is based on the work of A. J. Otter [17]. For treatment in greater detail, the excellent and comprehensive books by Baker, Ryder, and Baker should be consulted [18].

Some common methods of attachment of thermocouples to surfaces are shown in Figure 12.14. The junction can be held to the surface by solder, braze, weld, insulating cement, peen, or simply by pressing. The insulated thermocouple wires should be held in intimate contact with an isothermal portion of the surface for a length of at least 20 wire diameters to avoid steep temperature gradients in the vicinity of the thermocouple measuring junction.

The usual relation between indicated and true surface temperature is given by

$$z = \frac{T_{\text{surface}} - T_{\text{sensor}}}{T_{\text{surface}} - T_{\text{ambient}}}, \tag{12.38}$$

where the desired temperature difference, $T_{\text{surface}} - T_{\text{sensor}}$, is expressed as a fraction of the natural temperature difference, $T_{\text{surface}} - T_{\text{ambient}}$. The calibration factor z can be determined reliably only by experiment,

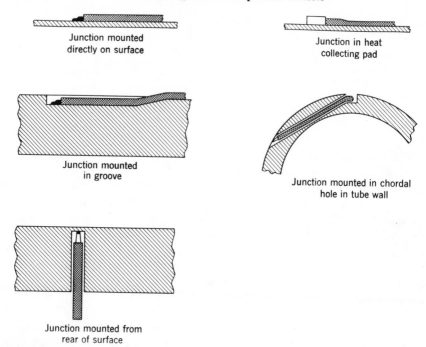

Figure 12.14 Some common methods of attaching thermocouples to surfaces. (After A. J. Otter, 1963.)

although many theoretical approximations are to be found in the literature. Typical calibration factors are given in Table 12.2, and the junction types for which these apply are shown in Figure 12.15. Naturally the results are applicable only for the conditions given in Table 12.2. It can be seen that the sensor size should be as small as possible and that, under the stated conditions, the junction types in order of preference are: separated, button, exposed, and (least desirable) grounded.

Certain general conclusions can be made as to minimizing surface temperature measurement errors.

1. Keep installation size as small as possible.

2. Keep sensor wires in an isothermal region for at least 20 wire diameters.

3. Locate sensor as close to surface as possible.

4. Disturb ambient conditions at the surface as little as possible by the sensor installation.

5. Reduce thermal resistance between sensor and surface to a minimum.

Table 12.2 Typical Calibration Factors for Thermocouple Junction Types of Figure 12.15[a]

Type of Junction	Calibration factor Z		
	100°F	200°F	300°F
$\frac{1}{8}$-in. grounded junction	0.168	0.155	0.148
$\frac{1}{16}$-in. grounded junction	0.059	0.057	0.057
$\frac{1}{8}$-in. exposed junction	0.128	0.113	0.113
$\frac{1}{8}$-in. button junction	0.051	0.045	0.038
$\frac{1}{16}$-in. button junction	0.016	0.016	0.016
$\frac{1}{8}$-in. separated junction	0.019	0.015	0.013

[a] Conditions: probes held normal to horizontal copper surface cooled by natural convection. (After A. J. Otter, 1963.)

   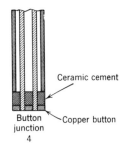

| Grounded junction 1 | Exposed junction 2 | Separated wire junction 3 | Button junction 4 |

Ceramic cement

Copper button

Figure 12.15 Various types of thermocouple junction. (After A. J. Otter, 1963.)

12.6 References

[1] W. M. Rohsenow, "A Graphical Determination of Unshielded-Thermocouple Thermal Correction," *Trans. ASME*, April 1946, p. 195.

[2] G. T. Lalos, "A Sonic-Flow Pyrometer for Measuring Gas Temperatures," *J. Res. Nat. Bur. Std.*, **47**, September 1951, p. 179.

[3] N. R. Johnson, A. S. Weinstein, and F. Osterle, "The Influence of Gradient Temperature Fields on Thermocouple Measurements," ASME Paper 1957-HT-18.

[4] R. P. Benedict, "High Response Aerosol Probe for Sensing Gaseous Temperature in a Two-Phase, Two-Component Flow," *Trans. ASME, J. Eng. Power*, July 1963, p. 245.

[5] J. E. Roughton, "Design of Thermometer Pockets for Steam Mains," *Proc. Inst. Mech. Eng.* **180**, Pt. 1, No. 39, 1965–1966, p. 907.

[6] J. E. Roughton, "Experimental Studies on the Design of Thermometer Pockets for Steam Mains," *Proc. Inst. Mech. Eng.*, **187**, 1973, p. 583.

[7] M. Dutt and T. Stickney, "Conduction Error in Temperature Sensors," ISA Transactions, **9**, 2, 1970, p. 81.

[8] W. H. McAdams, *Heat Transmission*, 2nd ed. McGraw-Hill, New York, 1942, p. 222.

[9] H. Grober, *Fundamentals of Heat Transfer*, 3rd ed. McGraw-Hill, New York, 1961, p. 279.

[10] R. P. Benedict and J. W. Murdock, "Steady-State Thermal Analysis of a Thermometer Well," *Trans. ASME, J. Eng. Power*, July 1963, p. 235.

[11] R. P. Benedict, "Graphical Heat-Transfer Estimation," *Electro-Technol.*, March 1964, p. 93.

[12] M. D. Scadron and I. Warshawsky, "Experimental Determination of Times Constants and Nusselt Numbers for Bare-Wire Thermocouples in High-Velocity Air Streams and Analytic Approximation of Conduction and Radiation Errors," NACA TN 2599, January 1952.

[13] A. F. Wormser, "Experimental Determination of Thermocouple Time Constants with Use of a Variable Turbulence, Variable Density Wind Tunnel, and the Analytic Evaluation of Conduction, Radiation, and Other Secondary Effects," SAE National Aero Meeting Paper 158D, April 1960.

[14] R. J. Moffat, "Gas Temperature Measurement," General Motors Research Lab. Report GMR-329, March 1961.

[15] E. R. G. Eckert, *The Transfer of Heat and Mass*, McGraw-Hill, New York, 1950, p. 142.

[16] H. Schenck, Jr., *Heat Transfer Engineering*, Prentice-Hall, Englewood Cliffs, N.J., 1959.

[17] A. J. Otter, "Thermocouples and Surface Temperature Measurement," Atomic Energy of Canada Limited. Report AECL-3062, 1963.

[18] H. D. Baker, E. A. Ryder, and N. H. Baker, *Temperature Measurement In Engineering*, Wiley, New York, Vol. 1, 1953, Vol. 2, 1961.

Nomenclature

Roman

a, b, c	coefficients
A	area
c_p	specific heat at constant pressure
d	inside diameter; also exact differential
D	outside diameter
e	base of natural logarithm
f	a function of
F	adjusted emissivity
F_A	area factor
F_ϵ	emissivity factor
g_c	gravitational constant $\left(= 32.174 \dfrac{\text{lbm ft}}{\text{lbf sec}^2} \right)$
G	flow per unit area
h	heat transfer coefficient
J	mechanical equivalent of heat
k	thermal conductivity

L immersion length
m coefficient
Nu   Nusselt number
Pr   Prandtl number
q rate of heat transfer
r frictional recovery factor
R overall recovery factor
Re   Reynolds number
T temperature
V directed velocity
$\mathring{w}$ weight rate of flow
x axial position
z calibration factor

Greek

α recovery factor; also absorptivity
Δ finite difference
ϵ emissivity
μ dynamic viscosity
σ Stefan-Boltzmann constant

Subscripts

adi   adiabatic
c convection
g gas
k conductivity
N number of divisions
r radiation
s static
w wall
x sensor
$0, 1$ axial positions

Problems

1. Find the forced convection film coefficient inside a pipe of 1 foot I.D. for air flowing at $G = 10$ lb/sec ft^2, and at $T = 800°F$.

 Ans. $h_c = 20.6$ Btu/hr ft^2°F.

2. Find the forced convection film coefficient for flow across a $\frac{1}{4}$ in. O.D. cylindrical probe if $Re > 40,000$ and at the same G and T as in Problem 1.

   Ans. $h_c = 53$ Btu/hr ft^2°F.

3. Find the forced convection film coefficient for the probe of Problem 2 if steam is flowing at $G = 10$ lb/sec ft^2, $T = 800$°F, and the pressure is 1000 psia.

   Ans. $h_c = 82$ Btu/hr ft^2°F.

4. Determination if the following thermometer well installation is satisfactory to yield fluid temperature to ±5°F. Air flows at $G = 2$ lb/sec ft^2, well O.D. $= \frac{1}{4}$ in., I.D. ≈ 0, indicated temperature $= 500$°F, $T_{wall} = 300$°F, immersion length $= 4$ in., $k_{well} = 15$ Btu/hr ft°F, $\epsilon_{well} = 0.95$, $\epsilon_{air} = 0$, at low Reynolds number.

   Ans. $T_{adi} = 540.1$°F; therefore, not satisfactory.

5. Same as Problem 4 except the flowing fluid is steam at an emissivity of 0.35 and at a pressure of 14.7 psia.

   Ans. $T_{adi} = 520.8$°F; therefore, not satisfactory.

Chapter 13

TRANSIENT TEMPERATURE MEASUREMENT

"... the true meaning of a term is to be found by observing what a man does with it, not what he says about it..."

Percy Bridgman (1946)

13.1 General Remarks

Because of inertia, no instrument (or anything else for that matter) responds instantly or with perfect fidelity to a change in its environment. In mechanical systems, mass is the familiar measure of inertia, whereas in electrical and thermal systems inertia is characterized by capacitance. We are concerned here with the response of a temperature sensor to a change in its environmental temperature. The simplified, hence manageable, temperature changes considered here are (*a*) the ramp change, in which the environment temperature shifts linearly with time from T_1 to T_2; (*b*) the step change, in which the temperature of the sensor environment shifts instantaneously from T_1 to T_2; and (*c*) the periodic change, in which the environment temperature alternates sinusoidally with time between $+T_2$ and $-T_2$ (see Figure 13.1). We seek answers to the following questions: What is the speed or measure of the sensor response? What is the fidelity or faithfulness of the sensor response?

Thermal response belongs, fundamentally, in the realm of transient heat transfer. The rate of response of a temperature sensor clearly depends on the physical properties of the sensor, the physical properties of its environment, as well as on the dynamical properties of its environment. Amplifying on this, we note that, because physical properties

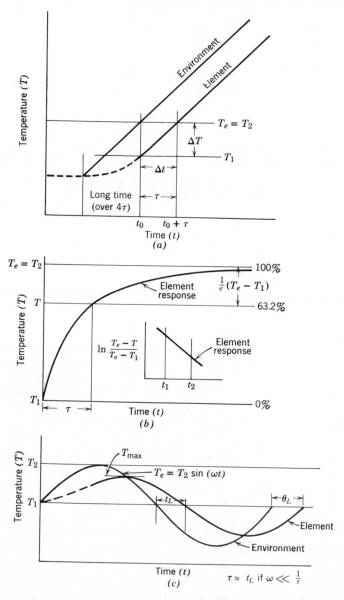

Figure 13.1 Various simplified environmental changes: (*a*) ramp change; (*b*) step change; and (*c*) periodic change.

normally change with temperature, it follows that the response time of a sensor will vary with temperature level; because heat transfer coefficients are strongly dependent on the Reynolds number, it follows that sensor response will vary with the mass velocity of its environment.

It is common practice to characterize the response of a temperature sensor to a nonisothermal change of state of its environment by a thermal time constant. Although no one time constant can exactly describe the response behavior of any but the simplest of systems, it is often quite adequate to consider first-order response only. In the next section, 13.2, we confine our attention to systems in which the sensor exhibits a rate of change in temperature that is exactly proportional to the temperature difference between the sensor and its environment. We defer until Section 13.6 the complications that arise from the additional considerations of conduction, radiation, temperature level, turbulence, and distributed thermal capacities.

13.2 Mathematical Development of First-Order Response

A simplified heat balance can be written for a temperature sensor subjected to a time-varying environmental temperature. We assume that all the heat transferred to the sensor is by convection, and that all this heat is retained by the sensor; that is, the thermal resistance of the system is lumped in the convective heat transfer film around the sensor, and the thermal capacity of the system is lumped in the sensor itself (see Figure 13.2). Thus the heat transfer rate through the film to the sensor exactly equals the rate of heat storage in the sensor. Expressed in terms of Newton's law of cooling and Black's heat capacity equation, we have

$$hA(T_e - T) = Mc\frac{dT}{dt}, \qquad (13.1)$$

where h is the convective heat transfer coefficient of the fluid film surrounding the sensor, A is the surface area of the sensor through which heat is transferred, T_e is the environment temperature at time t, T is the sensor temperature at time t, M is the mass of the sensing portion (it can also be expressed as ρV, i.e., density times volume), and c is the specific heat capacity of the sensing portion.

Separating the variables in (13.1) yields

$$\frac{dt}{(\rho Vc/hA)} = \frac{dT}{T_e - T}, \qquad (13.2)$$

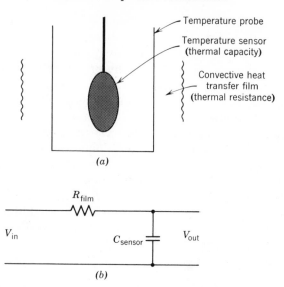

(a)

(b)

Figure 13.2 First-order system. (a) Temperature-sensing system. (b) Equivalent electrical circuit.

where the quantity in parentheses is taken to be a lumped constant, given the symbol τ, seen to have the dimensions of time, and called the *time constant*. Thus

$$\tau = \frac{\rho \mathbf{V} c}{hA} = \frac{\text{thermal capacitance of sensor}}{\text{thermal conductance of film}}. \qquad (13.3)$$

The first-order, first-degree, linear, differential equation expressed by (13.2) has the general solution [1], [2], [3]:

$$T = Ce^{-t/\tau} + \frac{1}{\tau} e^{-t/\tau} \int_0^t T_e e^{t/\tau} \, dt, \qquad (13.4)$$

where C is a constant of integration that is determined by inserting the proper boundary conditions.

Ramp Change

Under this condition at $t = 0$, $T = T_1 = C$, although in general, $T_e = T_2 + Rt$, where R represents the rate of change of environment temperature, $\Delta T/\Delta t$. Insertion of these boundary values in (13.4) yields

$$T = T_1 e^{-t/\tau} + \frac{1}{\tau} e^{-t/\tau} \left(T_2 \int_0^t e^{t/\tau}\, dt + R \int_0^t t e^{t/\tau}\, dt \right) \tag{13.5}$$

where
$$\int_0^t t e^{t/\tau}\, dt = t\tau e^{t/\tau} \Big|_0^t - \tau \int_0^t e^{t/\tau}\, dt.$$

Evaluating (13.5), we have

$$T = T_1 e^{-t/\tau} + T_2 - T_2 e^{-t/\tau} + Rt - R\tau + R\tau e^{-t/\tau}, \tag{13.6}$$

which, expressed in terms of the temperature difference, becomes

$$(T_e - T) = R\tau - R\tau e^{-t/\tau} + (T_2 - T_1)e^{-t/\tau}. \tag{13.7}$$

When $t \gg \tau$ the terms involving $e^{-t/\tau}$ approach zero, and (13.4) through (13.7) reduce to

$$(T_e - T) = R\tau. \tag{13.8}$$

According to (13.8), the time constant for a ramp change can be defined as follows:

> If an element is immersed for a long time in an environment whose temperature is rising at a constant rate (i.e., a ramp change), τ is the interval between the time when the environment reaches a given temperature and the time when the element indicates this temperature; that is, τ is the number of seconds the element lags its environment (see Figure 13.1).

Step Change

Under this condition at $t = 0$, $T = T_1 = C$, although, in general, $T_e = T_2$. Insertion of these boundary values in (13.4) yields

$$T = T_1 e^{-t/\tau} + T_2 - T_2 e^{-t/\tau}, \tag{13.9}$$

which, expressed in terms of the temperature difference, becomes

$$(T_e - T) = (T_e - T_1)e^{-t/\tau}. \tag{13.10}$$

According to (13.10), the time constant for a step change can be defined as follows:

> If an element is plunged into a constant-temperature environment (i.e., a step change), τ is the time required for the temperature difference

between the environment and the element to be reduced to $1/e$ of the initial difference; that is, τ is the number of seconds for the element to reach 63.2% of the initial temperature difference (see Figure 13.1).

Periodic Change

Under this condition at $t = 0$, $T = T_1 = C$, although in general $T_e = T_2 \sin(\omega t)$, where ω represents the frequency of the forcing oscillations of the environment in radians per unit of time. Insertion of these boundary values in (13.4) yields

$$T = T_1 + \frac{T_2}{(1+(\omega\tau)^2)^{1/2}} \sin\left[\omega t - \tan^{-1}(\omega\tau)\right] + \frac{T_2\omega\tau}{1+(\omega\tau)^2} e^{-t/\tau}. \quad (13.11)$$

When $t \gg \tau$, the last term above approaches zero and the sensor response will lag the environment by the phase angle

$$\theta_L \tan^{-1}(\omega\tau), \quad (13.12)$$

which in time units corresponds to a lag of

$$t_L = \frac{\tan^{-1}(\omega\tau)}{\omega}, \quad (13.13)$$

whereas the ratio of the sensor amplitude to that of the environment is given by

$$\frac{T_{max}}{T_2} = \frac{1}{(1+(\omega\tau)^2)^{1/2}}. \quad (13.14)$$

Although no general definition of the time constant for a periodic change is forthcoming, a restricted definition can be given as follows:

If an element is immersed for a long time in an environment whose temperature is varying sinusoidally (i.e., a periodic change), and if the frequency is much less than $1/\tau$, then, to a close approximation, τ is the number of seconds the element lags its environment (see Figure 13.1).

In (13.4) and in all equations thereafter, the time constant τ consistently has represented the lumped constant $(\rho Vc/hA)$; thus it follows that $\tau_{ramp} = \tau_{step} = \tau_{periodic}$ whenever all conditions stated under the three separate sections on the forcing functions are met.

13.3 Second-Order Response

When several thermal resistances are combined along with several thermal capacities, as in the case of a temperature sensor inserted in a thermometer well, an equivalent electric circuit more complex than that shown in Figure 13.2 must be considered. Also, solutions more complex than those given for first-order systems, namely, (13.8) and (13.10), must be used.

Step Change

The equivalent circuit of a second-order system, as a thermometer well-sensor system, is given in Figure 13.3. An appropriate solution describing the temperature-time response of the thermal system of Figure 13.3 to a *step change* in temperature can be given [4]–[7] as

$$T = \Delta T \left[1 - \left(\frac{r_1}{r_1 - r_2} \right) e^{-r_2 t} + \left(\frac{r_2}{r_1 - r_2} \right) e^{-r_1 t} \right] \tag{13.15}$$

where T is the sensor temperature at any time t, ΔT is the step change in temperature, with the initial temperature normalized at zero,

$$r_1, r_2 = (a \pm \sqrt{a^2 - 4b})/2b, \tag{13.16}$$

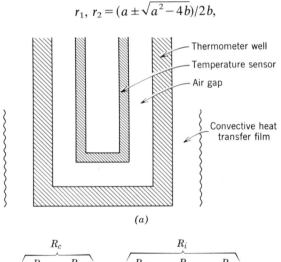

(a)

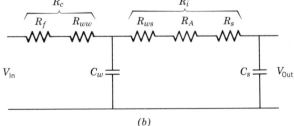

(b)

Figure 13.3 Second-order system. (*a*) Temperature-sensing system. (*b*) Equivalent electrical circuit.

and represent the roots of the second-order quadratic,

$$a = C_s(R_o + R_i) + C_W R_o \qquad (13.17)$$

and
$$b = C_W C_s R_o R_i. \qquad (13.18)$$

The thermal resistances and capacities, as shown in Figure 13.3, can be defined mathematically as:

$$C_s = \rho_s c_{ps} \pi (d_o^2 - d_i^2)/4, \qquad (13.19)$$

and represents the sensor capacity;

$$C_W = \rho_w c_{pw} \pi (D_o^2 - D_i^2)/4, \qquad (13.20)$$

and represents the well capacity;

$$R_o = R_f + R_{WW}, \qquad (13.21)$$

and represents the sum of the well-side resistances, where

$$R_f = 1/\pi h_f D_o, \qquad (13.22)$$

and represents the heat transfer film outside the well, and

$$R_{WW} = \ln (D_o/D_M)/2\pi k_W, \qquad (13.23)$$

and represents the well-to-capacity resistance;

$$R_i = R_{Ws} + R_A + R_s, \qquad (13.24)$$

and represents the sum of the sensor-side resistances, where

$$R_{Ws} = \ln (D_o/D_i)/2\pi k_W, \qquad (13.25)$$

and represents the well-to-sensor resistance, and

$$R_A = \ln (D_i/d_o)/2\pi k_A, \qquad (13.26)$$

and represents the air film between well and sensor, and

$$R_s = \ln (d_o/d_M)/2\pi k_s, \qquad (13.27)$$

and represents the sensor-to-capacity resistance.
The geometric definitions are as follows:

D_o = outer diameter of well,
D_i = inner diameter of well,
$D_M = \sqrt{\tfrac{1}{2}(D_o^2 + D_i^2)}$ = mean diameter of well,
d_o = outer diameter of sensor,
d_i = inner diameter of sensor,
$d_M = \sqrt{\tfrac{1}{2}(d_o^2 + d_i^2)}$ = mean diameter of sensor.

Equation 13.15 is the second-order counterpart of the first-order solution given by (13.10). The roots r_1, r_2 of (13.15) are reciprocals of the time constants, that is,

$$r_1 = \frac{1}{\tau_1}, \qquad r_2 = \frac{1}{\tau_2}. \tag{13.28}$$

When, as is often the case, one time constant dominates the other, that is, when $\tau_2 \gg \tau_1$, (13.15) can be reduced to the form of a first-order solution, namely

$$T = \Delta T \left[1 - \left(\frac{r_1}{r_1 - r_2} \right) e^{-t/\tau_2} \right]. \tag{13.29}$$

Of course, (13.10), the first-order solution, can be written

$$T = \Delta T [1 - e^{-t/\tau}]. \tag{13.10'}$$

The conclusion expressed by (13.29) is in agreement with Looney and Coon [5], [6] to the effect that in most cases a single τ, representing the 63.2% definition of the time constant, is adequate to represent even the more complex temperature-sensing systems.

Ramp Change

The same equations for r_1, r_2 hold for the ramp change as for the step change, that is, (13.16, 13.17, 13.18, and 13.28). Hence the same time constants apply as well. Just as stated after (13.7), after a time lapse of about five dominant time constants, the temperature-sensing system will follow a temperature-time ramp of the same slope as its environment (see Figure 13.1), and the lag for the second-order system will be

$$L = \tau_1 + \tau_2, \tag{13.30}$$

and the temperature error will be

$$T_e - T = RL \tag{13.31}$$

as compared to the first order (13.8).

Periodic Change

When the environment temperature is varying sinusoidally, at a frequency below $\frac{1}{4}\pi L$, the temperature-sensing system behaves essentially as a single time constant system with an effective τ equaling the lag of (13.30).

13.4 Experimental Determination of Time Constant

The ramp change definition of the time constant provides one method for determining τ. The sensor, initially at some uniform temperature, is inserted into an environment whose rate of change of temperature with time is fixed and known. However, several problems are encountered. The environment temperature T_e must be known as a function of time, and this requires a sensor of known τ or a sensor having an insignificantly small τ. In addition, and this is a problem common to all methods of determining τ, the film coefficient of the environment-sensor interface must be known [4], [8], [9]. By the Nusselt equation for heat transfer by forced convection

$$Nu = \phi_1(Re, Pr), \tag{13.32}$$

where Nu = Nusselt number = hD/k,
  Re = Reynolds number = DG/μ,
  Pr = Prandtl number = $c_p\mu/k$.

It follows that, along with the physical properties of the film and sensor, a knowledge of the mass velocity G of the environment relative to the sensor is required if the film coefficient is to be determined within a reasonable uncertainty.

The step change definition of the time constant provides the usual method of determining τ. The sensor is plunged from an initially different temperature into a constant-temperature bath. It does not matter if the sensor is heated or cooled by the step change, but we should not start timing the response at the instant of the plunge. At least four τs should elapse first to allow stabilization of the fluid film on the sensor, since during this initial time the response of the sensor is not well approximated by the first-order equation. As in the ramp change, the film coefficient must be known. In a stagnant bath, reliable, repeatable values for the film coefficient cannot be obtained because of the variableness of the natural convection currents set up in the bath by the temperature gradients.

Murdock, Foltz, and Gregory [10] have discussed a practical method for determining response times of thermometers in stirred liquid baths. Their detailed method is somewhat as follows: Noting that the fluid properties k, μ, and c_p can be taken as constants for a given liquid bath, (13.32) reduces to

$$hD = \phi_2(DG)^m. \tag{13.33}$$

Evaluation of the effective mass velocity around the sensor is complicated by the fact that the liquid swirls in a three-dimensional flow pattern. It varies not only with rate of agitation but also with the physical location of the sensor in the bath. By fixing the sensor location in the bath, we can express the effective value of G in terms of the stirrer speed as

$$G = \phi_3 N^b. \tag{13.34}$$

Hence (13.33) can also be given as

$$hD = \phi_4 D^m N^c. \tag{13.35}$$

Murdock et al. [10] summarize their experimental data for cylindrical sensors in stirred liquid baths with variable agitation by the empirical equation

$$hD = 3.11(DN)^{0.6}, \tag{13.36}$$

where the film coefficient h is in Btu per h-ft^2-°F, the sensor diameter D is in feet, and the stirrer speed N is in rpm. Equation 13.36 indicates that the exponents m and c of (13.35) are equal.

If the stirrer speed is also fixed, (13.35) reduces further to

$$hD = \phi_5 D^m. \tag{13.37}$$

By expressing (13.3) for cylindrical sensors in the form of (13.37), the unknown function of D can be obtained as

$$hD = \left(\frac{\rho c}{4\tau}\right) D^2. \tag{13.38}$$

According to the method under discussion, the physical properties of the sensor in (13.38) vary almost linearly with temperature and are to be evaluated at the average temperature corresponding to 36.8% of the difference between the initial and final temperatures. Thus this method provides an experimental determination of the effective film coefficient of a stirred liquid bath according to (13.36) and (13.38).

To determine τ, one starts with a liquid bath of known physical properties, stirred at known speed, held at a known temperature. A sensor of known physical properties is plunged into a fixed location in the bath and, after waiting an appropriate time, one starts timing its response. The timing is stopped at a predetermined percent of the temperature difference from start to final temperature, where the 63.2% mark yields the first-order time constant directly.

As indicated in Figure 13.1, a first-order response plots as an exponential curve on linear coordinates and as a straight line on a semilog grid. Since it is usually desirable to come closer to the final temperature than 63.2%, (13.10) can be rewritten as

$$a = \frac{T_e - T}{T_e - T_1} = e^{-t/\tau}, \tag{13.39}$$

where a indicates how close the sensor temperature is to the final temperature. Equation 13.39 yields, for a few points, the table

a	0.5	0.368	0.2	0.1	0.05	0.01
t	0.7τ	τ	1.6τ	2.3τ	3τ	4.6τ

13.5 Applying the Time Constant

Several examples will best illustrate the use of the concepts discussed thus far.

Example 1. Determine the first-order time constant in seconds of a thermocouple embedded in a steel cylinder of 1 in. O.D. when determined in a salt bath that is stirred at 800 rpm. Assume $\rho_{steel} = 488$ lb/ft^3 and $c_{steel} = 0.11$ Btu/lb°F.

By (13.19):

$$hD = 3.11(800 \times \tfrac{1}{12})^{0.6} = 38.5 \text{ Btu/h-ft}^2\text{-}°F,$$

$$h_{salt} = 38.5/\tfrac{1}{12} = 462 \text{ Btu/h-ft}^2\text{-}°F.$$

By (13.21):

$$\tau = \frac{\rho c D^2}{4(hD)} = \frac{488 \times 0.11}{4 \times 38.5}(\tfrac{1}{12})^2 = 0.00242 \text{ h} = 9.7 \text{ sec.}$$

Example 2. If the above sensor is to be used in a steam turbine where the film coefficient is estimated to be 1200 Btu/h-ft^2-°F, what will be the approximate time required to reach 99% of a given step change in temperature?

$$\tau_{steam} = \tau_{salt}\left(\frac{h_{salt}}{h_{steam}}\right) = 9.7\left(\frac{462}{1200}\right) = 3.7 \text{ sec,}$$

$$t = 4.6\tau = 4.6 \times 3.7 = 17 \text{ sec.}$$

Example 3. If the above sensor is to be used in an air furnace that has a film coefficient of 46.2 Btu/h-ft^2-°F and is cycling periodically every 10 min, what are the phase angle lag, the time lag, and the amplitude ratio of the sensor response with respect to the air furnace?

$$\tau_{air} = \tau_{salt}\left(\frac{h_{salt}}{h_{air}}\right) = 9.7\left(\frac{462}{46.2}\right) = 97 \text{ sec,}$$

$$\omega = \frac{2\pi}{10} \text{ rad/min,}$$

$$\theta_L = \tan^{-1}\left(\frac{2\pi}{10} \times \frac{97}{60}\right) \approx 45°$$

$$t_L = 45° \times \left(\frac{2\pi}{360} \text{ rad/deg}\right) \times \frac{10}{2\pi} \text{ min/rad} \approx 1.25 \text{ min}$$

$$\frac{a}{A} = \frac{1}{[1 + (2\pi/10 \times 97/60)^2]^{1/2}} \approx \frac{1}{1.414} \approx 0.7.$$

Example 4.   A copper temperature probe of $d_0 = 0.05193$ ft and $d_i = 0.0078$ ft is inserted in a copper thermometer well of $D_0 = 0.0833$ ft and $D_i = 0.05307$ ft. Using physical properties from Table 13.1, and with $h_f = 241$ Btu/hr ft^2°F, find the sensor temperature after 100 sec if the system is subjected to a step change from 0 to 200°F.

By (13.19) $C_s = 578 \times 0.1003 \times \pi/4(0.05193^2 - 0.0078^2) = 0.120018$ Btu/°F.

By (13.20) $C_W = 578 \times 0.1003 \times \pi/4(0.0833^2 - 0.05307^2) = 0.187705$ Btu/°F.

By (13.22) $R_f = (1/\pi \times 241 \times 0.0833) \times 3600$ sec/hr $= 57.08088$ sec°F/Btu.

By (13.23) $R_{WW} = (\ln (0.0833/0.06984)/2\pi \times 212) \times 3600 = 0.47628$ sec°F/Btu.

By (13.21) $R_o = 57.08088 + 0.47628 = 57.5572$ sec°F/Btu.

By (13.25) $R_{Ws} = (\ln (0.0833/0.05307)/2\pi \times 212) \times 3600 = 1.21824$ sec°F/Btu.

By (13.26) $R_A = (\ln (0.05307/0.05193)/2\pi \times 0.02777) \times 3600 = 448.02$ sec°F/Btu.

By (13.27) $R_s = (\ln (0.05193/0.03713)/2\pi \times 212) \times 3600 = 0.90648$ sec°F/Btu.

By (13.24) $R_i = 1.21824 + 448.02 + 0.90648 = 450.145$ sec°F/Btu.

By (13.17) $a = 0.120018 \times 507.702 + 0.187705 \times 57.5572 = 71.737$ sec.

By (13.18) $b = 0.120018 \times 0.187705 \times 57.5572 \times 450.145 = 583.679$ sec^2.

By (13.16) $r_1 = (71.737 + 53.0236)/1167.3582 = 0.10687$ 1/sec.

            $r_2 = (71.737 - 53.0236)/1167.3582 = 0.01603$ 1/sec.

By (13.15) $T \doteq 200[1 - 1.1765e^{-0.01603t} + 0.1765e^{-0.10687t}]$.

At $t = 100$ sec, (13.15) becomes

$$T = 200(1 - 0.23682 + 0.000004) = 152.64°F.$$

Note that the time constants, according to (13.28), are

$$\tau_1 = 9.357163 \text{ sec}, \qquad \tau_2 = 62.383032 \text{ sec}, \quad \text{hence} \quad \tau_2 \gg \tau_1$$

and the system can be represented by (13.29), namely,

$$T = 200(1 - 0.23682).$$

Table 13.1   *Physical Properties of Selected Materials (after [7])*

Properties	Copper	Stainless Steel	Air
Thermal conductivity k, Btu/hr ft°F	212	11.2	0.0277
Specific heat capacity C_p, Btu/lb$_m$°F	0.1003	0.135	—
Density ρ, lb$_m$/ft^3	578	500	—

Example 5.   For the same temperature-sensing system of Example 4, find the temperature error after 5 min if the environment is changing at an average rate of 0.6°F/sec.

By (13.30) $L = 9.357 + 62.383 = 71.74$ sec.

By (13.31) $T_{error} = T_e - T = 0.6 \times 71.74 = 43°F.$

13.6 Modifying Considerations

Many experimental and analytical studies have been made concerning factors that influence the time constants of temperature sensors in addition to the factors ρ, c, V, A, and h previously discussed. These include Mach number, size of temperature change, axial conduction, radiation, fluid turbulence, and installation.

Below a Mach number of 0.4, the time constant is hardly affected by the Mach number, and this effect can be minimized by basing physical properties on the total temperature and total pressure of the fluid.

The size of the temperature change affects τ because physical properties are not necessarily linear functions of temperature. Wormser [11] notes a 25% variation in τ for a 400% change in the size of ΔT.

Axial conduction, for example, along bare thermocouple wires from the measuring junction to the supporting probe, definitely affects the time constant. Under usual conditions, in which $T_{support}$ approaches $T_{junction}$ in the steady state, conduction effects cause an increase in τ. The thermal linkage of junction and support by conduction causes an increase in the effective mass of the junction. Wormser [11] indicates that τ can increase by as much as 80% from this source at the lower mass velocities. He presents the equation

$$\tau_{k,c} = \tau_c\left(1+\frac{\pi^2\tau_c\alpha}{4L^2}\right) \tag{13.40}$$

to calculate this effect where $\tau_{k,c}$ signifies the time constant for heat transfer by conduction and convection, τ_c is the time constant for convection alone, α is the thermal diffusivity of the wires (i.e., $k/\rho c$), and L is the distance from the support to the junction.

Example 6. The convective time constant of a Chromel-Alumel thermocouple that has $\frac{1}{8}$ in. of bare wire from junction to support is 0.5 sec. What is the percentage increase in the time constant because of conduction effects?

From Table 13.2

$$\bar{\alpha} = \frac{\alpha_{Ch}+\alpha_{Al}}{2} = \frac{0.0077+0.0103}{2} = 0.009 \text{ in.}^2/\text{sec.}$$

By (13.40), $\tau_{k,c} = 0.5\left(1+\frac{\pi^2\times0.5\times0.009}{4\times(\frac{1}{8})^2}\right) = 0.86$ sec.

Thus   percentage increase in $\tau = \frac{\tau_{k,c}-\tau_c}{\tau_c}\times100 = 72\%$.

Under some setups, $T_{support}$ remains constant, independent of $T_{junction}$. In such cases, a continuous conductive heat transfer takes place between the thermocouple junction and the support. This causes a reduction in

Table 13.2 Some Physical Properties of Thermocouple Materials[a]

Material	α (in.^{2}/sec)	k (Btu/sec-ft-°R)	c (Btu/lb°R)	ρ (lb/ft^3)
Alumel	0.0103	0.0048	0.124	537
Chromel	0.0077	0.0031	0.106	545
Constantan	0.0101	0.0038	0.099	553
Iron	0.0268	0.0096	0.107	491
Copper	0.1720	0.0616	0.093	555

[a] From NACA TN 2599, at 530 to 660°R.

both the junction temperature and the junction time constant. Scadron and Warshawsky suggest the following relations [8]:

$$\tau_{r,c} = \frac{\tau_c}{1 + (4\beta\epsilon_w/T_g)}, \qquad (13.41)$$

where $\tau_{r,c}$ represents the time constant for heat transfer by radiation and convection, and

$$\tau_{k,r,c} = \tau_{r,c}(1 - \psi), \qquad (13.42)$$

where $\tau_{k,r,c}$ represents the time constant in the presence of conduction, radiation, and convection. Scadron and Warshawsky present convenient nomographs (in TN 2599) for determining the factors β and ψ. An example given by Scadron [12] illustrates the use of these equations.

Example 7. The convective time constant of a $\frac{1}{8}$ in. long bare Chromel-Alumel thermocouple of $D = 0.04$ in. and $\epsilon_w = 0.9$ at $Ma = 0.2$ and $p = 2$ atm is 0.83 sec. If T_{gas} is 2000°R, $T_{support}$ is 1800°R, β is 120°R, and ψ is 0.12, what is the percentage decrease in the time constant because of conduction effects?

By (13.41)
$$\tau_{r,c} = \frac{0.83}{1 + (4 \times 120 \times 0.9/2000)} = 0.68 \text{ sec.}$$

By (13.42)
$$\tau_{k,r,c} = 0.68(1 - 0.12) = 0.60 \text{ sec.}$$

Thus
$$\text{percentage decrease in } \tau = \frac{\tau_c - \tau_{k,r,c}}{\tau_c} \times 100 = 28\%.$$

These opposing results for the effects of conduction, as illustrated by examples 6 and 7, indicate that great care must be exercised in choosing a mathematical model; it must match the existing conditions. In general, the conduction effect will be minimized by exposing a greater wire length to the fluid temperature.

Moffat [13] indicates that, under the most severe conditions of his tests, radiation increased the time constant only 10%, which is on the same order as the 20% increase in Scadron's example.

Fluid turbulence tends to reduce the time constant by increasing the film coefficient. Wormser [11] states that τ will be reduced by as much as 25% by a turbulence level increase of 1.5%.

13.7 Methods of Improving Response

In general, the response of a temperature sensor can be improved by concentrating on the variables that make up the first-order time constant, as given by (13.3). Also, the response of a second-order system, as the thermometer well-temperature sensor combination, can be improved by reducing the air gap between the sensor O.D. and the well I.D. (see Figure 13.4), by choosing the well material to be of a higher thermal conductivity (as copper instead of stainless steel), or by filling the air gap with a highly thermal conducting material (e.g., a soft solid metal can be pounded into this space).

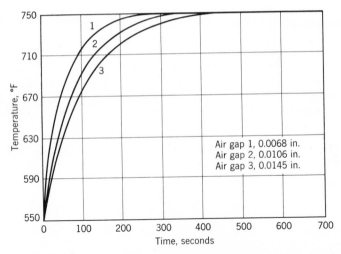

Figure 13.4 Effect of varying air gap on second-order response (after Reference 7).

When efforts have been exhausted along these lines, we still have recourse to multijunctions and electrical compensation. These are discussed briefly.

Multijunction Effect

When three thermocouple junctions are differentially connected, the net response R_n of the system is given by an equation analogous to the single junction equation 13.39, namely,

$$R_n = 1 - e^{-t/\tau_1} + r^{-t/\tau_2} - e^{-t/\tau_3}. \tag{13.43}$$

If it is mechanically arranged that $\tau_1 = \tau_3$, the net response is given by

$$R_n = 1 - 2e^{-t/\tau_1} + e^{-t/\tau_2}. \tag{13.44}$$

If it is further mechanically arranged that τ_2 be *less* than τ_1, the net temperature recovery will be *greater* than that of a single junction of response τ_1. For $\tau_2 = 2\tau_1$, the response of the system will be almost twice as fast as the response of a single junction.

Longer time constants are achieved by using junctions of a larger mass, by embedding junctions in insulating materials, and so on. A transient heat transfer analysis, such as presented in [14], for example, demonstrates the unusual fact that, when the direction of heat transfer is from insulator to metal, the metal has a much slower response than the insulator, when both are at a given distance from the heat source. Thus by embedding one junction of a multijunction system in the surface of a metal plug and by then embedding all three junctions of the system in an insulator such that all junctions are in a plane at the same distance from the heat source (see Figure 13.5), the net response of the device can be more than twice as fast as a single junction.

The performance of a single-junction and several multijunction thermocouples is shown in Figure 13.6, where we also note that, in the

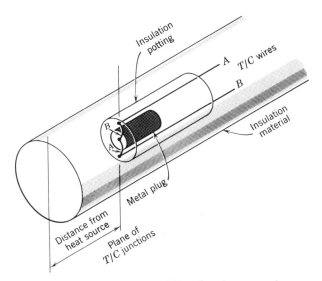

Figure 13.5 High-response multijunction thermocouple.

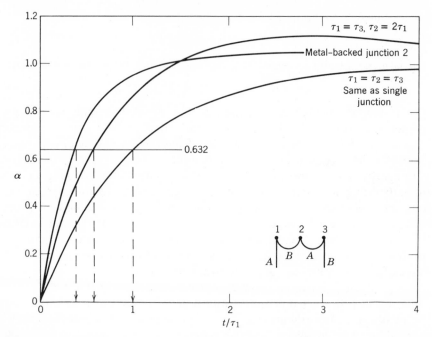

Figure 13.6 The response of several multijunction thermocouples compared with that of a single junction thermocouple.

steady state (over 4.6τ), there is no error in temperature level caused by the use of multijunctions.

Electrical Compensation

An NACA report [15] by Shepard and Warshawsky presents a method for providing up to a thirtyfold reduction in the effective time lag by the use of passive electrical networks. They indicate that this improvement is obtained without attenuation of the voltage signal, but does result in a large reduction in the amount of electric power available. Three basic forms of compensators are used, namely, the *RC*, the *RL*, and the transformer types. For further detail, the original references should be consulted.

13.8   References

[1] D. R. Harper III, "Thermometric Lag," NBS Scientific Paper 185, March 1912.
[2] R. Beck, "Thermometric Time Lag," *Trans. ASME*, August 1941, p. 531.

[3] A. J. Hornfeck, "Response Characteristics of Thermometer Elements," *Trans. ASME*, February 1949, p. 121.

[4] T. C. Linahan, "The Dynamic Response of Industrial Thermometer Wells," *Trans. ASME, May* 1956, p. 759.

[5] R. Looney, Method for Presenting the Response of Temperature-Measuring Systems," *Trans. ASME*, November 1957, p. 1851.

[6] G. A. Coon, "Response of Temperature-Sensing-Element Analogs," *Trans. ASME*, November 1957, p. 1857.

[7] D. R. Keyser, "The Dynamic Response of Thermometer-Well Assemblies, "R and D Report NAVSEC PHILA. DIV. Project A-669, March 15, 1968.

[8] M. D. Scadron and I. Warshawsky, "Experimental Determination of Time Constants and Nusselt Numbers for Bare-Wire Thermocouples in High-Velocity Air Streams and Analytic Approximation of Conduction and Radiation Errors," NACA TN 2599, January 1952.

[9] F. R. Caldwell, L. O. Olsen, and P. D. Freeze, "Intercomparison of Thermocouple Response Data," SAE National Aero Meeting Paper 158F, April 1960.

[10] J. W. Murdock, C. J. Foltz, and C. Gregory, "A Practical Method of Determining Response Time of Thermometers in Liquid Baths," *Trans. ASME, J. Eng. Power,* January 1963, p. 27.

[11] A. F. Wormser, "Experimental Determination of Thermocouple Time Constants with Use of a Variable Turbulence, Variable Density Wind Tunnel, and the Analytic Evaluation of Conduction, Radiation, and Other Secondary Effects," SAE National Aero Meeting Paper 158D, April 1960.

[12] M. D. Scadron, "Time Response Characteristics of Temperature Sensors," SAE National Aero Meeting Paper 158H, April 1960.

[13] R. J. Moffat, "Designing Thermocouples for Response Rate," *Trans. ASME*, February 1958, p. 257.

[14] H. H. Lowell and N. Patton, "Response of Homogeneous and Two-Material Laminated Cylinders to Sinusoidal Environmental Temperature Change, with Applications to Hot-Wire Anemometry and Thermocouple Pyrometry," NACA TN 3514, September 1955.

[15] C. E. Shepard and I. Warshawsky, "Electrical Techniques for Compensation of Thermal Time Lag of Thermocouples and Resistance Thermometer Elements," NACA TN 2703, January 1952.

Nomenclature

Roman

a amplitude of test sensor response; also $e^{-t/\tau}$

A area

b exponent

c specific heat capacity; also exponent

C constant of integration; also capacity

d exact differential

D diameter

e base of natural logarithm

G   mass velocity
h   convective film coefficient
k   thermal conductivity
N   stirrer speed
Nu   Nusselt number
Pr   Prandtl number
R   rate of change of temperature; also response; also resistance
Re   Reynolds number
t   time
T   empirical temperature
V   volume
M   mass

Greek

α thermal diffusivity
β conduction factor
Δ finite difference
ϵ emissivity
θ phase angle
μ dynamic viscosity
ρ density
τ time constant
ϕ a function of
ω frequency

Subscripts

1, 2   initial, final state
c   convection
e   environment
k   conduction
L   lag
n   net
p   at constant pressure
r   radiation

Problems

1. A temperature-sensing probe is inserted in a thermometer well. Find the temperature 2 min after the system is subjected to a step change of 300°F if conditions are such that $r_1 = 0.2$ and $r_2 = 0.02/sec$.

   Ans. $T = 269.76$°F.

2. What temperature would be predicted for the conditions of Problem 1 if only the dominant time constant were applied in a first-order solution?

    Ans. $T = 272.78°F$.

3. For the sensing system of Problem 1, find the temperature error after 3 min if the environment is changing at a rate of 0.5°F/sec.

    Ans. $\Delta T = 27.5°F$.

4. For the first-order system of Problem 2, find the temperature error as in Problem 3.

    Ans. $\Delta T = 25°F$.

5. The time constant of a thermometer is determined to be 10 sec in water where the film coefficient is 100 Btu/hr ft²°F. Find the time to reach 95% of a step change in steam where the film coefficient is 1500 Btu/hr ft²°F.

    Ans. $t_{95\%} = 2$ sec.

Part II
Pressure and Its Measurement

"... in the interior of the liquid the stress, which in this case is usually called the pressure, is everywhere normal to the surface on which it acts..."

Ludwig Prandtl (1949)

INTRODUCTION TO PART II

Part II concerns pressure: its early history, the various standards, and many of the conventional transducers used to measure it.

Like temperature, pressure is a fundamental design parameter. Rare is the engineer or scientist who at some time is not faced with the problem of controlling or measuring pressure. Thus ever since Evangelista Torricelli conceived the first mercury barometer more than 300 years ago, a complex arsenal of instruments, standards, and techniques for measuring pressure has evolved. Selecting the appropriate ones for a given problem is a matter of familiarity with the capabilities of each.

Our purposes in Part II are therefore: to give the basic definitions of pressure; to describe the standards of pressure and give the pertinent correction terms; to consider some of the conventional mechanical and electrical pressure transducers; to discuss the measurement of pressure in moving fluids (including definitions of static, total, and dynamic pressures, and the means of measuring them), and to give a brief review of transient pressure measurements.

Chapter 14

THE CONCEPTS OF PRESSURE

"... to make an instrument which would show the changes in the air, which is at times heavier and thicker, and at times lighter and more rarefied..."

Evangelista Torricelli (1644)

14.1 The Concepts

For a fluid at rest, pressure can be defined as the force exerted perpendicularly by the fluid on a unit area of any bounding surface [1], that is,

$$p \equiv \frac{dF}{dA}.$$ (14.1)

Thus pressure is seen to be basically a mechanical concept (in the field of mechanics it is called "compressive stress") that can be fully described in terms of the primary dimensions of mass, length, and time (see Figure 14.1).

This definition and the following three observations encompass the whole of pressure measurement.

1. It is a familiar fact that pressure (since it is a local property of the fluid) is strongly influenced by position within a static fluid but, at a given position, it is quite independent of direction (see Figure 14.2). Thus we note the expected variation in fluid pressure with elevation

$$dp \equiv -w \, dh$$ (14.2)

and account for the usefulness of manometry.

2. It has also been amply demonstrated that pressure is unaffected by

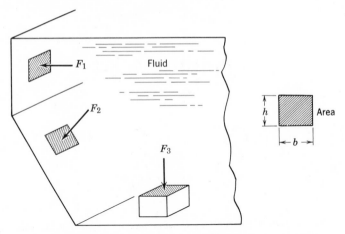

Figure 14.1 Basic definition of fluid pressure. By definition $p = F/A$ where $F = Ma$ and $A = bh$. Typical units of pressure are lb/in.^2, lb/ft^2, and dyn/cm^2. Primary dimensions: since $F = MLT^{-2}$ and $A = L^2$, $p = ML^{-1}T^{-2}$. (In dimensional analysis T is the symbol for the primary dimension of time.)

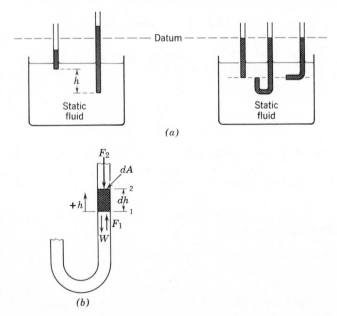

Figure 14.2 Static fluid pressure and position. (*a*) *Qualitatively*: fluid pressure varies with depth but is the same in all directions at a given depth. (*b*) *Quantitatively*: variation in fluid pressure with elevation is obtained by balancing forces on a static fluid element. The following equations may be used: $\sum F_h = 0$; $F_1 = F_2 + W$; $(p - d_p) \, dA = p \, dA + w \, dA \, dh$; $dp = -w \, dh$. For a constant density fluid $(p_1 - p_2) = w(h_2 - h_1)$.

290

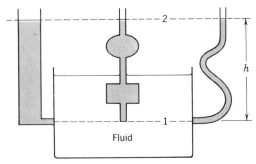

Figure 14.3 Pressure is independent of the size and shape of its confining boundaries: $(p_1 - p_2) = wh$ (where w is specific weight of manometer fluid).

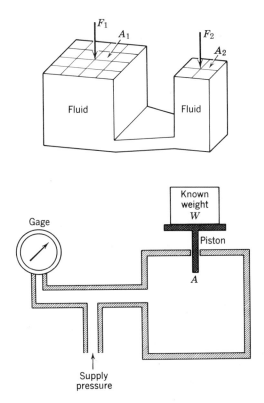

Figure 14.4 (a) Hydraulic lift uses force multiplication based on undiminished pressure transmission in a confined fluid. $p = F_1/A_1 = F_2/A_2$, (in this case $F_1 = 4F_2$, since $A_1 = 4A_2$). (b) Basic principle of deadweight testing: at balance of known weight the gauge pressure is $p = W/A$.

the shape of the confining boundaries. Thus a great variety of fluid pressure transducers is available (see Figure 14.3).

3. Finally, it is well known that a pressure applied to a confined fluid via a movable surface is transferred undiminished throughout the fluid to all bounding surfaces. Thus hydraulic lifts and deadweight testers make their appearance (see Figure 14.4).

14.2 Historical Resume

These basic concepts, which today are generally taken for granted, emerged only slowly over the years. A short historical review concerning the development of the pressure principle is given next. Much can be learned from it, for it involves some of the great names in the physical sciences [2], [3], [4].

Evangelista Torricelli (born October 15, 1608), briefly a student of Galileo, induced his friend Viviani (another pupil of Galileo, see p. 5 Ch. 1) to experiment with mercury and atomspheric pressure in 1643. Inverting a glass tube initially filled with mercury into a shallow dish also filled with mercury, the mercury in the tube was found to sink to a level of about 30 in. above the mercury level in the dish. Torricelli realized that the atmosphere exerted a pressure on the earth which maintained the mercury column in equilibrium. He reasoned further that the height of mercury thus supported varied from day to day, and he also concluded that this height would decrease with altitude. Incidentally, by this method of inverting the filled mercury tube, Torricelli successfully pulled a *vacuum* (i.e., a region essentially devoid of matter).

Torricelli's letter of June 11, 1644 to another friend, Michelangelo Ricci, described this first measurement of atmospheric pressure as follows [5], [6]:

> "We have made many glass vessels · · · with necks two ells long. When these were filled with quicksilver, their mouths stopped with the finger, and then turned upside-down in a vase which had some quicksilver in it, they were seen to empty themselves, and nothing took the place of the quicksilver · · · Nevertheless the neck always remained full to the height of an ell and a quarter and a finger more."

Blaise Pascal, in 1647, in addition to first stating observation 3 given above (now known as Pascal's law), got his brother-in-law, Perier, to measure the height of an inverted mercury column on the top of a mountain (the Puy de Dôme) and at its base. Of this experiment they wrote: "... thus between the heights of the quicksilver in these two experiments there was a difference of three inches and one and a half lines; which ravished us all with admiration and astonishment . . ." That is, a difference of about 3 in. was recorded for the difference of about 3200 feet between these two locations, and this is the basis for the general rule-of-thumb "1 in./1000 ft," to express change in atmospheric pressure with altitude. Pascal named the mercury-under-vacuum instrument they used to sense atmospheric pressure the *barometer*.

In 1660 Robert Boyle discerned that, "whatsoever is performed in the material world, is really done by particular bodies, acting according to the laws of motion." He further stated the now-famous relation: "the product of the measures of pressure and volume is constant for a given mass of air at fixed temperature."

And it was Boyle who first used the word *barometer* in print [7]: ". . . consulting the barometer (if to avoid circumlocutions I may so call the whole instrument wherein a mercurial cylinder of 29 or 30 inches is kept suspended after the manner of the Torricellian experiment) I found . . ."

Robert Hooke, at one time Boyle's assistant, considered the pressure of an enclosed gas as resulting from the continuous *impact* of large numbers of hard, independent, fast moving particles on the container walls. However, it remained for Daniel Bernoulli, in 1738, to develop the *impact theory* of gas pressure to the point where Boyle's law could be deduced analytically. Bernoulli also anticipated the Charles–Gay-Lussac law by stating that pressure is increased by heating a gas at constant volume.

In 1811, Amedeo Avagadro at Turin declared that equal volumes of pure gases, whether elements or compounds, contained equal numbers of molecules at equal temperatures and pressures. The number, later determined to be 2.69×10^{19} molecules/cm^3 at 0°C and 1 atm, attests well to the insight of Hooke and Bernoulli as to the very large numbers of particles involved in a gas sample under usual conditions.

In rapid succession James Prescott Joule, Rudolf Clausius, and James Clerk Maxwell, in the years 1847–1859, developed the *kinetic theory* of gas pressure in which pressure is viewed as a measure of the total kinetic energy of the molecules; that is,

$$p \equiv \frac{2}{3} \frac{\mathrm{KE}}{\mathbf{V}} = \tfrac{1}{3}\rho C^2 = NRT. \tag{14.3}$$

Since kinetic energies are additive, so are pressures, and this leads to Dalton's law (to the effect that the pressure of a mixture is made up of the sum of the partial pressures exerted separately by the constituents of the mixture). If volume changes when kinetic energy (i.e., temperature) is held constant, pressure is seen to vary inversely with volume, which is

Boyle's law. Alternatively, if pressure is held constant, Charles' law follows. Such were the immediate successes of the kinetic theory of gas pressure.

Although we have confined our discussion so far to fluid pressures, the concept of pressure as a result of impacts is not so restricted. As Sir James Jeans [8] has pointed out, these laws are also found to hold for (a) osmotic pressure of weak solutions, (b) pressure exerted by free electrons moving about in the interstics of a conducting solid, and (c) pressure exerted by the atmosphere of electrons surrounding a hot solid. In the case involving electrons, parameter N of (14.3) represents the number of free electrons per unit volume.

Still another viewpoint of pressure is gained from macroscopic thermodynamics [9]. The reversible work done by a closed system (i.e., $\delta W_{closed, reversible}$), is given by the product $p\,d\mathbf{V}$, where the δ denotes that work is a path function, that is, one strongly dependent on the process joining the end states. For the more realistic irreversible process, the internal heat generated (i.e., the friction) is defined as

$$\delta F \equiv p\,d\mathbf{V} - \delta W_{closed}, \tag{14.4}$$

where friction, like work and heat, is also inherently a path function. However, when (14.4) is rewritten as

$$d\mathbf{V} = \frac{\delta W + \delta F}{p}, \tag{14.5}$$

we see that pressure can be viewed as an integrating factor that transforms the "path function" $(\delta W + \delta F)$ into the "point function" $d\mathbf{V}$. This relationship is analogous to the familiar second law of thermodynamics

$$dS = \frac{\delta Q + \delta F}{T}, \tag{14.6}$$

where temperature serves as the integrating factor for heat absorbed (δQ) and heat generated (δF).

14.3 Brief Summary

In *mechanics,* pressure is force per unit area

$$p \equiv \frac{dF}{dA}. \tag{14.1}$$

In *hydraulics,* pressure is specific weight times height

$$dp \equiv -w\,dh. \tag{14.2}$$

In *kinetics*, pressure is molecular kinetic energy per unit volume

$$p \equiv \frac{2}{3}\frac{\mathrm{KE}}{\mathbf{V}}.$$ (14.3)

In *thermodynamics*, pressure is work per unit volume

$$p \equiv \frac{\delta \mathbf{W} + \delta F}{d\mathbf{V}}.$$ (14.4)

All basic pressure measurements are made in accord with (14.1) or (14.2). Equations 14.3 and 14.4, however, serve to broaden our understanding of the pressure principle.

14.4 References

[1] L. Prandtl, *Essentials of Fluid Dynamics*, Hafner, 1952.
[2] G. Holton, *Introduction to Concepts and Theories in Physical Science*, Addison-Wesley, Reading, Mass., 1952.
[3] H. Rouse and S. Ince, *History of Hydraulics*, Dover, New York, 1963.
[4] F. W. Sears, *Principles of Physics, Mechanics Heat and Sound*, Addison-Wesley, Reading, Mass., 1944.
[5] W. E. Knowles Middleton, *The History of the Barometer*, The Johns Hopkins Press, Baltimore, 1964.
[6] P. L. M. Heydemann, "Pressure Measurement and Calibration," ASME Paper 74-PVP-43, 1974.
[7] R. Boyle, "New Experiments and Observations Touching Cold," London, 1665 (reported in [5]).
[8] Sir James Jeans, *An Introduction to the Kinetic Theory of Gases*, Cambridge, 1952.
[9] R. P. Benedict, "Essentials of Thermodynamics," *Electro-Technol.*, July 1962, p. 107.

Nomenclature

Roman

A area
C^2 average value of the square of the molecular velocities.
d exact differential
F force; also friction
h height
KE kinetic energy
N number of molecules per unit volume
p absolute pressure
Q quantity of heat

R	specific gas constant
S	entropy
T	absolute temperature
V	volume
w	specific weight
W	work

Greek

δ	inexact differential
ξ	fluid density

Chapter 15

PRESSURE STANDARDS

"... we took then a long glass tube, which, by a dexterous hand and the help of a lamp, was in such a manner crooked at the bottom, that the part turned up was almost parallel to the rest of the tube ..."

Robert Boyle (1660)

No definition of pressure is really useful to the engineer until it is translated into measurable characteristics. Pressure standards are the basis of all pressure measurements. Those generally available are deadweight piston gauges, manometers, barometers, and McLeod gauges. Each is discussed briefly as to its principle of operation, its range of usefulness, and the more important corrections that must be applied for its proper interpretation.

15.1 Deadweight Piston Gauge

Use of the deadweight, free-piston gauge (see Figure 15.1) for the precise determination of steady pressures was reported as early as 1893 by E. H. Amagat [1]. The gauge serves to define pressures in the range from 0.01 to upwards of 10,000 psig, in steps as small as 0.01% of range within a calibration uncertainty of from 0.01 to 0.05% of the reading.

Principle

The gauge consists of an accurately machined piston (sometimes honed to microinch tolerances) that is inserted into a close-fitting cylinder, both of known cross-sectional areas. In use [2], [3], [4], a number of masses of known weight are first loaded on one end of the free piston. Fluid

297

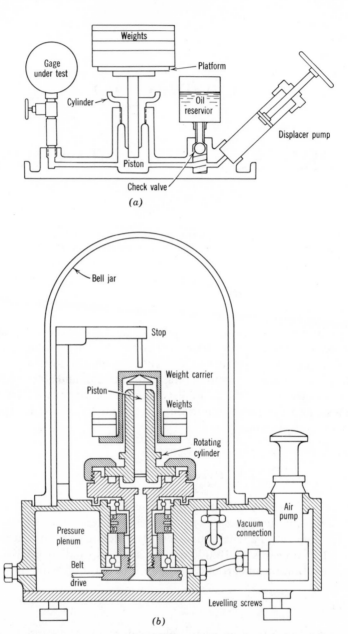

Figure 15.1 Various deadweight piston gauges. (*a*) High-pressure hydraulic gauge. (*b*) Low-pressure gas gauge. (After ASME PTC Supplement 19.2, 1964).

pressure is then applied to the other end of the piston until enough force is developed to lift the piston-weight combination. When the piston is floating freely within the cylinder (between limit stops), the piston gauge is in equilibrium with the unknown system pressure and hence defines this pressure in terms of (14.1) as

$$p_{DW} = \frac{F_E}{A_E},\tag{15.1}$$

where F_E, the equivalent force of the piston-weight combination, depends on such factors as local gravity and air buoyancy, whereas A_E, the equivalent area of the piston-cylinder combination, depends on such factors as piston-cylinder clearance, pressure level, and temperature. (The subscript DW indicates deadweight.)

There will be fluid leakage out of the system through the piston-cylinder clearance. Such a fluid film provides the necessary lubrication between these two surfaces. The piston (or, less frequently, the cylinder) is also rotated or oscillated to further reduce the friction. Because of fluid leakage, system pressure must be continuously trimmed upward to keep the piston-weight combination floating. This is often achieved in a gas gauge by decreasing system volume by a Boyle's-law apparatus (Figure 15.2). As long as the piston is freely balanced, the system pressure is defined by (15.1).

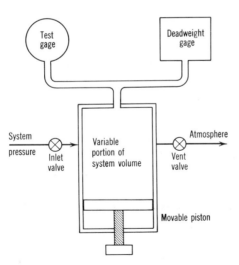

Figure 15.2 Pressure-volume regulator to compensate for gas leakage in a deadweight gauge. As gas leaks, the mass and hence the pressure decrease. As the system volume is decreased, the pressure is reestablished according to $p\mathbf{V} = MRT$.

Corrections

The two most important corrections to be applied to the deadweight piston gauge indication p_I to obtain the system pressure of (15.1) concern air buoyancy and local gravity [5]. According to Archimedes' principle, the air displaced by the weights and the piston exerts a buoyant force that causes the gauge to indicate too high a pressure. The correction term for this effect is

$$C_{tb} = -\left(\frac{w_{air}}{w_{weights}}\right). \tag{15.2}$$

Weights are normally given in terms of the standard gravity value of 32.1740 ft/sec^2 at sea level. Whenever the gravity value differs because of latitude or altitude variations, a gravity correction term must be applied. It is given according to [2] and [10] as

$$C_g = \left(\frac{g_{local}}{g_{standard}} - 1\right)$$
$$= -(2.637 \times 10^{-3} \cos 2\phi + 9.6 \times 10^{-8} h + 5 \times 10^{-5}), \tag{15.3}$$

where ϕ is the latitude (in degrees) and h is the altitude above sea level (in feet). The corrected deadweight piston gauge pressure is given in terms of (15.2) and (15.3) as

$$p_{DW} = p_I(1 + C_{tb} + C_g). \tag{15.4}$$

The effective area of the deadweight piston gauge is normally taken as the mean of the cylinder and piston areas, but temperature affects this dimension. The effective area increases between 13 and 18 ppm/°F for commonly used materials, and a suitable correction for this effect may also be applied [6].

Example. In Philadelphia, at a latitude of 40°N and an altitude of 50 ft above sea level, the *indicated* piston gauge pressure was 1000 psig. The specific weights of the ambient air and the piston weights were 0.076 and 494 lb/ft^3, respectively. The dimensions of the piston and cylinder were determined at the temperature of use (75°F) so that no temperature correction was required for the effective piston gauge area. The *corrected* pressure, according to (15.2), (15.3), and (15.4), was therefore

$$P_{DW} = 1000\left(1 - \frac{0.076}{494} - 2.637 \times 10^{-3} \cos 80° - 9.6 \times 10^{-8} \times 50 - 5 \times 10^{-5}\right) \text{psig},$$

$$= 1000(1 - 0.000154 - 0.000458 - 0.000005 - 0.000050) \text{ psig},$$

$$= 1000(0.999333) = 999.333 \text{ psig}.$$

Nomographs are often used to simplify the correction procedure (see Figure 15.3).

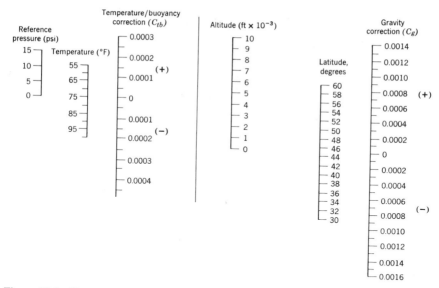

Figure 15.3 Nomographs for temperature/air-bouyancy correction C_{tb} and gravity correction C_g for deadweight gauge measurements.

A variation on the conventional deadweight piston gauges of Figure 15.1 is given in Figure 15.4. Here a force balance system with a binary-coded decimal set of deadweights is used in conjunction with two free pistons moving in two cylinder domes. The highly sensitive equal arm force balance indicates when the weights plus the reference pressure times the piston area on one arm are precisely balanced by the system

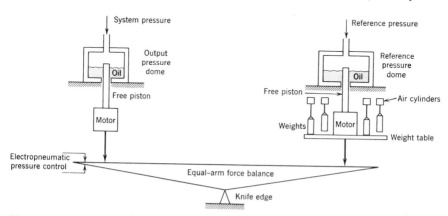

Figure 15.4 An equal-arm force balance piston gauge. (After Gilmore Industries Bulletin PC-100.)

pressure times the piston area on the other arm. This system provides convenient facilities for establishing vacuum reference pressures, controlling and maintaining system pressures, and selecting and loading required deadweights [7]. The pistons in this system are continuously rotated by electric motors, which are integral parts of the beam balance, thus eliminating mechanical linkages. The corrections of (15.4) apply equally well to the force balance piston gauge apparatus.

15.2 Manometer

The manometer (Figure 15.5) was used as early as 1662 by R. Boyle [8] for the precise determination of steady fluid pressures. Because it is founded on a basic principle of hydraulics, and because of its inherent simplicity, the U-tube manometer serves as a pressure standard in the range from 0.1 in. of water to 100 psig, within a calibration uncertainty of from 0.02 to 0.2% of the reading.

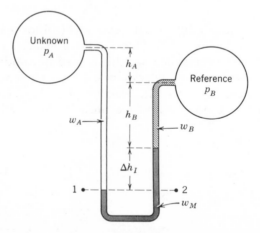

Figure 15.5 Hydraulic correction factor (C_h) for a generalized manometer (capillarity neglected). Given $p_1 = p_2$ (i.e., same level in same fluid at rest has same pressure), then $p_A + w_A(h_A + h_B + \Delta h_I) = p_B + w_B h_B + w_M \Delta h_I$ (all specific weights are those corrected for temperature and gravity), and

$$(p_A - p_B) = w_M\{1 + (w_B/w_M)(h_B/\Delta h_I) - (w_A/w_M)[(h_A + h_B)/\Delta h_I + 1]\}\,\Delta h_I.$$

Thus, in general, $C_h = \{1 + (w_B/w_M)(h_B/\Delta h_I) - (w_A/w_M)[(h_A + h_B)/\Delta h_I + 1]\}$. If $w_A = w_B$, $C_h = 1 - (w_A/w_M)(h_A/\Delta h_I + 1)\}$. If, in addition, $h_A = 0$, $C_h = 1 - w_A/w_M$.

Principle

The manometer consists of a transparent tube (usually of glass) bent or otherwise constructed in the form of an elongated U, and partially filled with a suitable liquid. Mercury and water are the most commonly preferred manometric fluids because detailed information is available on their specific weights. To measure the pressure of a fluid that is less dense than and immiscible with the manometer fluid, it is applied to the top of one of the tubes of the manometer while a reference fluid pressure is applied to the other tube. In the steady state, the difference between the unknown pressure and the reference pressure is balanced by the weight per unit area of the equivalent displaced manometer liquid column according to a form of (14.2)

$$\Delta p_{\mathrm{man}} = w_M \Delta h_E, \tag{15.5}$$

where w_M, the corrected specific weight of the manometer fluid, depends on such factors as temperature and local gravity, and Δh_E, the equivalent manometer fluid height, depends on such factors as scale variations with temperature, relative specific weights and heights of the fluids involved, and capillary effects. As long as the manometer fluid is in equilibrium (i.e., exhibiting a constant manometer fluid displacement, Δh_I), the applied pressure difference is defined by (15.5).

Corrections

The variations of specific weights of mercury and water with temperature in the range of manometer usage are well described by the relations

$$(w_{s,t})_{\mathrm{mercury}} = \frac{0.491154}{1 + 1.01(t-32)10^{-4}} \; \mathrm{lb/in}^3, \tag{15.6}$$

and

$$(w_{s,t})_{\mathrm{water}}$$
$$= \frac{(62.2523 + 0.978476 \times 10^{-2}t - 0.145 \times 10^{-3}t^2 + 0.217 \times 10^{-6}t^3)}{1728} \; \mathrm{lb/in}^3, \tag{15.7}$$

where the subscript s, t signifies evaluation at the standard gravity value and at the Fahrenheit temperature of the manometric fluid. These equations are the basis of the tabulations given in Table 15.1. The specific weight called for in (15.5), however, must be based on the local value of gravity. Hence it is clear that the gravity correction term of (15.3) or

Table 15.1 The Specific Weights of Mercury and Water[a]

Temperature, °F	Specific weight ($w_{s,t}$)	
	Mercury (lb/in.³)	Water (lb/in.³)
32	0.491154	0.036122
36	0.490956	0.036126
40	0.490757	0.036126
44	0.490559	0.036124
48	0.490362	0.036120
52	0.490164	0.036113
56	0.489966	0.036104
60	0.489769	0.036092
64	0.489572	0.036078
68	0.489375	0.036062
72	0.489178	0.036045
76	0.488981	0.036026
80	0.488784	0.036005
84	0.488588	0.035983
88	0.488392	0.035958
92	0.488196	0.035932
96	0.488000	0.035905
100	0.487804	0.035877

[a] At standard gravity value of 32.1740 ft/sec².

Figure 15.3, introduced for the deadweights of the piston gauge, also applies to the specific weight of any fluid involved in manometer usage according to the relation

$$w_c = w_{s,t}(1 + C_g), \qquad (15.8)$$

where w_c is the corrected specific weight of any fluid of standard specific weight w_{st}. Temperature gradients along the manometer can cause local variations in the specific weight of the manometer fluid, and these are to be avoided in any reliable pressure measurement because of the uncertainties they necessarily introduce. Evaporation of the manometer fluid will cause a shift in the manometer zero, but this is easily accounted for and is not deemed a problem. However, it is not good practice to use a mixture for the manometer fluid since selective distillation may cause an unknown change in the specific weight of the mixture.

The most important correction [9], [10] to be applied to the manometer indication Δh_I is that associated with the relative specific weights and heights of the fluids involved. According to the notation of Figure 15.5, the hydraulic correction factor is, in general,

$$C_h = \left\{1 + \left(\frac{w_B}{w_M}\right)\left(\frac{h_B}{\Delta h_I}\right) - \left(\frac{w_A}{w_M}\right)\left[\left(\frac{h_A + h_B}{\Delta h_I}\right) + 1\right]\right\}, \quad (15.9)$$

where all specific weights in (15.9) are to be corrected in accordance with
(15.8). The effect of temperature on the scale calibration is not considered significant in manometry, since these scales are usually calibrated
and used at near-room temperatures. As for capillary effects, it is well
known that the shape of the interface between two fluids at rest depends
on their relative gravity, and on cohesion and adhesion forces between
the fluids and the containing walls. In water-air-glass combinations, the
crescent shape of the liquid surface (called the meniscus) is concave
upward, and water is said to wet the glass. In this situation, adhesive
forces dominate, and water in a tube will be elevated by capillary action.
Conversely, for mercury-air-glass combinations, cohesive forces dominate, the mercury meniscus is concave downward, and the mercury level
in a tube will be depressed by capillary action (see Figure 15.6). From
elementary physics, the capillary correction factor for manometers is

$$C_c = \frac{2\cos\theta_M}{w_M}\left(\frac{\sigma_{A-M}}{r_A} - \frac{\sigma_{B-M}}{r_B}\right), \quad (15.10)$$

where θ_M is the angle of contact between the manometer fluid and the
glass, σ_{A-M} and σ_{B-M} are the surface tension coefficients of the manometer fluid M with respect to the fluids A and B above it, and r_A and r_B are

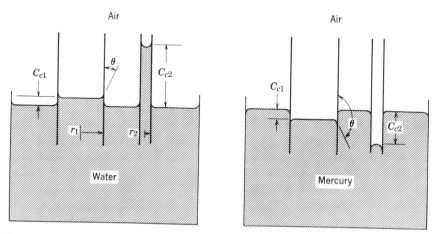

Figure 15.6 Capillary effects in water and mercury. Pressure difference inside and outside
tubes is zero so that variations in liquid heights because of capillarity must be accounted for
in pressure measurements. The single tube correction factor is $C_c = 2\sigma\cos\theta/w_M r$.

the radii of the tubes containing fluids A and B. Typical values for these capillary variables are given in Table 15.2. When both hydraulic and capillary corrections are taken into account, the equivalent manometer fluid height is given in terms of (15.9) and (15.10) as

$$\Delta h_E = C_h \Delta h_I \pm C_c. \tag{15.11}$$

Table 15.2   Capillary Effects

Combination	Surface Tension σ,		Contact Angle θ, degrees
	dyn/cm	lb/in.	
Mercury-vacuum-glass	480	2.74×10^{-3}	140
Mercury-air-glass	470	2.68×10^{-3}	140
Mercury-water-glass	380	2.17×10^{-3}	140
Water-air-glass	73	0.416×10^{-3}	0

For mercury manometers, C_c is positive when the larger capillary effect occurs in the tube showing the larger height of manometer fluid. For water manometers under the same conditions, C_c is negative. When the same fluid is applied to both legs of the manometer, the capillary effect is often neglected. This can be done because the tube bores are approximately equal in standard U-tube manometers, and hence capillarity in one tube just counterbalances that in the other. The capillary effect can be extremely important, however, and must always be considered in manometer-type instruments.

To minimize the effect of a variable meniscus, which can be caused by the presence of dirt, the method of approaching equilibrium, the tube bore, and so on, the tubes are always tapped before reading, and the measured liquid height is always based on readings taken at the center of the meniscus in each leg of the manometer. To reduce the capillary effect itself, the use of large-bore tubes (over $\frac{3}{8}$ in. diameter) is most effective.

Example. In Denver, Colorado, at a latitude of 39°40'N, at an altitude of 5380 ft above sea level, and at a temperature of 76°F, the *indicated* manometer fluid height was 50 in. of mercury in uniform $\frac{1}{8}$ in. bore tubing. The reference fluid was air at atmospheric pressure, and the higher pressure fluid was water at an elevation of 10 ft above the water-mercury interface. According to (15.6) and (15.8), the corrected specific weight of mercury was

$$w_{\text{mercury}} = \left[\frac{0.491154}{1+1.01(76-32)10^{-4}}\right](1 - 2.637 \times 10^{-3} \cos 79°20' - 9.6 \times 10^{-8} \times 5380$$
$$-5 \times 10^{-5}),$$

or $w_{\text{mercury}} = (0.488981)(0.998945) = 0.488465 \text{ lb/in.}^3.$

The corrected specific weight of water was similarly

$$w_{\text{water}} =$$

$$\left\{ \frac{(62.2523 + 0.978476 \times 10^{-2} \times 76 - 0.145 \times 10^{-3} \times (76)^2 + 0.217 \times 10^{-6} \times (76)^3)}{1728} \right\}$$

$$\times (0.998945),$$

$$w_{\text{water}} = (0.036026)(0.998945) = 0.035988 \text{ lb/in.}^3,$$

and, following the notation of Figure 15.5, $w_{\text{water}} = w_A$. Because of the extremely small specific weight of air (w_B) compared to that of the manometer fluid (w_M), the air column effect in (15.9) is neglected. Hence according to (15.9), (15.10), and (15.11), the equivalent manometer fluid height was

$$\Delta h_E = \left\{ 1 - \left(\frac{0.035988}{0.488465} \right) \left[\left(\frac{10 \times 12 - 50}{50} \right) + 1 \right] \right\} 50 + \frac{2(-0.7660)}{0.488465}$$

$$\times \left[\frac{2.17 \times 10^{-3}}{1/16} - \frac{2.68 \times 10^{-3}}{1/16} \right]$$

or $$\Delta h_E = (0.82318)50 + (0.02559) = 41.1846 \text{ in.}$$

The *corrected* manometer pressure difference, according to (15.5), was

$$\Delta p_{\text{man}} = (0.488465)(41.1846) = 20.1172 \text{ psig.}$$

15.3 Micromanometers

Extending the capabilities of conventional U-tube manometers are various types of micromanometers that serve as pressure standards in the range from 0.0002 to 20 in. of water at pressure levels from 0 absolute to 100 psig. Three of these micromanometer types are discussed next. These have been chosen on the basis of simplicity of operation. A very complete and authoritative survey of micromanometers is given in [11].

Prandtl-Type

In the Prandtl-type micromanometer (see Figure 15.7), capillary and meniscus errors are minimized by returning the meniscus to a reference null position (within a transparent portion of the manometer tube) before measuring the applied pressure difference. A reservoir, which forms one side of the manometer, is moved vertically with respect to an inclined portion of the tube, which forms the other side of the manometer, to achieve the null position. This position is reached when the meniscus falls within two closely scribed marks on the near-horizontal portion of the micromanometer tube. Either the reservoir or the inclined tube is moved

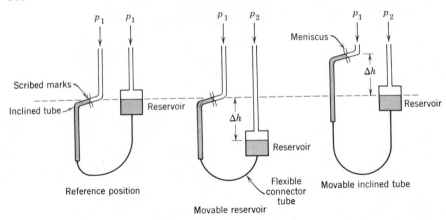

Figure 15.7 Two variations of the Prandtl-type manometer. (After application of pressure difference, either the reservoir or the inclined tube is moved by a precision lead screw to achieve the null position of the meniscus.)

by means of a precision lead screw arrangement, and the micromanometer liquid displacement (Δh), corresponding to the applied pressure difference, is determined by noting the rotation of the lead screw on a calibrated dial. The Prandtl-type micromanometer is generally accepted as a pressure standard within a calibration uncertainty of 0.001 in. of water.

Micrometer-Type

Another method of minimizing capillary and meniscus effects in manometry is to measure liquid displacements with micrometer heads fitted with adjustable, sharp index points located at or near the centers of large-bore transparent tubes that are joined at their bases to form a U. In some commercial micromanometers [12], contact with the surface of the manometric liquid may be sensed visually by dimpling the surface with the index point, or by electrical contact. Micrometer-type micromanometers also serve as pressure standards within a calibration uncertainty of 0.001 in. of water. (See Figure 15.8)

Air Micromanometer

An extremely sensitive, high-response micromanometer uses air as its working fluid, and thus avoids all capillary and meniscus effects usually encountered in liquid manometry. Such an instrument has been described by J. F. Kemp [13]. (See Figure 15.9.) In this device, the reference

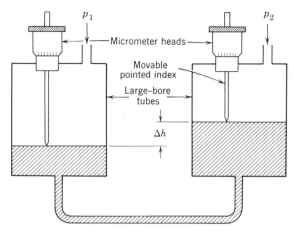

Figure 15.8 A micrometer-type manometer.

pressure is mechanically amplified by centrifugal action in a rotating disk. This disk speed is adjusted until the amplified reference pressure just balances the unknown pressure. This null position is recognized by observing the lack of movement of minute oil droplets sprayed into a glass indicator tube located between the unknown and amplified pressure lines. At balance, the air micromanometer yields the applied pressure difference through the relation

$$\Delta p_{\text{micro}} = K\rho n^2, \tag{15.12}$$

where ρ is the reference air density, n is the rotational speed of the disk, and K is a constant that depends on disk radius and annular clearance

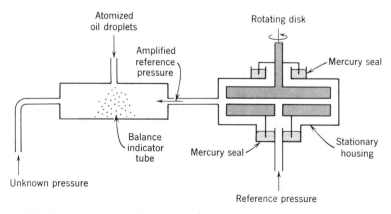

Figure 15.9 An air-type centrifugal micromanometer. (After J. F. Kemp, 1959.)

between the disk and housing. Measurements of pressure differences as small as 0.0002 in. of water can be made with this type of micromanometer, within an uncertainty of 1%.

REFERENCE PRESSURES. A word on the reference pressure employed in manometry is pertinent at this point in the discussion. If atmospheric pressure is used as a reference, the manometer yields *gauge pressures.* Such pressures vary with time, altitude, latitude, and temperature, because of the variability of air pressure (see Figure 15.10). If, however, a

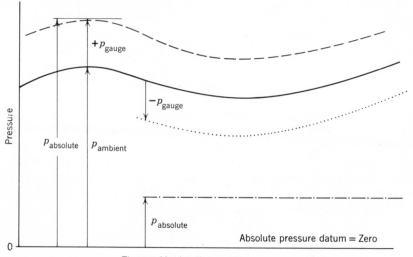

Figure 15.10 Relations among terms used in pressure measurements: (– – –) locus of a constant positive gauge pressure; (———) gauge pressure datum = ambient; (. . . .) locus of a constant negative gauge pressure; (—.—.) locus of a constant absolute pressure.

vacuum is used as reference, the manometer yields *absolute pressures* directly, and it could serve as a barometer (which is considered in the next section). In any case, the obvious but important relation between gauge and absolute pressures is

$$p_{absolute} = p_{gauge} + p_{ambient},$$ (15.13)

where by ambient pressure we mean pressure surrounding the gauge. Most often, ambient pressure is simply the atmospheric pressure.

15.4   Barometer

As already indicated, the barometer was used as early as 1643 by Torricelli for the precise determination of steady atmospheric pressures and continues today to define such pressures within a calibration uncertainty of from 0.001 to 0.03% of the reading.

Principle

The cistern barometer consists of a vacuum-referred mercury column immersed in a large-diameter ambient-vented mercury column that serves as a reservoir (the cistern). The most common cistern barometer in general use is the Fortin type [after Nicolas Fortin (1750–1831)] in which the height of the mercury surface in the cistern can be adjusted. (See Figure 15.11.) The cistern in this case is essentially a leather bag supported in a bakelite housing. The cistern level adjustment provides a fixed zero reference for the plated brass, mercury height scale that is adjustably attached, but fixed at the factory during calibration, to a metal tube. The metal tube, in turn, is rigidly fastened to the solid parts of the cistern assembly and, except for reading slits, surrounds the glass tube containing the barometer mercury. A short tube, which is movable up and down within the first tube, carries a vernier scale and a ring used for sighting on the mercury meniscus in the glass tube.

In use, the datum-adjusting screw is turned until the mercury in the cistern just makes contact with the ivory index, at which point the mercury surface is aligned with zero on the instrument scale. Next, the indicated height of the mercury column in the glass tube is determined. The lower edge of the sighting ring is lined up with the top of the meniscus in the tube. A scale reading and vernier reading are taken and combined to yield the indicated mercury height h_{ti} at the barometer temperature t.

The atmospheric pressure exerted on the mercury in the cistern is just balanced by the weight per unit area of the vacuum-referred mercury column in the barometer tube according to a form of (14.2), that is,

$$p_{baro} = w_{Hg}h_{t0}, \qquad (15.14)$$

where w_{Hg}, the referred specific weight of mercury, depends on such factors as temperature and local gravity, whereas h_{t0}, the referred height of mercury, depends on such factors as thermal expansions of the scale and of mercury.

Corrections

When reading mercury height on a Fortin barometer with the scale zero adjusted to agree with the level of mercury in the cistern, the correct height of mercury at temperature t (called h_t) will be greater than the

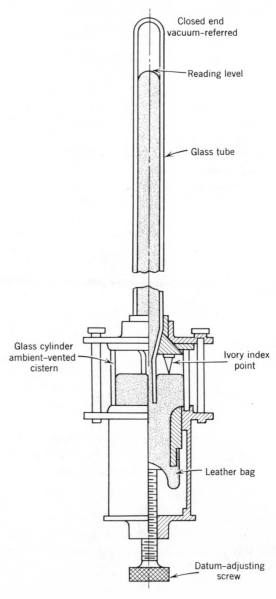

Figure 15.11 A Fortin-type barometer. (After ASME PTC Supplement 19.2, 1964.)

indicated height of mercury at temperature t (called h_{tI}) whenever $t > t_s$ (where t_s is the temperature at which the scale was calibrated). This difference can be expressed in terms of the scale expansion in going from t_s to t as

$$h_t = h_{tI}[1 + S(t - t_s)], \tag{15.15}$$

where S is the linear coefficient of thermal expansion of the scale per degree [10]. If, as is usual, the height of mercury is desired at some reference temperature t_0, the correct height of mercury at t will be greater than the referred height of mercury at t_0 (called h_{t0}) whenever $t > t_0$. This difference can be expressed in terms of the mercury expansion in going from t_0 to t as

$$h_t = h_{t0}[1 + m(t - t_0)], \tag{15.16}$$

where m is the cubical coefficient of thermal expansion of mercury per degree. A temperature correction factor can be defined in terms of the indicated reading at t and the referred height at t_0 as

$$C_t = h_{t0} - h_{tI}. \tag{15.17}$$

In terms of (15.15) and (15.16), this correction is

$$C_t = \left(\frac{1}{1 + m(t - t_0)} - \frac{1}{1 + S(t - t_s)} \right) h_t. \tag{15.18}$$

Replacement of h_t of (15.18) by (15.15) results in

$$C_t = \left[\frac{S(t - t_s) - m(t - t_0)}{1 + m(t - t_0)} \right] h_{tI}. \tag{15.19}$$

When standard values of $S = 10.2 \times 10^{-6}/°F$, $m = 101 \times 10^{-6}/°F$, $t_s = 62°F$, and $t_0 = 32°F$ are substituted in (15.19), the result is:

$$C_t = -\left[\frac{9.08(t - 28.63)(10^{-5})}{1 + 1.01(t - 32)(10^{-1})} \right] h_{tI}. \tag{15.20}$$

This temperature correction is zero at a barometer temperature of 28.63°F for all values of h_{tI}. For usual values of h_{tI} and t, the algebraically additive temperature correction factor of (15.20) is presented in Table 15.3. The temperature correction factor of (15.20) can be approximated with very little loss of accuracy as

$$C_{t\ approx} = -9(t - 28.6)(10^{-5}) h_{tI}. \tag{15.21}$$

This equation is useful for both hand and machine calculations. The uncertainty introduced in h_{t0} is always less than 0.001 in. of mercury for all values of h_{tI} and t presented in Table 15.3.

Temperature, °F	\multicolumn{7}{c}{Observed Barometer Height (h_{tI}), in. Hg}
	28.5	29.0	29.5	30.0	30.5	31.0	31.5
60	−0.081	−0.082	−0.084	−0.085	−0.086	−0.088	−0.089
61	−0.083	−0.085	−0.086	−0.088	−0.089	−0.091	−0.092
62	−0.086	−0.087	−0.089	−0.090	−0.092	−0.093	−0.095
63	−0.088	−0.090	−0.091	−0.093	−0.095	−0.096	−0.098
64	−0.091	−0.093	−0.094	−0.096	−0.097	−0.099	−0.101
65	−0.093	−0.095	−0.097	−0.098	−0.100	−0.102	−0.103
66	−0.096	−0.098	−0.099	−0.101	−0.103	−0.104	−0.106
67	−0.099	−0.100	−0.102	−0.104	−0.106	−0.107	−0.109
68	−0.101	−0.103	−0.105	−0.107	−0.108	−0.110	−0.112
69	−0.104	−0.106	−0.107	−0.109	−0.111	−0.113	−0.115
70	−0.106	−0.108	−0.110	−0.112	−0.114	−0.116	−0.117
71	−0.109	−0.111	−0.113	−0.115	−0.116	−0.118	−0.120
72	−0.111	−0.113	−0.115	−0.117	−0.119	−0.121	−0.123
73	−0.114	−0.116	−0.118	−0.120	−0.122	−0.124	−0.126
74	−0.117	−0.119	−0.121	−0.123	−0.125	−0.127	−0.129
75	−0.119	−0.121	−0.123	−0.125	−0.127	−0.130	−0.132
76	−0.122	−0.124	−0.126	−0.128	−0.130	−0.132	−0.134
77	−0.124	−0.126	−0.129	−0.131	−0.133	−0.135	−0.137
78	−0.127	−0.129	−9.131	−0.133	−0.136	−0.138	−0.140
79	−0.129	−0.132	−0.134	−0.136	−0.138	−0.141	−0.143
80	−0.132	−0.134	−0.136	−0.139	−0.141	−0.143	−0.146
81	−0.134	−0.137	−0.139	−0.141	−0.144	−0.146	−0.149
82	−0.137	−0.139	−0.142	−0.144	−0.147	−0.149	−0.151
83	−0.140	−0.142	−0.144	−0.147	−0.149	−0.152	−0.154
84	−0.142	−0.145	−0.147	−0.150	−0.152	−0.155	−0.157
85	−0.145	−0.147	−0.150	−0.152	−0.155	−0.157	−0.160
86	−0.147	−0.150	−0.152	−0.155	−0.157	−0.160	−0.163
87	−0.150	−0.152	−0.155	−0.158	−0.160	−0.163	−0.165
88	−0.152	−0.155	−0.158	−0.160	−0.163	−0.166	−0.168
89	−0.155	−0.158	−0.160	−0.163	−0.166	−0.168	−0.171
90	−0.157	−0.160	−0.163	−0.166	−0.168	−0.171	−0.174
91	−0.160	−0.163	−0.166	−0.168	−0.171	−0.174	−0.177
92	−0.162	−0.165	−0.168	−0.171	−0.174	−0.177	−0.180
93	−0.165	−0.168	−0.171	−0.174	−0.177	−0.179	−0.182
94	−0.168	−0.170	−0.173	−0.176	−0.179	−0.182	−0.185
95	−0.170	−0.173	−0.176	−0.179	−0.182	−0.185	−0.188
96	−0.173	−0.176	−0.179	−0.182	−0.185	−0.188	−0.191
97	−0.175	−0.178	−0.181	−0.184	−0.187	−0.191	−0.194
98	−0.178	−0.181	−0.184	−0.187	−0.190	−0.193	−0.196
99	−0.180	−0.183	−0.187	−0.190	−0.193	−0.196	−0.199
100	−0.183	−0.186	−0.189	−0.192	−0.196	−0.199	−0.202

[a] Based on $C_t = -\left[\dfrac{9.08(t - 28.63)(10^{-5})}{1 + 1.01(t - 32)(10^{-4})}\right] h_{tI}$. For use with brass scales correct at 62°F, for obtaining barometer heights referred to 32°F according to $h_{t0} = h_{tI} + C_t$.

The specific weight called for in (15.14) must be based on the local value of gravity and on the reference temperature (t_0). Thus once again the gravity correction term of (15.3) or Figure 15.3 must be applied. This time it is according to the relation

$$w_{\text{Hg}} = (w_{s,t0})(1 + C_g),$$

(15.22)

where, from Table 15.1, $w_{s,t0}$ is specifically 0.491154 lb/in.^{3}. Atmospheric pressure is now obtained in straightforward manner by combining (15.14), (15.17), and (15.22). (The gravity correction is sometimes applied instead to the referred height of mercury, and in this role it is often fallaciously looked on as a gravity correction to height.)

Several other factors that could contribute to the uncertainty in h_{tI} are detailed in NBS Monograph 8 [10] and are discussed briefly here. These factors introduce no additional correction terms.

1. *Lighting.* Proper illumination is essential to define the location of the crown of the meniscus. Precision meniscus sighting under optimum viewing conditions can approach ±0.001 in. Contact between index and mercury surface in the cistern, judged to be made when a small dimple in the mercury first disappears during adjustment, can be detected with proper lighting to much better than ±0.001 in.

2. *Temperature.* To keep the uncertainty in h_{tI} within 0.01% (0.003 in. Hg), the mercury temperature must be known within ±1°F. Although it is generally assumed that the scale and mercury temperatures are identical, the scale temperature need not be known to better than ±10°F for comparable accuracy. Uncertainties caused by nonequilibrium conditions could be avoided by installing the barometer in a uniform temperature room.

3. *Alignment.* Vertical alignment of the barometer tube is required for an accurate pressure determination. The Fortin barometer, designed to hang from a hook, does not of itself hang vertically. This must be accomplished by a separately supported ring encircling the cistern; adjustment screws control the horizontal position.

4. *Capillary Effects.* Depression of the mercury column in commercial barometers is accounted for in the initial calibration setting at the factory, since such effects could not be applied conveniently during use. The quality of the barometer is largely determined by the bore of the glass tube. Barometers with a bore of $\frac{1}{4}$ in. are suitable for readings of 0.01 in., whereas barometers with a bore of $\frac{1}{2}$ in. are suitable for readings of 0.002 in.

Finally, whenever the barometer is read at an elevation other than that of the test site, an altitude correction factor must be applied to the local absolute barometric pressure of (15.14). This is necessary because of the variation in atmospheric pressure with elevation as expressed by the relation

$$\int_{\text{baro}}^{\text{site}} \frac{dp}{w_{\text{air}}} = -\int_{\text{baro}}^{\text{site}} dz. \tag{15.23}$$

An altitude correction factor similar to C_t of (15.17), which can be added directly to the local barometric pressure, can be defined as

$$C_z = p_{\text{site}} - p_{\text{baro}}. \tag{15.24}$$

Using (15.23) and applying the realistic isothermal assumption between barometer and test sites, the altitude correction factor amounts to

$$C_z = p_{\text{baro}} \left[\exp\left(\frac{Z_{\text{baro}} - Z_{\text{site}}}{(RT)_{\text{air}}} \right) - 1 \right], \tag{15.25}$$

where Z is altitude in feet, R is the gas constant in feet per degree Rankine, and T is the absolute temperature.

Example. A Fortin barometer indicates 29.52 in. Hg at 75.2°F at Troy, New York, at a latitude of 42°41′N, and at an altitude of 945 ft above sea level. The test site is 100 ft below the barometer location. Equations 15.20, 15.21, and Table 15.3 all agree that $C_t = -0.124$ in. Hg. Equations 15.3 and 15.22 indicate that $w_{\text{Hg}} = 0.490980$ lb/in.^{3}. Hence, according to (15.14) and (15.17), the *corrected* barometric pressure was

$$p_{\text{baro}} = (0.490980)(29.52 - 0.124) = 14.433 \text{ psia.}$$

However, following (15.24) and (15.25), with $R = 53.35$ ft/°R, the barometric pressure at the test site was

$$C_z = (14.433) \left[\exp\left(\frac{100}{53.35 \times 534.8} \right) - 1 \right] = 0.051 \text{ psia.}$$

$$p_{\text{site}} = C_z + p_{\text{baro}} = 0.051 + 14.433 = 14.484 \text{ psia.}$$

15.5 McLeod Gauge

A special mercury-in-glass manometer was described by H. McLeod [14] in 1874 for the precise determination of very low absolute pressures of permanent gases. Based on an elementary principle of thermodynamics (Boyle's law), the McLeod gauge serves as the pressure standard in the range from 1 mm Hg above absolute zero to about 0.01 μ (where 1 $\mu = 1$ micron $= 10^{-3}$ mm Hg), with a calibration uncertainty of from 0.5% above 1 μ to about 3% at 0.1 μ.

Principle

The McLeod gauge consists of glass tubing arranged so that a sample of gas at the unknown pressure can be trapped and then isothermally compressed by a rising mercury column [15]. This amplifies the unknown pressure and allows measurement by conventional manometric means. The apparatus is illustrated in Figure 15.12. All of the mercury is initially contained in the volume below the cutoff level. The McLeod gauge is first exposed to the unknown gas pressure p_1; the mercury is then raised in tube A beyond the cutoff, trapping a gas sample of initial volume $\mathbf{V}_1 = \mathbf{V} + ah_c$, where $a =$ area of the measuring capillary. The mercury is continuously forced upward until it reaches the zero level in the reference capillary B. At this time, the mercury in the measuring capillary C reaches a level h where the gas sample is at its final volume $\mathbf{V}_2 = ah$, and at the final amplified manometric pressure $p_2 = p_1 + h$. The relevant equations at these pressures are

$$p_1\mathbf{V}_1 = p_2\mathbf{V}_2 \tag{15.26}$$

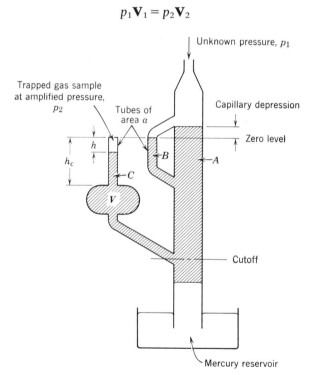

Figure 15.12 The McLeod gauge. Before gas compression takes place the mercury is contained in the reservoir. The cross-hatched area indicates the location of the mercury after the trapped gas is compressed.

and
$$p_1 = \frac{ah^2}{\mathbf{V}_1 - ah}.$$
(15.27)

If $ah \ll \mathbf{V}_1$, as is usually the case, then

$$p_1 = \frac{ah^2}{\mathbf{V}_1}.$$
(15.28)

It is clear from (15.26)–(15.28) that the larger the volume ratio $(\mathbf{V}_1/\mathbf{V}_2)$ the greater will be the magnification of the pressure p_1 and of the manometer reading h. Hence it is desirable that measuring tube C have a small bore. Unfortunately, for tube bores under 1 mm, the compression gain is offset by the increased reading uncertainty caused by capillary effects (see 15.10). In fact, the reference tube B is introduced just to provide a meaningful zero for the measuring tube. If the zero is fixed, (15.28) indicates that manometer indication h varies nonlinearly with initial pressure p_1. A McLeod gauge with such an expanded scale at the lower pressures will naturally exhibit a higher sensitivity in this region. The McLeod pressure scale, once established, serves equally well for all the permanent gases.

Corrections

There are no corrections to be applied to the McLeod gauge reading, but certain precautions should be taken [16], [17]. Moisture traps must be provided to avoid taking any condensable vapors into the gauge because such condensable vapors occupy a larger volume when in the vapor phase at the initial low pressures than they occupy when in the liquid phase at the high pressures of reading. Thus the presence of condensable vapors always causes pressure readings to be too low. Capillary effects, although partly counter-balanced by using a reference capillary, can still introduce significant uncertainties, since the angle of contact between mercury and glass can vary $\pm 30°$ depending on how the mercury approaches its final position. Finally, since the McLeod gauge does not give continuous readings, steady-state conditions must prevail for the measurements to be useful.

The mercury piston of the McLeod gauge can be motivated a number of ways. A mechanical plunger can force the mercury up tube A. A partial vacuum over the mercury reservoir can hold the mercury below the cutoff until the gauge is charged, and then the mercury can be allowed to rise by bleeding dry gas into the reservoir. There are also several types of swivel gauges [18] in which the mercury reservoir is located above the gauge zero during charging. A $90°$ rotation of the gauge causes the mercury to rise in tube A by the action of gravity alone. In a variation of the McLeod principle, the gas sample is compressed between two mercury columns, thus avoiding the need for a reference capillary and a sealed-off measuring capillary. A McLeod gauge with an automatic zeroing reference capillary has also been described recently [19].

A summary of the characteristics of the various pressure standards is given in Figure 15.13.

Type	Range	Uncertainty
Deadweight piston gage	0.01 to 10,000 psig	0.01 to 0.05% of reading
Manometer	0.1 to 100 psig	0.02 to 0.2% of reading
Micromanometer	0.0002 to 20 in. H_2O	1% of reading to 0.001 in. H_2O
Barometer	27 to 31 in. Hg	0.001 to 0.03% of reading
McLeod gage	0.01 μ to 1 mm Hg	3 to 0.5% of reading

Figure 15.13 Characteristics of various pressure standards.

PRESSURE SCALES AND UNITS. A word on pressure scales seems necessary here, since there is indeed a confusing array of scales and units to choose from in expressing pressures. Some of the more common include:

pascals	(N/m^2)
pounds per square inch	(psi)
inches of water	(in. H_2O)
inches of mercury	(in. Hg)
atmospheres	(atm)
microbars	(μbar)
millimeters of mercury	(mm Hg)
microns	(μ)

Conversion factors between these units are given in Table 15.4. Two useful approximations, to help sense order of magnitudes are:

$$1 \text{ mm} = 1000 \ \mu \sim 0.04 \text{ in. Hg,}$$

$$1 \ \mu = 0.001 \text{ mm} \sim 0.00004 \text{ in. Hg.}$$

Other units that have also been used to express pressures are: torrs (1 torr $\sim$ 1 mm Hg), pascals, deciboyles, stress-presses, to mention but a few.

Table 15.4 Pressure Unit Conversion Factors[a]

Pressure Unit	psi	in. H$_2$O	in. Hg	atm	μbar	mm Hg	μ
1 psi	1.000	27.730	2.0360	6.8046 $\times 10^{-2}$	68947.6	51.715	51715.0
1 in. H$_2$O (68°F)	0.036063	1.000	0.073424	2.4539 $\times 10^{-3}$	2486.4	1.8650	1865.0
1 in. Hg (32°F)	0.49115	13.619	1.000	3.3421 $\times 10^{-2}$	33864.0	25.400	25400.0
1 atm	14.69595	407.513	29.9213	1.000	1.01325 $\times 10^{6}$	760.000	7.6000 $\times 10^{5}$
1 μbar (dyn/cm^2)	1.4504 $\times 10^{-5}$	4.0218 $\times 10^{-4}$	2.9530 $\times 10^{-5}$	9.8692 $\times 10^{-7}$	1.000	7.5006 $\times 10^{-4}$	0.75006
1 mm Hg (32°F)	0.019337	0.53620	0.03937	1.3158 $\times 10^{-3}$	1333.2	1.000	1000.0
1 μ (32°F)	1.9337 $\times 10^{-5}$	5.3620 $\times 10^{-4}$	3.9370 $\times 10^{-5}$	1.3158 $\times 10^{-6}$	1.3332	0.0010	1.000

[a] Adopted from NBS Monograph 8, 1960.

In general, (14.1) and (14.2) serve to relate all these various pressure units, if proper attention is given to dimensional analysis. For further detail on the various systems of units, recent literature should be consulted [20]–[22].

15.6 References

[1] E. H. Amagat, "Mémoires sur l'élasticité et la dilatabilité des fluides jusqu'aux tres hautes pressions," *Ann. Chem. Phys.*, **29**, 1893, p. 68.

[2] D. P. Johnson and D. H. Newhall, "The Piston Gage as a Precise Pressure-Measuring Instrument," *Trans. ASME*, **75**, No. 3, April 1953, p. 301 (see also *Instr. Control Systems*, **35**, April 1962, p. 120).

[3] *Pressure Measurement*, ASME Power Test Codes, Supplement on Instruments and Apparatus, Pt. 2, 1964.

[4] J. O. Ess, "Generation of Precise Reference Pressures up to 15,750 psi," *Design News*, August 1965, p. 136.

[5] J. L. Cross, "Reduction of Data for Piston Gauge Pressure Measurements," *Nat. Bur. Std. Monograph 65*, June 1963.

[6] R. C. Dean *et al.*, *Aerodynamic Measurements*, M.I.T. Notes, September 1952.

[7] "Automatic Primary Pressure Standards," Gilmore Industries, Inc., Cleveland, Ohio, Bulletin PC-100.

[8] R. Boyle, "A Defence of the Doctrine Touching the Spring and Weight of the Air," in *New Physics-Mechanical Experiments*, Pt. II, Chapt. V, London, 1962.

[9] J. B. Meriam, "Manometers," *Instr. Control Systems*, **35**, February 1962, p. 114.

[10] W. G. Brombacher, D. P. Johnson, and J. L. Cross, "Mercury Barometers and Manometers," *Nat. Bur. Std. Monograph 8*, May 1960.

[11] W. G. Brombacher, "Survey of Micromanometers," *Nat. Bur. Std. Monograph 114*, June 1970.

[12] E. C. Hass and C. Hass, "Micrometer Standard Barometers," *Instr. Control Systems*, **35**, March 1962, p. 118.

[13] J. F. Kemp, "Centrifugal Manometer," *Trans. ASME, J. Basic Eng.*, **81**, September 1959, p. 341.

[14] H. McLeod, "Apparatus for Measurement of Low Pressure of a Gas," *Phil. Mag.*, **48**, 1874, p. 110.

[15] J. H. Leck, *Pressure Measurement in Vacuum Systems*, Reinhold, New York, 1957.

[16] S. Ruthbery (Chief, Vacuum Measurements Section Nat. Bur. Std.), Personal Communication, April 1966.

[17] C. F. Morrison, "Vacuum Gauge Calibration Today," *Research/Development*, September 1969, p. 54.

[18] R. Gilmont and M. C. Parkinson, "New Tilting Gauge Improves Accuracy," *Research/Development*, November 1962, p. 50.

[19] R. H. Work and C. E. Hawk, "McLeod Gauge with Automatic Zeroing," *Research/Development*, June 1966, p. 79.

[20] L. Tonks and G. C. Baldwin, "The Deciboyle: A Rational Scale of Pressure Designation," *Research/Development*, February 1965, p. 57.

[21] B. C. Wiggin, "Triple U: A Universal Unit System," *Research/Development*, February 1966, p. 29.

[22] G. C. Baldwin and L. Tonks, "How Shall We Measure Vacuum," *Research/Development*, March 1966, p. 66.

Nomenclature

Roman

a	capillary area
A	area
C_{tb}	bouyant correction factor
C_c	capillary correction factor
C_g'	gravity correction factor
C_h	hydraulic correction factor
C_t	temperature correction factor
C_z	altitude correction factor
F	force
h	altitude
m	cubical coefficient of expansion of mercury per degree, also metre
N	newton of force
p	pressure
r	radius
R	specific gas constant
S	linear coefficient of expansion of scale per degree
t	empirical temperature
T	absolute temperature
V	volume
w	specific weight
Z	elevation

Greek

Δ	finite difference
θ	angle of contact
σ	surface tension coefficient

Subscripts

c	corrected
E	equivalent
I	indicated
M	manometer
S	standard
t	at temperature t

Problems

1. A deadweight tester indicates 500 psig when located at latitude 25°N and at sea level. Using the specific weight of air as 0.075 lbf/ft³ and

the specific weight of the weights as 495 lbf/ft^3, find the corrected pressure.

Ans. $p_{DW} = 499.052$ psig.

2. The static pressure in a water-filled pipe is to be measured by a mercury manometer. Express the pipe pressure in terms of manometer deflection Δh if the zero pressure position of the manometer is A inches below the centerline of the pipe at which point the pressure tap is located. (Assume specific weights are in lbf/in.^{3}.)

    Ans. $p = p_{baro} + w_{H_2O}\left[\left(\dfrac{w_{Hg}}{w_{H_2O}} - \dfrac{1}{2}\right)\Delta h - A\right].$

3. A barometer indicates 30 in. of mercury at 80°F at a location at which $g = 32$ ft/sec^2. Express the corrected barometric pressure in psia.

    Ans. $p_{baro} = 14.5871$ psia.

4. If a test site is 100 ft above the barometer location of Problem 3, find the barometric pressure at the site.

    Ans. $p_{baro} = 14.5365$ psia.

5. A McLeod gauge of $V_1 = 350$ in.^{3} and bore $d = \frac{1}{8}$ in. indicates 3.2 in. of mercury when the temperature is 76°F and $g = g_{standard}$. Express the pressure (a) in psia, (b) in microns.

    Ans. (a) $p = 1.756 \times 10^{-4}$ psia, (b) $p = 9.12$ microns.

Chapter 16

PRINCIPLES OF CONVENTIONAL PRESSURE TRANSDUCERS

"... an understanding of the correct principle according to which an invention operates may follow, instead of precede, the making of the invention ..."

Clemens Herschel (1887)

16.1 Definitions

In general, a *transducer* is a device that, being actuated by energy from one system, supplies energy (in any form) to another system. In particular the essential feature of a conventional *pressure* transducer is an *elastic element* that converts energy from the pressure system under study to a *displacement* in the mechanical measuring system. An additional feature found in many pressure transducers is an *electric element* which, in turn, converts the displacement of the mechanical system to an *electrical signal*. The popularity of electric element pressure transducers derives from the ease with which electrical signals can be amplified, transmitted, controlled, and measured. Electrical pressure transducers can be delineated further as follows: an *active* transducer is one that generates its own electrical output as a function of the mechanical displacement, whereas a *passive* transducer (i.e., one dependent on a change in electrical impedance) requires an auxiliary electrical input which it modifies (modulates), as a function of the mechanical displacement for its electrical output (see Figure 16.1) [1]–[4].

Examples of mechanical pressure transducers having elastic elements

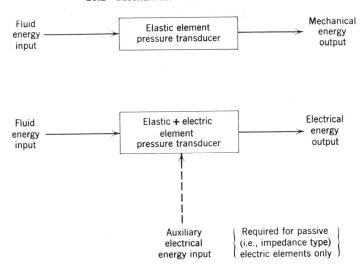

Figure 16.1 Block diagrams for pressure transducers.

only are: deadweight free-piston gauges, manometers, bourdon tubes, bellows, and diaphragm gauges. An example of an active electrical pressure transducer, combining in one the elastic and electric elements, is the piezoelectric pickup. Examples of electric elements employed in passive electrical pressure transducers include strain gauges, slide-wire potentiometers, capacitance pickups, linear differential transformers, variable reluctance units, and the like. Some of the more commonly used mechanical and electrical pressure transducers are considered next.

16.2 Mechanical Pressure Transducers

We have already described several types of mechanical pressure transducers in the discussion of manometer pressure standards. In addition, there are manometers not considered standards, yet used as conventional transducers. These include the well-, inclined-, and Zimmerli-types of manometers. In these, as in all manometers, the elastic element is the manometric fluid itself, which is moved by an applied pressure difference.

Well Type

The well-type manometer offers the advantage of a single scale reading for the pressure difference, the hope being that the level variation in the well is either negligible or can be accounted for in the construction of the

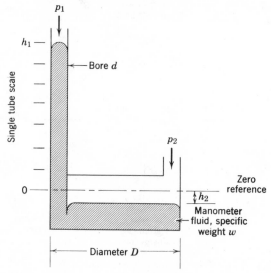

Figure 16.2   A well-type manometer pressure transducer: $p_2 - p_1 = w(h_1 + h_2)$; $h_1 d = h_2 D$; therefore $p_2 - p_1 = wh_1(1 + d/D)$.

single tube scale. Following the notation of Figure 16.2 the pertinent equation is

$$p_2 - p_1 = wh_1\left(1 + \frac{d}{D}\right). \tag{16.1}$$

If $D \gg d$ (say on the order of 500 to 1), variations in the well level can be neglected.

Inclined Type

The inclined-type manometer provides a single scale reading that is expanded along the single tube (i.e., the scale has more graduations per unit vertical height than the equivalent vertical scale of the well-type manometer). This allows for greater readability (on the order of ±0.01 in.) than in the *U*-tube manometer. The angle of incline (α) is generally about 10° from the horizontal (see Figure 16.3).

Zimmerli Type

The Zimmerli-type manometer [5] is another special form of manometer that features high readability at the lower absolute pressures (range is 0 to 100 mm Hg within 0.1 mm Hg). A mercury column is first separated by simultaneously applying the pressures to be measured to both sides of

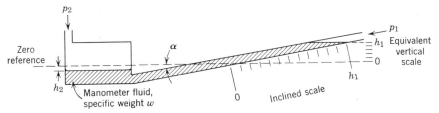

Figure 16.3 Inclined-type manometer pressure transducer.

the mercury. The resulting void between the two mercury columns (which occurs at an applied pressure of about 140 mm Hg) produces a near-absolute zero reference for the measurement. Any decrease in pressure beyond the separation point causes the mercury to drop in the reference leg and to rise in the measuring leg of the gauge until, at a pressure of about 0.1 mm Hg (i.e., 100 μ), no discernible difference in the elevations of mercury in the two legs is apparent. This, of course, represents the limit of usefulness of the Zimmerli manometer (see Figure 16.4).

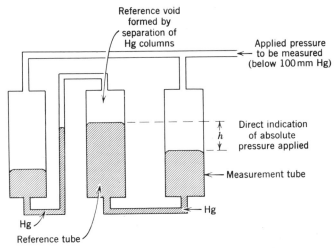

Figure 16.4 Zimmerli-type manometer pressure transducer.

Bourdon Tube

In the bourdon tube transducer, the elastic element is a small-volume tube, fixed at one end, which is open to accept the applied pressure, but free at the other end, which is closed to allow displacement under the deforming action of the pressure difference across the tube walls. In the most common model, a tube of oval cross section is bent in a circular arc.

Under pressure, the oval-shaped tube tends to become circular, with a subsequent increase in the radius of the circular arc. By an almost frictionless linkage, the free end of the tube rotates a pointer over a calibrated scale to give a mechanical indication of pressure (see Figure 16.5).

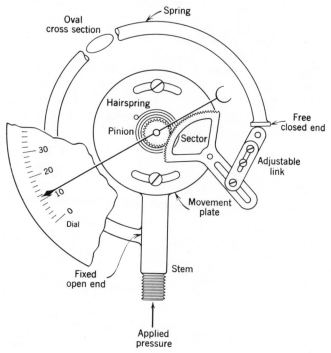

Figure 16.5 A common bourdon tube transducer. (After ASME PTC Supplement 19.2, 1964).

The reference pressure in the case containing the bourdon tube is usually atmospheric, so that the pointer indicates gauge pressures. Absolute pressures can be measured directly, without-evacuating the complete gauge casing, by biasing a sensing bourdon tube against a reference bourdon tube which is evacuated and sealed (Figure 16.6). Bourdon gauges are available for wide ranges of absolute, gauge, and differential pressure measurements within a calibration uncertainty of 0.1% of the reading. In contrast to the large angular displacements encountered in the mechanical-output bourdon gauges already described, the elastic element most often used in conjunction with electric elements (to yield electrical outputs) takes the form of a flattened tube that is twisted about its own

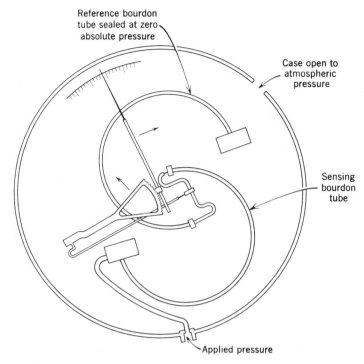

Reference bourdon
tube sealed at zero
absolute pressure

Case open to
atmospheric
pressure

Sensing
bourdon
tube

Applied pressure

Figure 16.6 Bourdon tubes arranged for absolute-pressure measurements. (After Wallace and Tiernan, Instruction Book F1A 101-1-1a.)

longitudinal axis and exhibits very small angular displacements (see Figure 16.7).

Bellows

Another elastic element used in pressure transducers takes the form of a bellows. In one arrangement, pressure is applied to one side of a bellows, and the resulting deflection is partly counterbalanced by a spring (see Figure 16.8). In another differential arrangement, one pressure is applied to the inside of one sealed bellows while the other pressure is led to the inside of another sealed bellows. By suitable linkages, the pressure difference is indicated by a pointer.

Diaphragm

A final elastic element to be mentioned because of its widespread use in pressure transducers is the diaphragm (see Figure 16.9). Such elements

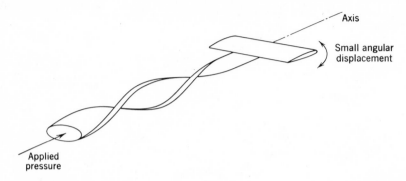

Axis

Small angular
displacement

Applied
pressure

Figure 16.7 Twisted bourdon tube for use with passive electrical elements.

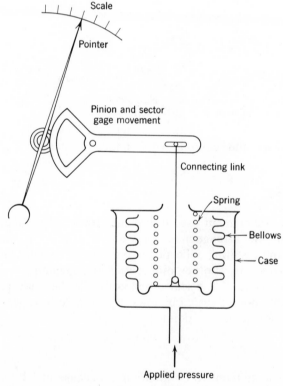

Scale

Pointer

Pinion and sector
gage movement

Connecting link

Spring

Bellows

Case

Applied pressure

Figure 16.8 Common bellows gauge. (After ASME PTC Supplement 19.2, 1964.)

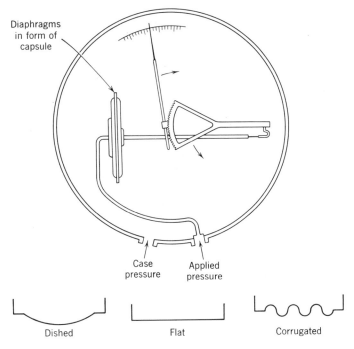

Figure 16.9   Precision capsule gauge. Diaphragm may be dished, flat, or corrugated. (After Wallace and Tiernan, Instruction Book F1A-101-1-1*a*.)

can appear in the form of flat, corrugated, or dished plates. The choice depends on the strength and amount of deflection desired. The literature on diaphragms is quite extensive and should be consulted for detailed information on diaphragm characteristics and on diaphragm-type pressure transducers [6], [7]. In high-precision instruments, a pair of diaphragms is used back-to-back to form an elastic capsule. One pressure is applied to the inside of the capsule which is surrounded on the outside by the other pressure. Such a differential pressure transducer exhibits the unique feature of a calibration that is almost independent (within 0.1%) of pressure level effects.

16.3   Electrical Pressure Transducers

An Active Electrical Pressure Transducer

A *piezoelectric* element provides the basis for the only active electrical pressure transducer in common use. It operates on a principle discovered in the 1880s by the Curie brothers that certain crystals (i.e., those not possessing a center of symmetry) produce a surface potential difference when they are stressed in appropriate directions [8], [9]. Quartz, Rochelle salt, barium-titanate, and lead-zirconate-titanate are some of the common crystals that exhibit usable piezoelectricity. Pressure pickups designed around such active elements have the crystal geometry oriented to give maximum piezoelectric response in a desired direction with little or no response in other directions. Sound pressure instrumentation makes extensive use of piezoelectric pickups in such forms as hollow cylinders, disks, and so on. Piezoelectric pressure transducers are also used in measuring rapidly fluctuating aerodynamic pressures or for short-term transients such as those encountered in shock tubes. Although the emf developed by a piezoelectric element may be proportional to pressure, it is nonetheless difficult to calibrate by normal static procedures. An attractive technique called "electrocalibration" has been described in the recent literature [10]. In this procedure, the piezoelectric pressure transducer is excited by an electric field rather than by an actual physical pressure to obtain the calibrations.

Passive Electrical Pressure Transducers

Of the passive electrical pressure transducers, none are more common than the variable resistance types.

STRAIN GAUGE TYPES. Electric elements of this type operate on the principle that the electrical resistance of a wire varies with its length under load (i.e., with strain). In the *unbounded* type, four wires run free between four electrically insulated pins located two on a fixed frame and two on a movable armature. The wires are installed under an initial tension and form the active legs of a conventional bridge circuit (see Figure 16.10). Under pressure, the elastic element (usually a diaphragm) displaces the armature, causing two of the wires to elongate while reducing the tension in the remaining two wires. This change in resistance causes a bridge imbalance proportional to the applied pressure, and these quantities can be related by calibration. The use of four wires in the manner indicated makes for increased bridge sensitivity, and allowing the wires to run free between the pins causes a high natural frequency for the transducer [11]. In the *bonded* type, the strain gauge takes the form of a fine wire filament, set in cloth, paper, or plastic, and fastened by a suitable cement to a flexible plate that takes the load of the elastic element (see Figure 16.11). Often two strain gauge elements are connected to the bridge in an attempt to nullify unavoidable temperature

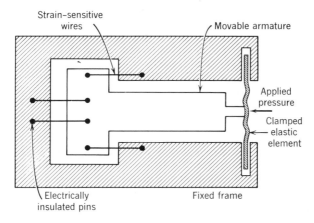

Figure 16.10 Typical unbonded strain gauge.

effects. The electrical energy input, required for all passive transducers, is in this case the excitation voltage of the bridge. The nominal bridge output impedance of most strain gauge pressure transducers is 350 Ω. The nominal excitation voltage is 10 V (ac or dc.) The natural frequency can be as high as 50,000 cps. Transducer resolution is infinite, and the usual calibration uncertainty of such gauges is within 1% of full scale.

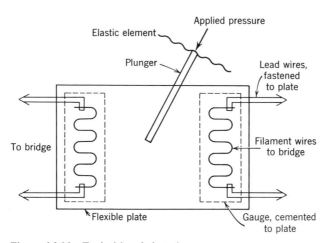

Figure 16.11 Typical bonded strain gauge.

POTENTIOMETRIC TYPE.   Other pressure transducers of the variable resistance type operate on the principle of movable contacts such as those found in slide-wire rheostats or potentiometers. In one arrangement, the elastic element is a helical bourdon tube, and a precision wire-wound potentiometer serves as the electric element. As pressure is applied to the open end of the bourdon, it unwinds and causes the wiper (which is connected directly to the closed end of the bourdon) to move over the potentiometer, thus varying the resistance of a suitable measuring circuit.

CAPACITANCE TYPE. In the variable capacitance-type pressure transducer, the elastic element is usually a metal diaphragm that serves as one plate of a capacitor. If pressure is applied, the diaphragm moves with respect to a fixed plate to change the thickness of the dielectric between the plates. By means of a suitable bridge circuit, the variation in capacitance can be measured and related to pressure by calibration.

Several variable inductance types of pressure transducers are considered next.

LINEAR VARIABLE DIFFERENTIAL TRANSFORMER TYPE (LVDT). The electric element in a LVDT is made up of three coils mounted in a common frame. A magnetic core centered in the coils is free to be displaced by an elastic element of either the bellows, bourdon, or diaphragm type (see Figure 16.12). The center coil is the primary winding of

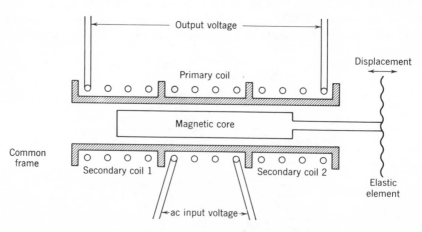

Figure 16.12   Linear variable differential transformer (LVDT.)

the transformer and as such has an ac excitation voltage impressed across it. The two outside coils form the secondaries of the transformer. When the core is centered, the induced voltages in these two outer coils are equal and 180° out of phase; this represents the zero pressure-position. However, when the core is displaced by the action of an applied pressure, the voltage induced in one secondary increases, whereas

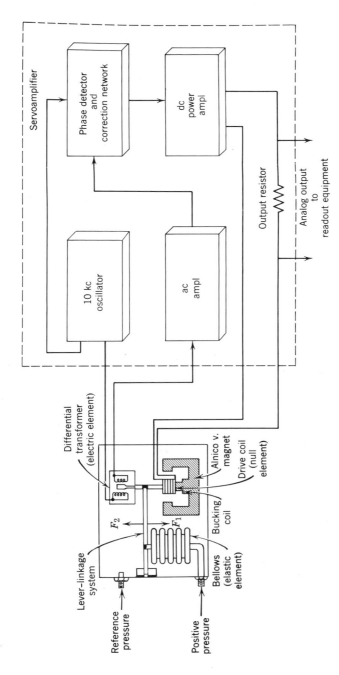

Figure 16.13 Block diagram and schematic of null-balance type LVDT pressure transducer. (From Consolidated Electrodynamics Corp., Bulletin 1016-21.)

335

that in the other decreases. This output voltage difference varies essentially linearly with pressure for the small core displacements allowed in LVDT pressure transducers. This voltage difference is measured and related to the applied pressure by calibration. In one variation of the above [12], a servoamplifier operates on the electrical output of the LVDT and causes the core to return to its null position for each applied pressure. Simultaneously it produces an appropriate electrical output signal (see Figure 16.13).

VARIABLE RELUCTANCE TYPES. Another class of pressure transducers whose electrical output signals are ultimately derived from variable inductances in the measuring circuits operates on the principle of a movable magnetic vane in a magnetic field. In one type, the elastic element is a flat magnetic diaphragm located between two magnetic output coils. Displacement of the diaphragm, caused by the applied pressure, changes the inductance ratio between the output coils and results in an output voltage proportional to the applied pressure (see Figure 16.14).

In a final type, the elastic element is a flat twisted tube such as already described in the section on bourdon tubes (see Figure 16.7). A flat

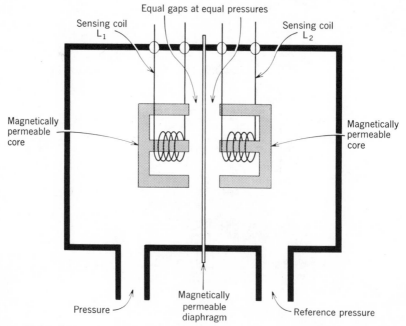

Figure 16.14 Magnetic reluctance differential pressure transducer (after Pace Wiancko literature.)

magnetic armature, connected directly to the closed end of the bourdon, rotates slightly when a pressure is applied. The accompanying small changes in the air gap between the armature and electromagnetic output coils alter the inductances in a bridge-type circuit. This variation in circuit inductance is used to modulate the amplitude or frequency of a carrier voltage, with the net result being an electrical response that is proportional to the applied pressure [13].

16.4 References

[1] J. H. Ruiter, Jr., and R. G. Murphy, "Transducers and What They Measure," *Machine Design*, November 1961, p. 139.

[2] D. P. Eckman, *Industrial Instrumentation*, Wiley, New York, 1950.

[3] R. J. Sweeney, *Measurement Techniques in Mechanical Engineering*, Wiley, New York, 1953.

[4] J. P. Holman, *Measurement Methods for Engineers*, McGraw-Hill, New York, 1966.

[5] "New, Improved Zimmerli Gauge," Scientific Glass Apparatus Co., Inc., Bloomfield, N.J.. Bulletin Z-251.

[6] A. M. Wahl, "Recent Research on Flat Diaphragms and Circular Plates with Particular Reference to Instrument Applications," *Trans. ASME*, **79,** January, 1957, p. 83.

[7] W. A. Wildhack, R. F. Dressler, and E. C. Lloyd, "Investigations of the Properties of Corrugated Diaphragms," *Trans. ASME*, **79,** January, 1957, p. 65.

[8] E. Eller, "Squeeze Electricity," *International Science and Technology*, July 1965, p. 32.

[9] W. G. Cady, *Piezoelectricity*, McGraw-Hill, New York, 1946.

[10] W. E. Anderson, "Calibrating Piezoelectric Transducers by Electrical Stimulation," *Electronic Instrument Digest*, September–October 1965, p. 61.

[11] "Statham Transducer Element," Statham Laboratories, Inc., Los Angeles, Bulletin 1.0.

[12] "Secondary Pressure Standard," Consolidated Electrodynamics Co., Pasadena, Calif., Sales Information Bulletin 1016-21.

[13] "System Engineer's Handbook," Daystrom Wiancko Engrg. Co., Pasadena, Calif., Application Notes AN 2.1.

Nomenclature

Roman

d small diameter
D large diameter
h height
p pressure
w specific weight

Greek

α angle of incline

Problems

1. Deduce the water pressure sensed by a pipe static pressure tap if the tap is located A in. above the reservoir of a single leg mercury manometer, and the mercury rises Δh in. above the reservoir level and is open to the atmosphere. (Assume ws are in $lbf/in.^3$)

   Ans. $p = p_{baro} + w_{H_2O} \left[\dfrac{w_{Hg}}{w_{H_2O}} \Delta h - A \right].$

2. A Zimmerli gauge indicates 2 in. of mercury. Express this as an absolute pressure measurement in millimeters of mercury and in microns.

   Ans. $p = 50.8$ mm Hg $= 50,800$ microns.

Chapter 17

PRESSURE MEASUREMENT
IN MOVING FLUIDS

". . . the idea of this machine is so simple and natural that the moment I conceived it I ran immediately to the river to make a first experiment with a glass tube . . ."

Henri de Pitot (1732)

17.1 Definitions

The pressure definitions and the measurements we have discussed thus far have concerned fluids in thermodynamic states of statistical equilibrium. Even for the very common case of a flowing fluid, however, there is a departure from the basic equilibrium condition, since a directed kinetic energy of flow is superposed on the random kinetic energy of the fluid molecules [1]. Three additional pressure definitions are usually offered to cover the fluid-flow situation.

Static (or *stream*) *pressure, p,* is the actual pressure of the fluid whether in motion or at rest. It can be sensed by a small hole drilled perpendicular to and flush with the flow boundaries so that it does not disturb the fluid in any way.

Dynamic (or *velocity*) *pressure, p_v,* is the pressure equivalent of the directed kinetic energy of the fluid when the fluid is considered as a continuum.

Total (*stagnation, impact,* or *pitot*) *pressure, p_t,* is the sum of the static and dynamic pressures. It can be sensed by a probe that is at rest with respect to the system boundaries when it locally stagnates the fluid isentropically (i.e., without losses and without heat transfer).

339

17.2 Mathematical Relations

The relation between these three pressures is

$$p_t = p + p_v. \tag{17.1}$$

This relationship is based on the general energy equation for steady flow and on the first law of thermodynamics. In a fluid continuum that is in directed motion with respect to the system boundaries, the energy relation at a single point where isentropic stagnation is postulated is

$$0 = \int_p^{p_{tP}} \frac{dp}{\rho} + \int_{V_P}^{0} \frac{V_P \, dV}{g_c}, \tag{17.2}$$

where the subscript P signifies a point quantity.

Here, it seems appropriate to discuss the relationship between specific weight, w, and density, ρ. These are related, through Newton's second law of motion [2], to the effect that

$$F = KMa$$
$$W = KMg$$

and
$$w = K\rho g$$

where $g =$ the *local* value of the acceleration of gravity. It remains to define the proportionality constant, K. In the SI system of units, where 1 newton is defined as that force which imparts to a mass of 1 kg an acceleration of 1 meter/sec/sec, it is clear that K must have the numerical value of unity. However, in the U.S. Customary system of units, where 1 pound force is defined as that force which imparts to a mass of 1 pound mass (i.e., 0.45359237 kg) an acceleration of 32.1740 ft/sec/sec, it is likewise clear that K must have the numerical value of $1/32.1740 = 1/g_c$. Thus,

$$w\,(\text{lbf/ft}^3) = \frac{1}{g_c}\, \rho\,(\text{lbm/ft}^3)\, g\,(\text{ft/sec}^2)$$

where
$$g_c = 32.1740 \text{ lbm ft/lbf sec}^2.$$

It follows that

$$\frac{w}{g} = \frac{\rho}{g_c}.$$

Now it is necessary to distinguish between the relatively incompressible liquids and the readily compressible gases (for simplicity, only constant density liquids and perfect gases are analyzed).

For a *constant density liquid*, (17.2) yields

$$(p_{tP} - p)_{inc} = (p_{vP})_{inc} = \frac{\rho V_P^2}{2g_c},$$ (17.3)

where the subscript "inc" means incompressible. Equation 17.3 bears out the familiar relation between total, static, and dynamic pressures in accord with (17.1), and identifies the dynamic pressure.

For a perfect gas undergoing an isentropic process p/ρ^γ is a constant, where γ is the ratio of specific heats. Equation 17.2 yields

$$\left(\frac{p_{tP}}{\rho_t} - \frac{p}{\rho}\right)_{comp} = \left(\frac{\gamma-1}{\gamma}\right)\frac{V_P^2}{2g_c}.$$ (17.4)

where the subscript "comp" means compressible. To put this relationship in a form comparable to (17.3), we introduce the Mach number, $M = V/(\gamma g_c RT)^{1/2}$, which accounts for the elastic nature of the gas. We then expand in a binomial series [3]:

$$(p_{tP} - p)_{comp} = (p_{vP})_{comp} = \frac{\rho V_P^2}{2g_c}\left(1 + \frac{M^2}{4} + (2-\gamma)\frac{M^4}{24} + \cdots\right),$$ (17.5)

which again confirms the simple relationship given in (17.1). The dynamic pressure of (17.5) approaches that of (17.3) as the Mach number approaches zero, that is, as compressibility effects become less important.

When an *effective* total pressure $(\bar{p}_t)$ at a *plane* is required, it becomes necessary to modify the *point* definitions of total pressure p_{tP} as given by (17.3) and (17.5). For the constant density liquid, for example, the effective total pressure can be taken as the flow-weighted-average total pressure. Thus from (17.3)

$$\int_A (p_{tP})_{inc}\, d\dot{m} = \int_A p\, d\dot{m} + \int_A \left(\frac{\rho V_P^2}{2g_c}\right)(\rho V_P\, dA),$$ (17.6)

where the integrals are taken over the area A of the plane. Integration yields

$$(\bar{p}_t)_{inc} = p + \frac{\rho \int V_P^3\, dA}{2g_c VA}.$$ (17.7)

The last term in (17.7) is simply the density ρ of the flowing fluid times the *actual* kinetic energy per pound of flowing fluid, for note

$$\left(\frac{K.E.}{M}\right)_{actual} = \frac{\int \frac{1}{2}V_P^2(\rho V_P\, dA)}{g_c(\rho VA)} = \frac{\int V_P^3\, dA}{2g_c VA}.$$ (17.8)

Sometimes it is more convenient to work with the *uniform* velocity of continuity (i.e., the effective one-dimensional velocity V) rather than with the point velocity V_P:

$$\left(\frac{\text{K.E.}}{\text{M}}\right)_{\text{uniform}} = \frac{\int \frac{1}{2} V^2 (\rho V \, dA)}{g_c (\rho V A)} = \frac{V^2}{2 g_c}. \tag{17.9}$$

The ratio of the actual to the uniform kinetic energies of (17.8) and (17.9) defines a kinetic energy coefficient α, which can also be used to express the effective total pressure at a plane as

$$(\bar{p}_t)_{\text{inc}} = p + \alpha \left(\frac{\rho V^2}{2 g_c}\right). \tag{17.10}$$

Representative values of the kinetic energy coefficient are $\alpha_{\text{uniform}} = 1.0$, $\alpha_{\text{laminar}} = 2.0$, and $1.02 < \alpha_{\text{turbulent}} < 1.15$. According to [4], α in a straight pipe for fully developed turbulent flow can be given by the simplified expression

$$\alpha = 1 + 2.7 f \tag{17.11}$$

where f is the familiar friction factor of equation 17.18.

For constant density fluids, effective total pressures defined in terms of flow-weighted-average total pressures provide direct measurement of the *head loss h* between planes [5]; that is,

$$(h_{\text{loss}})_{\substack{\text{inc} \\ 1 \text{ to } 2}} = \left(\frac{\bar{p}_{t1}}{\rho} - \frac{\bar{p}_{t2}}{\rho}\right)_{\text{inc}}. \tag{17.12}$$

where the units of each term in (17.12) are ft lbf/lbm, sometimes called feet of head, but this is only true if $g_{\text{local}} = g_c$.

For the compressible fluids, however, no simple counterparts of (17.7), (17.10), and (17.12) are available. For example, the difference in total pressure heads for a perfect gas flowing without heat transfer between planes gives no indication of head loss, since

$$\left(\frac{\bar{p}_{t1}}{\rho_{t1}} - \frac{\bar{p}_{t2}}{\rho_{t2}}\right)_{\substack{\text{comp} \\ \text{adiabatic}}} = 0 \tag{17.13}$$

This is in direct contrast to the incompressible head loss of (17.12). Nevertheless, basic expressions have been given for p_t, p and p_v, for both compressible and constant density fluids, and the next sections are concerned with the measurement of these moving fluid pressures.

17.3 Sensing Static Pressure

Static pressures are used for determining flow rates by head-type fluid meters (e.g., venturis, nozzles, orifices). They are required in velocity determinations, to establish thermodynamic state points, and are also useful for obtaining indications of flow direction. From its definition, static pressure can be sensed in at least three ways. Small holes can be drilled in the surface of the flow boundary in such a way that the streamlines of the flow are relatively undisturbed. The familiar *wall taps*, first used extensively by D. Bernoulli, function in this manner. Small holes can also be so located on probes that streamline curvatures and other effects caused by the probe presence in a flowing fluid stream are self-compensating. In this class belong the *static tubes*, such as those designed by L. Prandtl. Finally, small holes can be strategically located at critical points on *aerodynamic bodies* at which static pressures naturally occur. The sphere, cylinder, wedge, and cone probes are examples of this type. Briefly expanded, separate accounts of the more basic static pressure sensors will serve to broaden our understanding of some of the problems involved.

Wall Taps

It is generally assumed that infinitely small square-edged holes installed normal to flow boundaries give the correct static pressure. Now it seems a relatively simple matter to drill a small hole perpendicular to a flow boundary and hence to sense the static pressure. But in reality, small holes are difficult to machine, they are exceedingly difficult to keep burr-free, and small holes are slow to respond to pressure changes (see Chapter 18). Therefore, one usually settles for other than the infinitely small hole, and pays the price for compromise by applying a correction to the reading of a finite diameter pressure tap.

The problem of correcting pressure tap readings was discussed by Darcy as early as 1857. The American engineer Mills did an exceptionally precise experiment to determine hole size effects in 1879 [6]. Many others, including Allen and Hooper [7] in 1932, have continued this work.

The recommended geometry for a pipe wall tap is shown in Figure 17.1.

The effect of hole size is almost always evaluated experimentally with respect to an arbitrary small diameter reference tap (see, for example, Figure 17.2). Such results are then extrapolated to zero hole size to obtain absolute errors [8], [9].

Franklin and Wallace [10], in a very ingenious experiment based on the use of diaphragm pressure transducers set flush in the walls of a wind tunnel, avoided such extrapolations to zero hole size (which procedure they incidentally proved to be a questionable technique). All experimenters agree, however, that pressure tap errors *increase* with hole size.

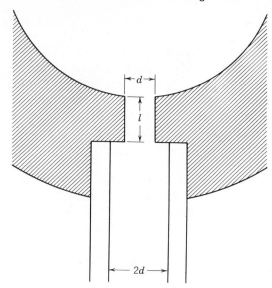

Figure 17.1 Recommended static-pressure wall-tap geometry $(0.5 < L/d < 6.0)$.

Although square-edged holes yield small positive errors, radius-edged holes introduce additional positive errors, and chamfer-edged holes introduce small negative effects (see Figure 17.3). These shape effects can be visualized as follows. At the upstream edge of a square tap, the fluid separates cleanly (in effect, the fluid fails to note the removal of the constraining boundary). Thus there are only minor deflections of the

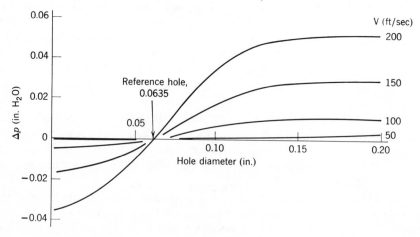

Figure 17.2 Typical experimental determination of hole size effect for $1.5 < L/d < 6.0$. (After Shaw, April 1960.)

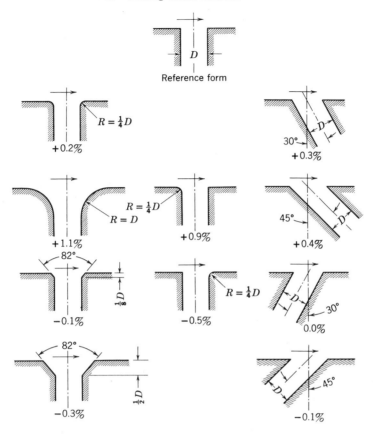

Figure 17.3 Effect of orifice edge form on static pressure measurement. Variation in percentage of dynamic pressure. (From Rayle, December 1959.)

streamlines into a square tap; however, by fluid viscosity, a slight forward motion is imparted to the fluid in the hole. It is the arresting of this motion by the downstream wall of the tap that accounts for the pressure rise. On the contrary, flow over a tap with a rounded edge does not immediately separate, but instead is guided into the hole with the resulting recovery of a portion of the dynamic pressure. Finally, although the flow does separate at the upstream edge of a countersunk tap, it also accelerates at the sloping downstream edge of the tap; this latter effect induces a sucking action that detracts from the stream pressure in the countersunk tap.

Static pressure error, Δp, for fully developed turbulent flow has been generalized in terms of $\Delta p/\tau_0$ versus a friction Reynolds number, R_d^*,

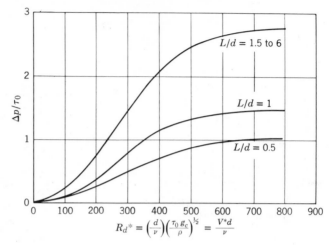

Figure 17.4 Wall tap errors for incompressible turbulent flows for $0.025 < d < 0.175$. (After Shaw, April 1960.)

based on the tap diameter d and the friction velocity,

$$V_* = \left(\frac{\tau_0 g_c}{\rho}\right)^{1/2} = \left(\frac{\tau_0}{2 p_v}\right)^{1/2} V \tag{17.14}$$

(see Figure 17.4). Note that the wall shear stress τ_0 enters the correlation since it characterizes the flow velocity gradient (i.e., $\tau_0 = \mu (dV_p/dy)_{y=0}$ which is $\propto \rho V^2/2 g_c$), and this has an important effect on the wall tap performance.

The curves of Shaw (Figure 17.4) indicate that L/d (of Figure 17.1) should be ≥ 1.5 to avoid tap error dependence on L/d. Thus most experimenters keep L/d between 1.5 and 15. Franklin and Wallace [10] extended Shaw's work (which was limited to $R_d^* = 800$) to $R_d^* = 2000$, and because of their careful experimental work, and because they avoid extrapolations to zero tap size, their results are recommended as definitive, and are given in Figure 17.5. It is interesting to observe that for R_d^* on the order of 2000, $\Delta p/\tau_0$ is about 3.75.

Another method for presenting static pressure tap error, and one that is felt to be more convenient to apply, has been suggested [11]. Here, Δp is nondimensionalized by the dynamic pressure p_v, rather than the wall shear stress τ_0, and this quantity is given in terms of the pipe Reynolds number, $R_D = VD/\nu$, instead of the friction Reynolds number, R_d^*. Thus

$$\frac{\Delta p}{p_v} = f(R_D). \tag{17.15}$$

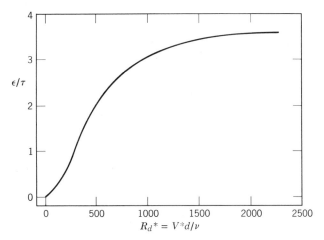

Figure 17.5 Nondimensional error curve for static pressure taps (after Franklin and Wallace, [10]). Hole diameters vary from 0.02 to 0.25 in.

Now, algebraically

$$\frac{\Delta p}{p_v} = \left(\frac{\Delta p}{\tau_0}\right)\left(\frac{\tau_0}{p_v}\right) \tag{17.16}$$

where $p_v = \rho V^2 / 2g_c$ in U.S. customary units. But, it is well established (e.g., [12]) that

$$\frac{\tau_0}{p_v} = \frac{f}{4} \tag{17.17}$$

where f is the Darcy friction factor defined in general, for smooth pipes, by the implicit empirical Prandtl equation [13]

$$\frac{1}{\sqrt{f}} = 2 \log\left(R_D \sqrt{f}\right) - 0.8. \tag{17.18}$$

It should be mentioned at this point that all the above work on static taps was done for smooth pipes. Static taps installed in rough pipes would yield questionable results, and hence this practice is not recommended.

Combining (17.16) and (17.17), there results

$$\frac{\Delta p}{p_v} = \left(\frac{f}{4}\right)\left(\frac{\Delta p}{\tau_0}\right). \tag{17.19}$$

The $(\Delta p/\tau_0)$ term is, of course, the very empirical information available from the Franklin and Wallace curve of Figure 17.5, once we can enter with the friction Reynolds number. This we can get from the pipe

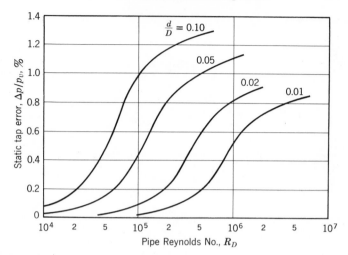

Figure 17.6 Static tap error as function of pipe Reynolds number (after Wyler, March 1976.)

Reynolds number via

$$R_d^* = \left(\frac{f}{8}\right)^{1/2}\left(\frac{d}{D}\right)R_D \tag{17.20}$$

(Where $R_d^* = V^* d/\nu$ and $\tau_0/p_v = f/4$).

It is now possible to plot general correction curves for static taps in smooth pipes in fully developed turbulent flows according to (17.15), based on (17.18), (17.19), and (17.20). Such a correction curve has been proposed by Wyler [14] and is presented in Figure 17.6.

***Example* 1.** Water at 70°F is flowing at a velocity of 10 ft/sec in a 4 in. smooth pipe. The density and kinematic viscosity of this water are: $\rho = 62.3$ lbm/ft^3 and $\nu = 1.11 \times 10^{-5}$ ft^2/sec. Find the absolute error in the static pressure measurement sensed by a $\frac{1}{4}$ in. square-edged tap.

Pipe Reynolds number:

$$R_D = \frac{VD}{\nu} = 10 \times \frac{4}{12} \times \frac{10^5}{1.11} = 3 \times 10^5.$$

Friction factor from (17.18):

$$f = 0.01447.$$

Friction Reynolds number from (17.20):

$$R_d^* = \left(\frac{0.01447}{8}\right)^{1/2} \times 3 \times 10^5 \times \frac{1}{16} = 797.$$

Tap error from Figure 17.5:

$$\Delta p/\tau_0 = 2.9.$$

Tap error from (17.19)

$$\Delta p/p_v = 0.01447 \times 2.9/4 = 1.05\%.$$

As a check, one could obtain the tap error directly from Figure 17.6 at $d/D = 0.0625$:

$$\Delta p/p_v = 1\%.$$

These slight differences can be attributed to inaccuracies in reading Figures 17.5 and 17.6.

Since
$$p_v = \frac{\rho v^2}{2g_c} = \frac{62.3 \times 100}{2 \times 32.174 \times 144} = 0.672 \text{ psi,}$$

it follows that the absolute error predicted for this tap when used under these conditions is

$$\Delta p = 0.00672 \text{ psi.}$$

In terms of inches of water at standard gravity and at a temperature of 76°F, Table 15.1 indicates $w_{H_2O} = 0.036026$ lbf/in.^{3},

thus
$$\Delta p = \frac{0.00672}{0.036026} = 0.186 \text{ in. } H_2O.$$

Such an error may or may not be significant in the overall measurement problem.

Example 2. Show that the static pressure error reaches about 1% of the dynamic pressure when $d/D = 0.1$ and $R_D = 2 \times 10^5$.

From (17.18):
$$f = 0.01564.$$

From (17.20):
$$R_d^* = \left(\frac{0.01564}{8}\right)^{1/2} \times 2 \times 10^5 \times 0.1 = 884.3.$$

From Figure 17.5:
$$\Delta p/\tau_0 = 3.$$

From (17.19):
$$\Delta p/p_v = 0.01564 \times \tfrac{3}{4} = 1.17\%.$$

As a check, one could obtain the tap error directly from Figure 17.6 at $d/D = 0.1 : \Delta p/p_v = 1.16\%$.

These two answers are within the accuracy of reading $\Delta p/\tau_0$ and $\Delta p/p_v$ from the figures.

Another correlation sometimes useful is that of Rayle's $\Delta p/p_v$ versus the Mach number (see Figure 17.7). Note that Rayle [8] states that a 0.030 in. hole with a 0.015 in. deep countersink should give nearly true static pressure. Actually, Mach number effects have not been considered by many experimenters because they are not usually important in pipe wall taps. On pressure probes, however, Mach number effects could be very important. Rainbird [15] has done some important work in this area.

As a final note on wall tap performance, Emmett and Wallace [16] have stated that the magnitude of errors found in *rough* channel tests is approximately one-half to two-thirds the value found in smooth channel tests. They are speaking of turbulence effects of which we usually have no firm quantitative idea, but their work suggests that static tap errors can

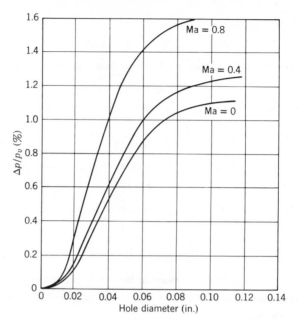

Figure 17.7   Wall tap errors for compressible flows. (After Rayle, December 1959.)

have a large uncertainty associated with them. A work on throat tap nozzles [11] does support the Emmett and Wallace contention that static taps in high turbulence regions may show only one-half the error of the usual low turbulence tests.

Static Tubes

The accuracy in static pressure measurement using static tubes depends mainly on the position of the sensing holes with respect to the nose of a tube and its supporting stem. Acceleration effects caused by the nose tend to lower the tap pressure; stagnation effects caused by the stem tend to raise the tap pressure. In a properly compensated tube, these two effects will just cancel at the plane of the pressure holes [17], [18]. A satisfactory arrangement uses four square-edged pressure holes (of 0.040 in. diameter) placed 90° apart in a plane that is located five tube diameters back from the nose and 15 tube diameters forward from the probe stem (see Figure 17.8). The *disk* probe [19] is another static pressure sensor that uses the compensation principle. Streamline curvature induced at the leading edge of the disk is countered by depressions in the plane of the pressure holes (see Figure 17.9).

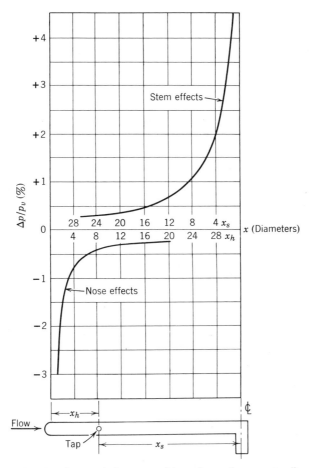

Figure 17.8   Static tube characteristics: x_s, position of taps from center line of stem; x_h, position of taps from base of nose. (From Dean, 1952.)

Aerodynamic Probes

The accuracy of static pressure measurement using pressure taps in aerodynamic bodies depends on the accuracy of location, the size of the holes, and the direction and variation in the direction of the flow [20], [21]. A *cylindrical* probe inserted normal to the flow is representative of such sensors. The actual pressure distribution over the surface of a cylinder is well known (see Figure 17.10), as is the variation in the critical angle with Mach number (the critical angle is the angle at which static

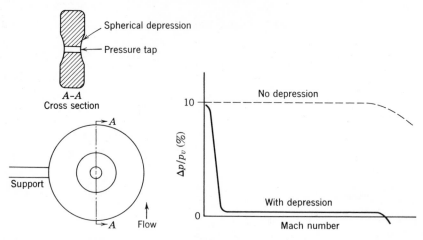

Figure 17.9 Disk probe and characteristics. (After Gilmer, 1952.)

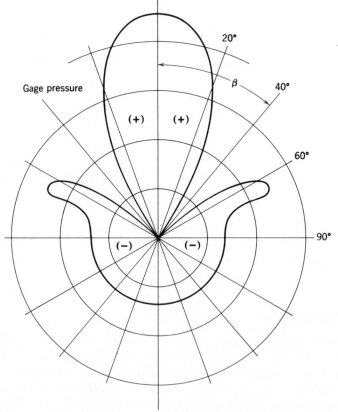

Figure 17.10 Pressure distribution on surface of cylinder inserted normal to flow (see also Figure 17.14).

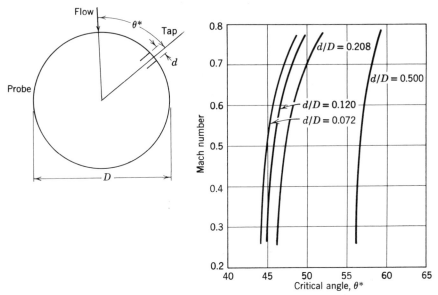

Figure 17.11 Critical angle. (*a*) Definition. (*b*) Typical variation with Mach number and diameter ratio (*d* = tap diameter; *D* = probe diameter; θ^* = critical angle).

pressure occurs on the cylinder surface) (see Figure 17.11.) Calibration characteristics of cylindrical probes depend primarily on tap-angle location with respect to flow direction, as indicated in Figure 17.12. If two pressure taps are located in the same plane on the cylindrical surface and separated by about twice the critical angle, the direction of flow normal to the cylinder can be obtained. This is accomplished by rotating the cylinder until the two pressure taps, connected across a manometer (or better yet a bubbler), sense identical pressures (see Figure 17.13). This operation is similar to that of range finding or, more basically, to sensing direction by our ears. In a variation of this idea, the two taps are manifolded rather than connected individually across a manometer [22]. The static pressure sensed by such a probe (faced in the nominal direction of flow) will be insensitive to flow-angle variations of as much as ±20°. The manifolded arrangement takes advantage of the linear portion of the pressure distribution curve in the vicinity of ±45° from the flow direction (see Figure 17.14). As the flow direction veers from nominal, one tap is exposed to a lower pressure, whereas the other tap experiences a higher pressure of the same magnitude.

The wedge probe has a less rapid change in tap pressure in the region of the pressure taps (as a function of tap position) than does the cylinder (see Figure 17.15). Unfortunately, the somewhat fragile knife edge of the

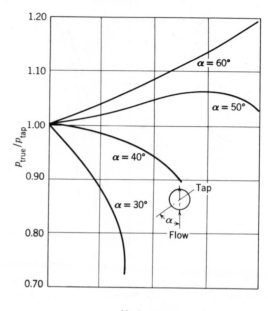

Mach number

Figure 17.12 Calibration curves showing effect of static hole angle α.

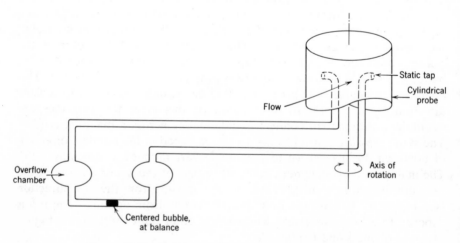

Figure 17.13 Bubbler method for detecting flow direction.

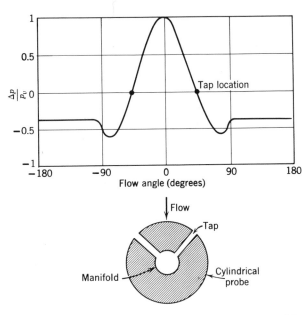

Figure 17.14 Performance of manifolded cylindrical probe with two taps. (Note that if flow direction veers pressure increase in one tap is just offset by decrease in the other tap.)

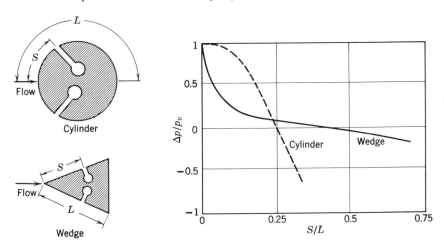

Figure 17.15 Comparison of performance of cylinder and wedge-shaped aerodynamic probes.

wedge makes it a less robust instrument than the cylinder for many applications.

Probe Blockage

A probe inserted in a flow passage changes the velocity in the near vicinity of the probe by an amount δV. This effect has been characterized by a probe blockage factor defined as

$$\epsilon = \frac{\delta V}{V} \tag{17.21}$$

This parameter has been shown [23] to influence the various thermodynamic quantities in the following ways

$$\frac{\delta p}{p} = -\gamma M^2 \epsilon, \tag{17.22}$$

$$\frac{\delta \rho}{\rho} = -M^2 \epsilon, \tag{17.23}$$

$$\frac{\delta T}{T} = -(\gamma - 1)M^2 \epsilon, \tag{17.24}$$

$$\frac{\delta p}{p_v} = -2\epsilon, \tag{17.25}$$

and

$$\frac{\delta p_v}{p_v} = (2 - M^2)\epsilon. \tag{17.26}$$

Thus, if we knew the magnitude and sign of ϵ, and the Mach number of the free stream flow, we could determine the changes in the various thermodynamic quantities as given above.

Two cases are considered here. The first concerns closed channel flow as typified by flow in a pipe or duct. The other can be described as free jet flow such as encountered in the potential cones of pressure probe calibration nozzles (see, for example, [24]).

PIPE FLOW. A probe inserted in a pipe causes *increases* in the local velocity of the fluid. Hence, the blockage parameter increases, and by (17.22) the static pressure *decreases*. This means that the probe will measure a static pressure lower than would have been in the pipe without the probe.

Wyler [23] has shown for subsonic pipe flow that

$$\epsilon = \left(\frac{C_D}{2}\right) \frac{(S/C)}{(1 - M^2)}, \tag{17.27}$$

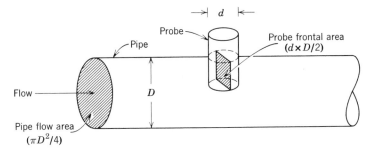

Figure 17.16 Geometric considerations in probe blockage.

where C_D is the drag coefficient, related to the pressure coefficient via

$$\frac{C_D}{\pi \bar{P}} = \frac{\int_0^\pi P_\theta \cos\theta\, d\theta}{\int_0^\pi P_\theta\, d\theta} \tag{17.28}$$

and to the Mach number, approximately, via

$$C_D = 1.15 + 0.75(M - 0.2). \tag{17.29}$$

The geometric factor, S/C, is the ratio of the probe frontal area to the pipe flow area (see Figure 17.16). Typically, for a cylindrical probe inserted one-half way in a pipe

$$\frac{S}{C} = d\left(\frac{D}{2}\right) \Big/ \frac{\pi D^2}{4} = \frac{2d}{\pi D} \tag{17.30}$$

where D is the pipe diameter and d is the probe diameter.

Combining equations (17.25), (17.27), (17.29), and (17.30), there results

$$\frac{\delta p}{p_v} = -2\epsilon = \frac{-2}{(1 - M^2)} \left\{ \frac{1.15 + 0.75(M - 0.2)}{2} \right\} \left(\frac{2d}{\pi D} \right). \tag{17.31}$$

This can be plotted as in Figure 17.17 to show acceptable blockages or resultant pressure errors in terms of d/D, M, and δ_p/p_v.

Example 3. If a $\frac{1}{4}$ in. diameter probe were installed to the midpoint in a 5 in. diameter pipe, compare the static pressure error as a percent of the dynamic pressure in water and in air at a Mach number of 0.5. By (17.31), in *water* (with $M = 0$)

$$\frac{\delta p}{p_v} = -2\left\{ \frac{1.15 - 0.15}{2} \right\} \left(\frac{2 \times \frac{1}{4}}{\pi \times 5} \right) = -3.2\%.$$

By (17.31), in *air* (with $M = 0.5$)

$$\frac{\delta p}{p_v} = \frac{-2}{(1 - 0.5^2)} \left\{ \frac{1.15 + 0.75(0.5 - 0.2)}{2} \right\} \left(\frac{2 \times \frac{1}{4}}{\pi \times 5} \right) = -5.8\%.$$

As a check, one could use Figure 17.17 directly.

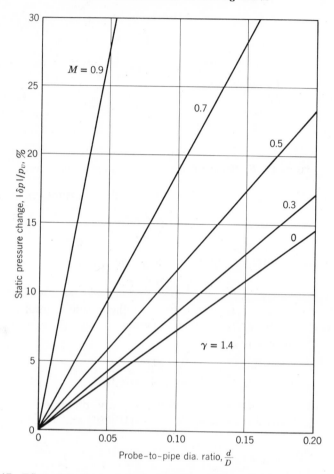

Figure 17.17 Effect of blockage on static pressure at various Mach numbers (after Wyler, October 1975).

The blockage effect decreases approximately linearly with distance upstream of the probe. For example, the effect is reduced about 40% at a distance upstream of $\frac{1}{2}D$.

FREE JET FLOW. A probe inserted in the potential cone of a free jet causes *decreases* in the local velocity of the fluid. Hence the blockage parameter decreases, and by (17.22) the static pressure *increases*. This means that the probe will measure a static pressure higher than the free jet pressure without the probe.

Wyler [23] has shown, for subsonic free jet flow, that (17.27) applies as to magnitude, but that the blockage effect is in the *opposite* direction. Thus, if Example 3 were to involve a 5 in. diameter free jet instead of the pipe the two results would be

$$\left(\frac{\delta p}{p_v}\right)_{\text{water}} = +3.2\%$$

$$\left(\frac{\delta p}{p_v}\right)_{\text{air}} = +5.8\%.$$

From Figure 17.11, we see that the critical angle θ^* varies with tap-to-probe diameter ratio and with Mach number. It also has been found to vary with blockage (i.e., probe-to-flow diameter ratio) and with turbulence. A useful relation that approximates some of these variations in free jets, when the tap-to-probe diameter ratio is about 0.12, can be given as:

$$\delta\theta^* \simeq \frac{22 \, d/D}{1 - M^2} \tag{17.32}$$

where D is the jet diameter and d is the probe diameter. Thus, when d/D is 0.1 and the Mach number is 0.4, the critical angle increases by 2.6° over the zero blockage case.

Pressure Tap Displacement

Even if a pressure tap were to cause no change in the flow pattern in its vicinity, a correction to the indicated pressure still would be required when the tap is in a region of a strong pressure gradient. Such gradients would be encountered routinely by pressure taps located on the surfaces of cylinders and spheres at other than the stagnation point.

A series of experiments by Pugh et al. [25] on spherical-shaped bodies indicate that the measured pressure is equal to the true surface pressure at a point about 0.35 d upstream, toward the stagnation point. At first thought this correction would appear to be a function of the strength of the pressure gradient, but it happens that the gradient is essentially independent of angle in the region on the probe surface where static taps are normally placed (i.e., near the critical angle) (see Figure 17.14). Hence the corrected pressure distribution is always slightly steeper than the measured distribution.

This constant correction to be applied to pressure taps, on cylinders and spheres, located near the critical angle is given by

$$\theta_{\text{correction}} = 0.35\left(\frac{360}{\pi}\right)\left(\frac{d}{D}\right) = 40.107\left(\frac{d}{D}\right), \text{ degrees} \tag{17.33}$$

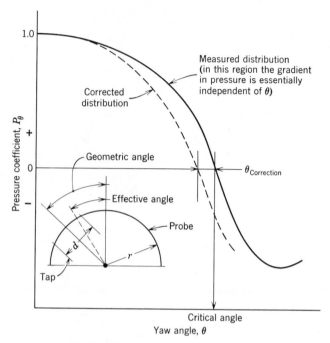

Figure 17.18 Pressure tap displacement.

where d is the tap diameter and D is the probe diameter. This correction is always applied to reduce the effective tap angle (see Figure 17.18).

Example 4. A cylindrical probe of $\frac{3}{16}$ in. O.D. is equipped with static taps of $\frac{1}{32}$ in. diameter located at 45° from the stagnation point. By how much is the effective tap angle reduced?

By (17.33)

$$\theta_{\text{correction}} = 40.107\left(\frac{1/32}{3/16}\right) = 6.68°,$$

which means that instead of sensing the pressure at 45°, the static taps indicate the pressure at 38.32°.

17.4   Sensing Total Pressure

Total pressures are used to determine head-loss data and to establish velocities, state points, and flow rates. By definition, total pressure can be sensed only by stagnating the flow isentropically. Such stagnation can be accomplished by a pitot tube as first described by Henri de Pitot in 1732. To obtain velocities, he made use of two open tubes immersed to the same depth in flowing water. The lower opening in one of the tubes was perpendicular to the flow, and the rise in water in this tube was taken to

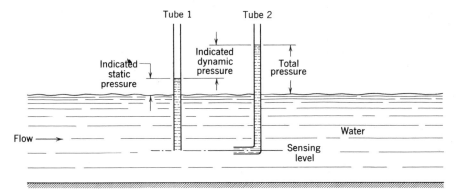

Figure 17.19 Basic Pitot-tube method of sensing static, dynamic, and total pressure.

be an indication of the static pressure. The other tube was bent 90° so that its lower opening faced into the flow direction, and the water rise in this tube was taken to be an indication of the total pressure. Pitot took the difference in water levels in these two tubes to be a measure of the velocity (see Figure 17.19). Although his static pressure tube would now be considered highly inadequate, his simple Pitot tube bent into the flow is, to this day, the basic method for sensing total pressure. Total pressure can also be sensed by holes located at stagnation points on aerodynamic bodies such as spheres and cylinders (see Figure 17.20). In all cases it is assumed that fluid stagnation takes place so rapidly in the vicinity of the pressure holes that heat transfer and frictional effects can be neglected. In other words, it is assumed that the fluid is brought to rest isentropically in the vicinity of the tap. Most references bear this out, indicating that the total pressure is remarkably easy to measure over wide variations in Reynolds number, Mach number, probe geometry, and alignment with flow direction. Thus the Pitot coefficient

$$C_P = \frac{p_{tI} - p}{p_v} \tag{17.34}$$

(where p_{tI} is the indicated total pressure) is essentially equal to one under usual flow conditions. Factors that influence the Pitot coefficient include geometry effects coupled with misalignment of the flow, viscous effects at the lower Reynolds numbers, and transverse pressure-gradient effects in the vicinity of the total pressure hole. These are discussed briefly below.

GEOMETRY EFFECTS. Since flow direction is not always precisely known, there is a definite advantage in using a total pressure tube that is relatively unaffected by alignment. A great deal of experimental work has been done on this effect [26], and some of the results for Pitot tubes are

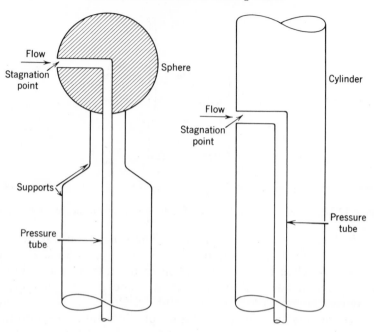

Figure 17.20   Aerodynamic total pressure probes.

summarized in Figure 17.21. Tubes having a cylindrical external shape are found to be less sensitive to misalignment than tubes having conical or ogival nose sections [27]. For simple, straight-bore cylindrical Pitot tubes, the range of insensitivity can be increased by increasing the relative size of the pressure hole and increased still more by incorporating a 15° internal bevel extending about 1.5 tube diameters to the rear of the nose. The flow angle range over which a Pitot tube remains insensitive to within 1% of the dynamic pressure can be as much as ±25°. Figure 17.22 indicates the sensitivity of aerodynamic total pressure probes to flow angle variations as a function of the relative size of the pressure hole.

VISCOUS EFFECTS.   The Pitot coefficient C_P of (17.34) has been given as having the value one for usual applications; however, detailed work indicates wide variations at the lower Reynolds number. Since the Reynolds number (Re) of fluid mechanics is the ratio of the inertia forces on the fluid to the viscous forces, a very low Re signifies relatively large viscous forces. Hence the pronounced effect on C_P in this flow regime is charged to viscosity. Although there is no exact theoretical treatment that describes the viscosity effect on a Pitot tube reading, experimental data

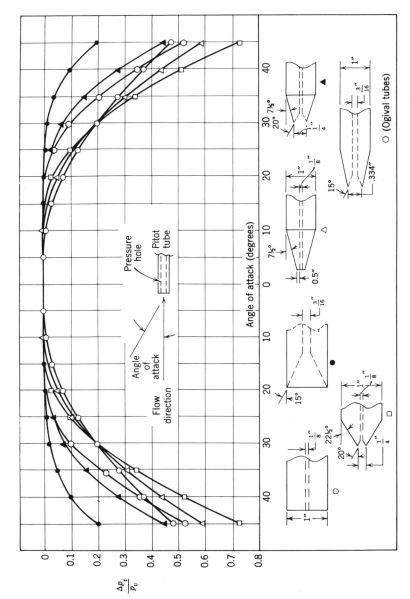

Figure 17.21 Variation of total pressure indication with angle of attack and geometry for Pitot tubes. (After NACA TN 2331, April 1951).

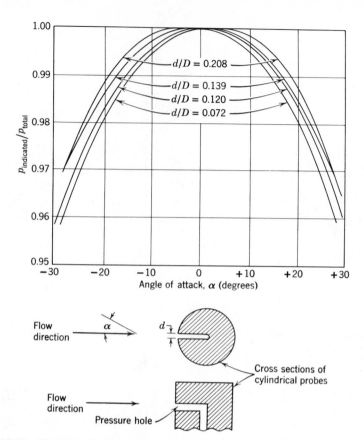

Figure 17.22 Variation of total pressure indication with angle of attack and hole size for cylindrical probes perpendicular to the flow.

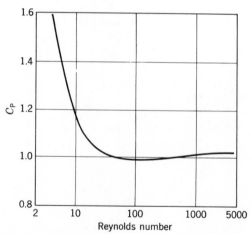

Figure 17.23 Viscosity effects on a Pitot tube. (After Hurd et al., 1953.)

indicate the overall variation in C_P with Re (Figure 17.23). The characteristic length used to form the Reynolds number is taken as the outside radius r of the Pitot tube. For Re > 1000, C_P can be taken as 1, as noted previously; for $50 < \text{Re} < 1000$, C_P falls between 0.99 and 1; if Re is below 10, C_P is always greater than 1, rapidly approaching the asymptotic value of 5.6/Re. Thus, if the Reynolds number (based on tube radius) is greater than 50, the Pitot tube is essentially unaffected by viscosity, but if the Reynolds number is below 10, viscous effects are extremely important [28], [29].

17.5 Transverse Gradient Effects

Streamline Displacement

When there is a gradient in total pressure (i.e., a gradient in velocity) upstream of a pitot tube, the pressure sensed will be greater than the total pressure on the streamline approaching the tube. That is, the effective center of a square-edged pitot tube will be displaced by an amount δ from its geometric center towards the region of higher velocity [30], [31].

The mechanism of this displacement has been explained [32] as the complex result of an induced secondary flow caused by the presence of the probe in a flow possessing vorticity.

Results from various theoretical and experimental analyses indicate that in shear flow the displacement of the velocity profile from its undisturbed pattern (see Figure 17.24) can be given in the dimensionless

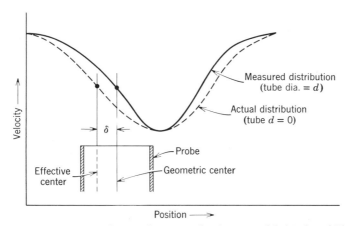

Figure 17.24 Displacement of stagnation streamline because of finite size of Pitot tube in gradient.

form as

$$\frac{\delta}{D} = f(K) \tag{17.35}$$

where D = probe outside diameter,

and K = shear parameter = $\dfrac{D(\Delta V/\Delta y)}{2V}$. $\tag{17.36}$

Young and Maas [33], in 1936, experimenting in airfoil wakes with square-edged *pitot tubes*, found

$$\frac{\delta}{D} = 0.18.$$

Hall [31], in 1956, neglected viscosity and deduced theoretically for a *sphere* that

$$\frac{\delta}{D} = 0.62K - 0.5876K^3. \tag{17.37}$$

He further indicated that the stagnation streamline would be shifted an amount

$$\theta_s = \sin^{-1}(0.9004K - 1.9357K^3) \tag{17.38}$$

for a *sphere*. For $K = 0.5$, this yields a shift of about 12°. Hall postulated that the displacement for a pitot tube would be similar to that of a sphere of slightly larger diameter. Finally, Hall declared that the displacement of a *cylinder* of like diameter would be only about one-fifth that of a sphere, specifically,

$$\frac{\delta}{D} = \frac{-1 + \sqrt{1 + K^2/2}}{2K}, \tag{17.39}$$

whereas the angular shift of the stagnation streamline for a cylinder would be

$$\theta_s = \sin^{-1}\left(\frac{K}{4}\right). \tag{17.40}$$

Livesey [32], in 1956, showed that the displacement effect encountered in transverse pressure gradients could be *minimized* by using conical-nosed pitot tubes with sharp lips rather than the conventional square-edged pitots (see Figure 17.25).

Davies [34], in 1957, presented important experimental data on square-edged pitot tubes of various diameters and demonstrated that the smaller tubes did indeed show less displacement for the same shear flow.

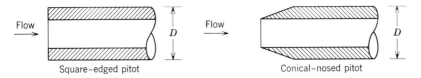

Figure 17.25 Several Pitot tubes with different displacement characteristics.

His results were not fit by empirical equations, but agree well with those of Sami which are discussed next.

Sami [35], in 1967, presented experimental results for square-edged pitot tubes in various shear flows and summarized his results by empirical equations patterned after Hall's theoretical formulations, namely,

$$\frac{\delta}{D} = 1.025K - 4.05K^3 \tag{17.41}$$

for $K < 0.3$, and by

$$\frac{\delta}{D} = 0.195 \tag{17.42}$$

for $K > 0.3$.

All of this information on streamline displacement for various probe geometries is summarized in Figure 17.26.

Flow Direction Error

As already mentioned, the flow direction in a *uniform* static and total pressure field can be obtained by balancing the pressures sensed by two separate taps located on the surface of a cylinder. However, when such a probe is exposed to a flow field that exhibits a velocity gradient, such as is always encountered in the wakes of airfoils or cylinders, then, as we have already seen, the streamlines in the vicinity of the probe will be displaced according to (17.39) and (17.40). If one attempts now to obtain flow direction by balancing the tap pressures on the cylinder, a correction must be applied to account not only for $\Delta\theta_s$, the displacement effect, but also for an *additional* error introduced by balancing two pressures in a shear flow. That is, one tap will be sensing a different dynamic pressure (p_v), and hence a different pressure coefficient (P_θ), than the other tap, at the balance condition, as shown in Figure 17.27. This latter effect can be summarized analytically, following Ikui and Inoue [36], as

$$\Delta\theta_T = 2\sin\left(\frac{\theta_{\text{taps}}}{2}\right)\frac{KP_{\theta 0}}{\left|\dfrac{\Delta P_\theta}{\Delta\theta}\right|_0}, \tag{17.43}$$

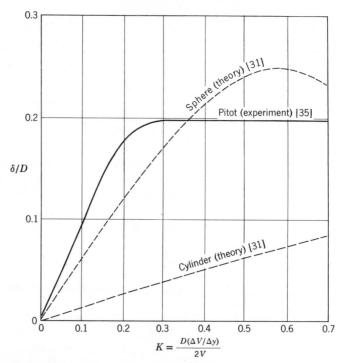

$$K = \frac{D(\Delta V/\Delta y)}{2V}$$

Figure 17.26   Stagnation streamline displacement in terms of shear parameter.

where θ_{taps} = total included angle between the two taps,

K = the shear parameter,

$P_{\theta 0}$ = pressure coefficient of the cylinder in *uniform* flow at the balance point,

$\left| \dfrac{\Delta P_\theta}{\Delta \theta} \right|_0$ = absolute slope of the pressure coefficient in *uniform* flow at the balance point.

Since $\Delta \theta_s$ is always in one direction, namely, toward the higher velocity region, and since $\Delta \theta_T$ can be in either direction, depending on the tap location with respect to the critical angle, it is possible to use the tap angle error to minimize the streamline displacement error by a judicious choice of tap locations.

Example 5.   A two-hole cylindrical probe is placed in a shear flow characterized by $K = 0.2$. If the taps are placed at 100° apart and the probe is such that in uniform flow $P_{\theta 0} = -0.2$ and $|\Delta P_\theta/\Delta \theta|_0 = 0.038$, find the net angle error at pressure balance.

By (17.40) $\Delta \theta_s = \sin^{-1}\left(\dfrac{0.2}{4}\right) = 2.87°$

because of streamline displacement.

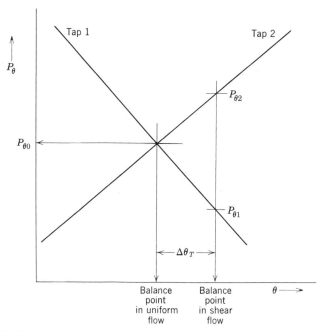

Figure 17.27 Showing that pressure taps on a cylinder sense different dynamic pressures at the balance point in shear flow.

By (17.43)

$$\Delta\theta_T = \frac{2\sin\left(\frac{100}{2}\right)0.2(-0.2)}{0.038} = -1.61°$$

because of tap location.
The net angle error

$$\Delta\theta_{\text{net}} = \Delta\theta_s + \Delta\theta_T = 2.87 - 1.61 = 1.26°,$$

toward the higher-velocity region.

17.6 Turbulence Effects

Turbulence is a random phenomenon concerning fluctuations in velocity (often characterized by the rms components of the turbulent velocity in three mutually perpendicular directions) that are superimposed on the mean velocity. Both total pressure sensors and static pressure sensors are affected by turbulence in a fluid stream as indicated by

$$p_{tI} = p + \frac{\rho V^2}{2} + \frac{\rho(v')^2}{2} \tag{17.44}$$

and
$$p_I = p + \frac{\rho(v')^2}{2},$$
(17.45)

where v' is the root mean square value of the axial component of the turbulent velocity, p_{tI} is the indicated total pressure, and p_I is the indicated static pressure. Turbulence levels have been established for only a few flow configurations; thus usually only qualitative effects of turbulence are considered. It has been noted, however, that the turbulent component v' must reach 20% of the mean velocity V before the turbulence terms in (17.44) and (17.45) amount to 4% of the mean dynamic pressure p_v. Since p_t and p are similarly affected by turbulence, velocity determinations should be almost independent of turbulence effects [37], [38].

17.7   Deducing Total Pressure

Because of blockage and displacement effects, it is not always convenient or judicious to install a total pressure tube into a flow duct. Also, it is often an effective total pressure that is required, as compared to a point total pressure. Analytic methods for determining effective total pressures have been given [39] as:

$$(p_t)_{\text{inc}} = p + \dot{m}^2 / 2 g_c \rho A$$
(17.46)

for liquids, and

$$(p_t)_{\text{comp}} = p \left\{ \frac{1}{2} + \left[\frac{1}{4} + \left(\frac{\gamma - 1}{2\gamma} \right) \left(\frac{\dot{m}}{Ap} \right)^2 \left(\frac{RT_i}{g_c} \right) \right]^{1/2} \right\}^{\gamma/\gamma - 1}$$
(17.47)

for gases.

17.8   References

[1] R. P. Benedict, "Temperature Measurement in Moving Fluids," ASME Paper 59A257, December 1959. (See also *Electro-Technol.*, October 1963, p. 56).

[2] R. P. Benedict, "The Force-Mass Dilemma," *Design News*, November 1968, p. 116.

[3] R. P. Benedict, "Some Comparisons Between Compressible and Incompressible Treatment of Compressible Fluids," *Trans. ASME, J. Basic Eng.*, **86,** September 1964, p. 527.

[4] R. P. Benedict and J. S. Wyler, "A Generalized Discharge Coefficient for Differential Pressure Type Fluid Meters," *Trans. ASME, J. Eng. Power*, October 1974, p. 440.

[5] W. G. Steltz and R. P. Benedict, "Thermodynamics of Constant-Density Fluids," *Electro-Techol.*, April 1963, p. 70. (See also *Trans. ASME, J. Eng. Power*, January 1962, p. 44).

[6] H. F. Mills, "Experiments upon Piezometers used in Hydraulic Investigations," *Proc. Am. Acad. Arts Sci. (New Series)*, Vol. 6, 1879, p. 26.

[7] C. M. Allen and L. J. Hooper, "Piezometer Investigation," *Trans. ASME*, Hydraulic Section, May 1932, p. 1.

[8] R. E. Rayle, "Influence of Orifice Geometry on Static Pressure Measurements," ASME Paper 59-A-234, December 1959. (See also M. I. T. Master's thesis, 1949).

[9] R. Shaw, "The Influence of Hole Dimensions on Static Pressure Measurements," *J. Fluid Mech.*, 7. Pt. 4, April 1960, p. 550.

[10] R. E. Franklin and J. M. Wallace, "Absolute Measurements of Static-Hole Error using Flush Transducers," *J. Fluid Mech.*, Vol. 42, Pt. 1, 1970, p. 33.

[11] J. S. Wyler and R. P. Benedict, "Comparisons Between Throat and Pipe Wall Tap Nozzles," *Trans. ASME, J. Eng. Power*, October 1975, p. 569.

[12] H. Schlicting, *Boundary-Layer Theory*, translated by J. Kestin, McGraw-Hill, New York, 1955, p. 401.

[13] R. P. Benedict, "Friction in Pipe Flow," *Instruments and Control Systems*, December 1969, p. 91.

[14] J. S. Wyler (STDE, Westinghouse Electric Corp.), Personal Communication, March 1976.

[15] W. J. Rainbird, "Errors in Measurement of Mean Static Pressure of a Moving Fluid Due to Pressure Holes," *Quart. Bull. Div. Mech. Engrg. Nat. Aero. Est.*, Nat. Res. Counc. Canada Rept DME/NAE 1967 (3).

[16] W. W. Emmett and J. R. Wallace, "Errors in Piezometric Measurement," *J. Hyd. Div*, Proc. A. Soc. Civil Engrs., November 1964, p. 45.

[17] L. N. Krause, "Effects of Pressure-Rake Design Parameters on Static-Pressure Measurement for Rakes Used in Subsonic Free Jets," NACA TN 2520, October 1951.

[18] W. M. Schulze, G. C. Ashby, Jr., and J. R. Erwin, "Several Combination Probes for Surveying Static and Total Pressure and Flow Direction," NACA TN 2830, August 1952.

[19] W. N. Gilmer (AGT Division, Westinghouse Electric Co.), Personal Communication, June 1952.

[20] R. E. Marris, "Multiple Head Instrument for Aerodynamic Measurements," *The Engineer*, August 1961, p. 315.

[21] C. A. Meyer and R. P. Benedict, "Instrumentation for Axial-Flow-Compressor Research," *Trans. ASME*, **74,** November 1952, p. 1327.

[22] A. H. Glaser, "The Pitot Cylinder as a Static Pressure Probe in Turbulent Flow," *J. Sci. Instr.*, **29,** July 1952, p. 219.

[23] J. S. Wyler, "Probe Blockage Effects in Free Jets and Closed Tunnels," *Trans. ASME, J. Eng. Power*, October 1975, p. 509.

[24] R. P. Benedict, "The Flow Field of a Free Jet," *Flow: Its Measurement and Control in Science and Industry*, Vol. 1, Pt. 1, ISA, Pittsburgh, 1974, p. 327.

[25] P. G. Pugh, J. W. Peto, and L. C. Ward, "Experimental Verification of Predicted Static Hole Size Effects on a Model with Large Streamwise Pressure Gradients," Aero Res. Counc., Current Paper 1139, 1971.

[26] R. G. Folsom, "Review of the Pitot Tube," *Trans. ASME*, October 1956, p. 1447.

[27] W. Gracey, W. Letko, and W. R. Russell, "Wind-Tunnel Investigation of a Number of Total-Preasure Tubes at High Angles of Attack," NACA TN 2331, April 1951.

[28] P. L. Chambre and S. A. Schaaf, "The Impact Tube," in *Physical Measurements in Gas Dynamics and Combustion*, Princeton University Press, Princeton, N. J., 1954, p. 111.

[29] C. W. Hurd, K. P. Chesky, and A. H. Shapiro, "Influence of Viscous Effects on Impact Tube," *Trans., ASME, J. Appl. Mech.*, June 1953, p. 253.

[30] M. J. Lighthill, "Contributions to the Theory of the Pitot-Tube Displacement Effect," *J. Fluid Mech.* Vol. 2, Pt. 2, 1957, p. 493.

[31] I. M. Hall, "The Displacement Effect of a Sphere in a Two-Dimensional Shear Flow," *J. Fluid Mech.*, Vol. 1, Pt. 2, July 1956, p. 142.

[32] J. L. Livesey, "The Behavior of Transverse Cylindrical and Forward Facing Total Pressure Probes in Transverse Total Pressure Gradients," *J. Aero. Sci.*, October 1956, p. 949.

[33] A. D. Young and J. N. Maas, "The Behaviour of a Pitot Tube in a Transverse Total-Pressure Gradient", Aero. Res. Counc., Lond., Rep. and Memo. No. 1770, 1936.

[34] P. O. A. L. Davies, "The Behaviour of a Pitot Tube in Transverse Shear," *J. Fluid Mech.*, Vol. 3, Pt. 2, 1957, p. 441.

[35] S. Sami, "The Pitot Tube in Turbulent Shear Flow," Proc. 11th Midwestern Mech. Conf., *Dev. in Mechanics*, Vol. 5, Paper 11, 1967, p. 191.

[36] T. Ikue and M. Inoue, "Pressure or Velocity Gradient Error in the Flow Direction Measurement," *Memoirs of the Faculty of Engrg.*, Kyushu Univ., Vol. 29, No. 3, March 1970, p. 121.

[37] R. C. Pankhurst and D. W. Holder, *Wind Tunnel Technique*, Pitman, London, 1952, p. 179.

[38] R. C. Dean et al., *Aerodynamic Measurements*, M. I. T. Notes, September 1952.

[39] R. P. Benedict, J. S. Wyler, J. A. Dudek, and A. R. Gleed, "Generalized Flow Across an Abrupt Enlargement," *Trans. ASME, J. Eng. Power*, July 1976, p. 327

Nomenclature

Roman

A	area
C_D	drag coefficient
C_P	Pitot coefficient
d	exact differential, diameter of static tap, diameter of probe
D	pipe diamerer, probe diameter, tap diameter
f	friction factor
g	acceleration of gravity
h_{loss}	head loss
KE	kinetic energy
L	depth of static tap
M	Mach number, Mass
p	absolute pressure
P_θ	pressure coefficient
R	specific gas constant
Re	Reynolds number
t	time
T	absolute temperature
v'	rms value of axial component of turbulent velocity

V	directed one-dimensional velocity
w	specific weight
W	weight

Greek

α	kinetic energy coefficient
Δ	finite difference
ϵ	blockage factor
γ	ratio of specific heats of perfect gas
ρ	fluid density
τ_0	shear stress at wall
θ	angle

Subscripts

I	indicated
P	point
t	total
v	dynamic

Problems

1. A submerged jet of water is discharged vertically downward into a reservoir of water. Tubing from a pitot tube, centered in the free jet, and located h inches below the reservoir surface, connects to a U-tube water manometer. Show that the water will rise in the open end of the manometer, above the reservoir surface, an amount X equivalent to the jet velocity.

    Ans. $X = V^2/2\,g$.

2. Find the error in a static pressure measurement made by a tap of $d = \frac{1}{8}$ in. in a 6 in. pipe if the fluid is water at 100°F, $\rho = 0.035877$ lbm/in.^{3}, $\nu = 0.74 \times 10^{-5}$ ft^2/sec, flowing at 15 ft/sec.

    Ans. $\Delta p/p_v = 0.84\%$.

3. Express the pressure error of Problem 2 in inches of water of a manometer at 72°F.

    Ans. $\Delta p = 0.35$ in. water.

4. A $\frac{1}{4}$ in. probe is inserted to the center of a 4 in. pipe. Find the blockage effect on the static pressure measurement if Mach number is 0.2.

    Ans. $\delta p/p_v = 4.77\%$ too low.

5. A 1 in. diameter spherical probe is used to measure pressure in an airfoil wake where the velocity gradient is 100 ft/sec/in. and the velocity level is 300 ft/sec. Find the displacement of the stagnation streamline in terms of probe diameter and in angle.

   Ans. $\delta = 0.1006D$, $\theta_s = 8.11°$.

6. Find the shift in angle measurement just because of balancing two pressure taps located 90° apart on a cylinder placed in a pressure gradient characterized by a shear parameter $K = 0.3$, where $P_{\theta 0}$ is -0.2, and $\left| \dfrac{\Delta P_\theta}{\Delta \theta} \right|_0 = 0.04$.

   Ans. $\Delta \theta_T = -2.12°$.

Chapter 18

TRANSIENT PRESSURE MEASUREMENT

"...the difference between the pressure at the mouth of the pitot tube and that in the manometer decreases exponentially with time..."

H. L. Weissberg (1953)

18.1 General Remarks

Most pressure-sensing systems involve small pressure holes that open to tubing of small diameter. These are connected by other tubing (usually flexible and of a larger diameter) to any of the mechanical or electrical pressure transducers.

Often these orifices and the associated tubing are subjected to time varying pressures. There may be a nonsteady flow, or the probe may be moving, perhaps rotating, through a pressure gradient field. Thus although there is usually no fluid flow in steady-state pressure-sensing systems, when transients occur, fluid flow generally accompanies them. At the least this causes some flow resistance in the orifices and tubing. Also, there are usually compression effects in the transducer. As a result, the pressure signal at the transducer tends to lag the impressed pressure at the orifice. The purpose of this Chapter is to indicate simple expressions that can serve to predict the response behavior of a pressure-sensing system when it is subjected to time varying pressure signals in both gases and liquids.

18.2   Mathematical Development for a Gas

A single time constant can describe the response behavior of only the simplest of systems. Fortunately, however, it is usually sufficient to consider only two regions in a pressure-sensing system. All the fluid resistance in the system is assigned (lumped) to one of these regions, called the *capillary portion*. If the system has more than one capillary in series, the equivalent length of the capillary portion is [1]:

$$L_e = L_1 + L_2 \left(\frac{d_1}{d_2}\right)^4 + \cdots + L_n \left(\frac{d_1}{d_n}\right)^4, \qquad (18.1)$$

where d_1 is taken as the capillary reference diameter. All the fluid mass in the system can be thought of as stored in a second region called the *transducer portion*. The volume of the transducer portion should include the volume of the equivalent capillary portion. This system is illustrated in Figure 18.1, in which a manometric-type transducer is included to

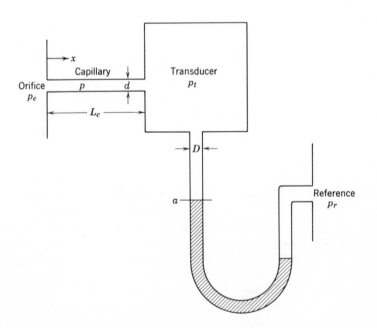

Figure 18.1   Schematic representation of transient pressure measurement; p_e is the environmental pressure.

represent the general case of time-varying system volume. The system can be characterized [2] by writing a flow balance at the capillary-transducer interface (i.e., at $x = L_e$). The mass flow rate into the transducer is equal to the rate of increase of mass in the transducer:

$$\int_0^{r_o} \rho_t V \, dA = \rho_t \frac{d\mathbf{V}}{dt} + \mathbf{V} \frac{d\rho_t}{dt}, \tag{18.2}$$

where the subscript t indicates time in this section, ρ_t is the fluid density at the interface, $\mathbf{V}$ is the system volume, r_o is the capillary radius from center to wall, and A is the interface area. The flow balance is localized in this manner to simplify the analysis, since at this interface the density term has consistent meaning. Appropriate expressions for the various terms in (18.2) are indicated next.

The fluid density at the capillary-transducer interface is defined as

$$\rho_t = \frac{p_t}{RT} = f(t), \tag{18.3}$$

where R is the specific gas constant and T is the absolute temperature of the fluid. Under isothermal conditions, the rate of change of density in the transducer is

$$\frac{d\rho_t}{dt} = \left(\frac{1}{RT}\right)\left(\frac{dp_t}{dt}\right). \tag{18.4}$$

The volume of the entire system, when considered as lumped in the transducer, is

$$\mathbf{V} = \mathbf{V}_\infty - \frac{A_e}{w}(p_f - p_t), \tag{18.5}$$

where $\mathbf{V}_\infty$ is the steady-state volume from the orifice to the surface of the manometric fluid; A_e is an equivalent area that depends on the geometry of the manometer (see Figure 18.2); p_f is the final environment pressure p_e at time $t = 0^+$, that is, after a step change takes place in pressure; p_t is the desired transducer pressure such that at $t = 0$, $p_t = p_i$, and at $t = \infty$, $p_t = p_f$; and w is the specific weight of the manometer fluid. The rate of change of volume is

$$\frac{d\mathbf{V}}{dt} = \left(\frac{A_e}{w}\right)\left(\frac{dp_t}{dt}\right). \tag{18.6}$$

If we assume laminar flow in the capillary, the instantaneous axial

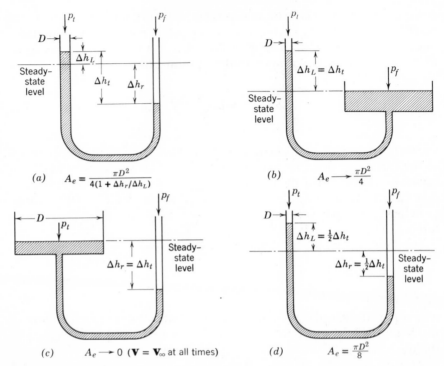

Figure 18.2 The definition of manometer equivalent area A_e for various types of manometer: (a) U-tube manometer, unequal diameters; (b) single leg on capillary side; (c) single leg on reference side; (d) U-tube manometer, equal diameters.

velocity at any radius r in the capillary is

$$V = -\frac{1}{4\mu}\left(\frac{\partial p}{\partial x}\right)(r_0^2 - r^2), \tag{18.7}$$

where μ is the dynamic viscosity of the fluid, r is the capillary radius, and x is the distance from the orifice.

With each of the terms of (18.2) now defined, the flow balance becomes

$$-\int_0^r \left(\frac{p_t}{RT}\right)\left(\frac{1}{4\mu}\right)\left(\frac{\partial p}{\partial x}\right)(r_0^2 - r^2)2\pi r\, dr$$

$$= \frac{1}{RT}\left[p_t\frac{A_e}{w} + \mathbf{V}_\infty - (p_f - p_t)\frac{A_e}{w}\right]\frac{dp_t}{dt}. \tag{18.8}$$

Equation 18.8 is a first-order, first-degree, nonlinear differential equation [3], and therefore cannot be used to define a conventional time constant.

In order to get a usable time constant, we must find a suitable linearized form of (18.8). As a first step, we impose the limitation that

$$(p_f - p_i) \ll p_i \tag{18.9}$$

so that the pressure difference is always much smaller than the initial pressure level p_i. This implies that p_t, which is bounded by p_i and p_f, can be well approximated by its average value

$$\bar{p}_t = \frac{\int_0^t p_t \, dt}{t} . \tag{18.10}$$

However, dp_t/dt cannot be replaced by $d\bar{p}_t/dt$ (since the latter, of course, equals zero), since system lag derives from the very fact that $p_t = f(t)$. Making use of the approximation $p_t \approx \bar{p}_t$, and separating the variables in (18.8), we obtain

$$-\left(\frac{\pi d_1^{\,4} p_t}{128\mu}\right) \int_{p_f}^{p_t} dp = \left[\frac{A_e}{w}(2\bar{p}_t - p_f) + \mathbf{V}_\infty\right]\frac{dp_t}{dt} \int_0^{L_e} dx. \tag{18.11}$$

On integration, we get the first-order, linear expression [4], [5], [6]:

$$\frac{dp_t}{dt} = \frac{1}{K}(p_f - p_t). \tag{18.12}$$

The constants of the system are lumped in K. The general solution for (18.12) is

$$p_t = Ce^{-t/K} + \frac{1}{K} e^{-t/K} \int_0^t p_f e^{t/K} \, dt. \tag{18.13}$$

For the case of a pressure-sensing system subjected to a step change in environment pressure, the initial condition is that $p_t = p_i$ at $t = 0$; thus $C = p_i$, and at any time p_f is a constant. The general solution then reduces to

$$(p_f - p_t) = (p_f - p_i)e^{-t/K}. \tag{18.14}$$

Thus K indicates the time required after the step change for the reduction of environment-transducer pressure difference to $1/e$ times the initial difference. In other words, K is the time required for the pressure-sensing system to indicate a pressure corresponding to approximately 63.2% of the initial pressure difference. Since this is exactly the meaning of the time constant of a first-order linear system, K is the time constant we seek. For higher-order, variously damped systems, the time constant has no such significance; that is, if the damping factor of a second-order system is very small, there is a decaying oscillatory response, and the 63% recovery has little meaning. However, if the damping factor is ≥ 1, the second-order response appears much the same as a first-order response, and the 63% recovery is significant (see Figure 18.3). For Reynolds numbers below 2000 a capillary acts like a highly damped system and therefore behaves like a first-order linear system.

The time required to reduce the pressure difference to any acceptable value can be predicted from the expression

$$a_t = \frac{p_f - p_t}{p_f - p_i} = e^{-t/K} \tag{18.15}$$

(compare with (13.39) for temperature response). Equation 18.15 tabulates as

a_t	0.5	0.368	0.2	0.1	0.05	0.01
t	$0.7K$	K	$1.6K$	$2.3K$	$3K$	$4.6K$

The average pressure level $\bar{p}_t$ can now be evaluated, and explicit expressions can be obtained for both the time constant K and the recovery time t. Inserting p_t from (18.14) and t from (18.15) into (18.10) results in

$$\bar{p}_t = p_f - (p_f - p_i)\left[\frac{1 - a_t}{\ln(1/a_t)}\right]. \tag{18.16}$$

Since p_t asymptotically approaches but can never equal p_f in any practical time, an arbitrary cutoff point must be chosen for a_t to allow a realistic evaluation of $\bar{p}_t$ (see Figure 18.4). For simplicity, an a_t corresponding to 98% recovery, is chosen, in which case

$$\bar{p}_t = p_f - \frac{(p_f - p_i)}{4}, \tag{18.17}$$

where $\bar{p}_t$ represents the average value of p_t from time zero to a time corresponding to the 98% recovery point. The time constant for a pressure sensing system is then

$$K = \frac{128\mu L_e}{\pi d_1^4 \bar{p}_t}\left[\frac{A_e}{w}\left(\frac{p_f + p_i}{2}\right) + \mathbf{V}_\infty\right], \tag{18.18}$$

and the general expression for recovery time is

$$t = K \ln\left(\frac{p_f - p_i}{p_f - p_t}\right). \tag{18.19}$$

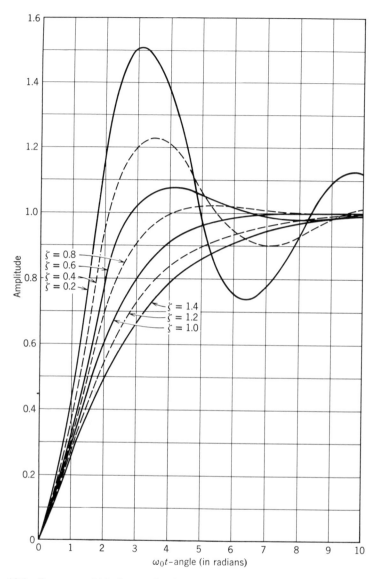

Figure 18.3 Response of ideal second-order system to step input of unit amplitude. (From ANSI B 88.1–1972, Reference [12]).

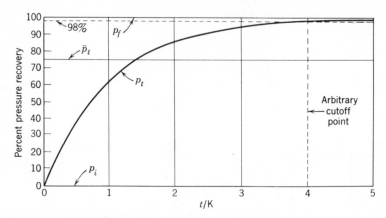

Figure 18.4 Approximation to p_t for a step change in environmental pressure. When $\bar{p}_t$ approximates p_t, then $p_t \approx \int_0^{t_{98\%}} p_t \, dt/t_{98\%}$.

These linear equations [(18.18) and (18.19)] reliably predict response time at various pressure levels for pressure differences on the order of 10% of the initial pressure level, as indicated by the excellent agreement between experimental points and the linear equations (see Figure 18.5).

Nonlinear Solution

There are many situations where the pressure response solution to (18.8) cannot be limited by the approximation of (18.9), wherein the step change in pressure was stated to be much smaller than the pressure level. This means that that linearized differential equation of (18.12), the linearized solution of (18.14), the linearized time constant of (18.18), and the linearized solution for time of (18.19) may not be adequate.

Very briefly, for the nonlinear case, the counterparts of the preceding linearized equations (for zero-volume-displacement systems such as typified by diaphragm-type transducers which do not allow system volume to vary with time) are for the nonlinear differential equation,

$$\frac{dp_t}{dt} = (2B)(p_f^2 - p_t^2), \tag{18.12'}$$

where
$$B = \frac{128\mu L_e \mathbf{V}_\infty}{\pi d_1^{\,4}} \, ;$$

and for the nonlinear solution

$$(p_f - p_t) = (p_f - p_i)e^{-[t/(B/p_f) + \ln b_r]}, \tag{18.14'}$$

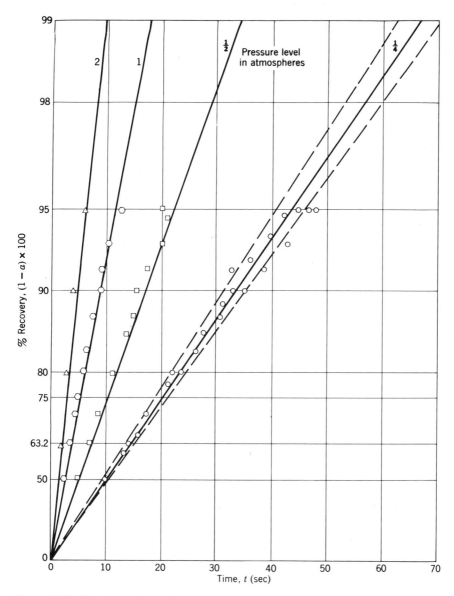

Figure 18.5 Experimental confirmation of linearized theory at various pressure levels. Notation: (———) linearized theory ($p_t \approx \bar{p}_t$); (— —) experimental uncertainty envelope ($\pm 6\%$); ($\triangle$, $\bigcirc$, $\square$, $\bigcirc$) experimental points.

where

$$b_t = \frac{p_f + p_i}{p_f + p_t} \; ;$$

and for the nonlinear (*NL*) time constant,

$$K_{NL} = (1 - \ln b_t)(B/p_f); \tag{18.18'}$$

and for the nonlinear solution in terms of recovery time,

$$t_{NL} = \left(\frac{B}{p_f}\right) \ln\left(\frac{1}{a_t b_t}\right). \tag{18.19'}$$

Example 1. An air-filled pressure-sensing system is subjected to a small step change in pressure of 1% at a pressure level of 15 psia. The temperature is such that the viscosity is 3.85×10^{-7} 1bf sec/ft^2. If $L_1 = 1$ in., $L_2 = 6$ ft, $d_1 = 0.02$ in., $d_2 = 0.10$ in., and $V_\infty = 8$ in.^{3}, find the time for the pressure to reach within 1% of the final pressure by the linear and nonlinear solutions.

Linear

By (18.1):
$$L_e = 1 + 72\left(\frac{0.02}{0.10}\right)^4 = 1.1152 \text{ in.}$$

By (18.18):
$$K_{\text{linear}} = \frac{128(3.85 \times 10^{-7})(1.1152)8}{\pi(16 \times 10^{-8})(15 \times 144)} = \frac{B}{\bar{p}_t}$$

$$= \frac{874.6687 \text{ 1bf sec/ft}^2}{15 \times 144 \text{ 1bf/ft}^2} = 0.405 \text{ sec.}$$

Since it takes about $4.6 K_L$ to be within 1% of the final pressure, $t_{99\%} = 4.6 \times 0.405 = 1.86$ sec.

Nonlinear

$a_t = (p_f - p_t)/(p_f - p_i) = 0.01$ for 99% recovery. But p_f, for a 1% step change, is 15.15 psia, and $p_t = p_f - (p_f - p_i)\, a_t$ is 15.1485 psia, thus

$$b_t = \frac{15.15 + 15.00}{15.15 + 15.1485} = 0.99509877.$$

By (18.18'): $K_{NL} = (1 - \ln b_t)\left(\dfrac{B}{p_f}\right) = (1 + 0.00491)\left(\dfrac{874.6687}{15.15 \times 144}\right) = 0.403$ sec.

By (18.19'):
$$t_{NL} = \left(\frac{B}{p_f}\right) \ln\left(\frac{1}{a_t b_t}\right)$$

$$= \left(\frac{874.6687}{15.15 \times 144}\right) \ln\left(\frac{1}{0.01 \times 0.99509877}\right)$$

$$= 0.40093 \times 4.61008$$

$$t_{NL} = 1.85 \text{ sec.}$$

Note how close are t_L and t_{NL} for the small pressure step of 1%.

Example 2. For the same pressure system of Example 1, find the linear and nonlinear solutions for 99% pressure recovery when the step change in pressure is 50% of the initial

15 psia. $a_t = 0.01$ for 99% recovery.

$$p_i = 15 \text{ psia}, \ p_f = 1.5 p_i = 22.5 \text{ psia},$$

$$p_t = p_f - (p_f - p_i) a_t = 22.5 - (7.5) \times 0.01 = 22.425 \text{ psia},$$

$$\bar{p}_t = p_f - \left(\frac{p_f - p_i}{4} \right) = 22.5 - 7.5/4 = 20.625 \text{ psia}.$$

$$B/p_f = 874.6687/(22.5 \times 144) = 0.27 \text{ sec}.$$

$$b_t = (22.5 + 15)/(22.5 + 22.425) = 0.83472.$$

By (18.18): $\qquad\qquad\qquad K_L = B/\bar{p}_t = 0.2945 \text{ sec}$

By (18.19): $\qquad\qquad t_L = (B/\bar{p}_t) \ln (1/a_t) = 0.2945 \times 4.605 = 1.356 \text{ sec}.$

By (18.18′): $\qquad\quad K_{NL} = (1 - \ln b_t)(B/p_f) = 1.18065 \times 0.27 = 0.3188 \text{ sec}.$

By (18.19′): $\qquad\quad t_{NL} = (B/p_f) \ln (1/a_t b_t) = 0.27 \times 4.78583 = 1.292 \text{ sec}.$

It must be emphasized that this approach, particularized by equations (18.18) and (18.19) for linear solutions and by (18.18′) and (18.19′) for nonlinear solutions, is valid only when there are two distinct regions in the pressure-sensing system, namely, a *capillary portion* (representing the inlet orifice, the probe, and the pressure tubing) and a relatively *large volume portion* (representing the transducer).

Systems made up of tubing only, without dominating volume portions, must be handled by different methods. One of these is the so-called *transmission line* approach [7]–[11]. Numerical solutions to such pneumatic pressure systems subjected to *large* transient pressure changes are presented and compared with experimental data in [7]. A simplified solution for transmission line response has been developed by Karam [10], [11]. For the high percentage pressure recovery of interest here, this solution can be given as

$$\frac{p_t}{p_i} = \text{erfc} \left[\frac{1}{2} \left(\frac{8 \gamma \tau_0^2}{\tau - \tau_o} \right)^{1/2} \right] \tag{18.20}$$

where $\tau = $ dimensionless time $= tv/r^2$,

$\quad \tau_0 = $ dimensionless delay time $= Tv/r^2$,

$\quad v = $ kinematic fluid viscosity,

$\quad r = $ pressure tubing radius,

$\quad T = $ tubing length/acoustic velocity $= L/a$,

$\quad \gamma = $ specific heat ratio,

and erfc = the complementary error function $= 1 - $ erf, as tabulated in many mathematical handbooks (see Table 18.1).

Example 3. An air filled pressure transmission line is subjected to a step change in pressure. The fluid kinematic viscosity is $0.18 \times 10^{-3} \text{ ft}^2/\text{sec}$, the acoustic velocity is 1200 ft/sec, and the ratio of specific heats is 1.4. If the tube length is 20 ft and the tube radius is 0.125 in., find the 99% pressure recovery time.

By (18.20) $\qquad\qquad\qquad\qquad \dfrac{p_t}{p_i} = 1 - \text{erf}(x) = 0.99$

Transient Pressure Measurement

Table 18.1 Error Function (Adapted From NBS AMS-55 [13])

x	erf x	x	erf x	x	erf x	x	erf x
0.00	0.00000	0.25	0.27632	0.50	0.52049	0.75	0.71115
0.01	0.01128	0.26	0.28689	0.51	0.52924	0.76	0.71753
0.02	0.02256	0.27	0.29741	0.52	0.53789	0.77	0.72382
0.03	0.03384	0.28	0.30788	0.53	0.54646	0.78	0.73001
0.04	0.04511	0.29	0.31828	0.54	0.55493	0.79	0.73610
0.05	0.05637	0.30	0.32862	0.55	0.56332	0.80	0.74210
0.06	0.06762	0.31	0.33890	0.56	0.57161	0.81	0.74800
0.07	0.07885	0.32	0.34912	0.57	0.57981	0.82	0.75381
0.08	0.09007	0.33	0.35927	0.58	0.58792	0.83	0.75952
0.09	0.10128	0.34	0.36936	0.59	0.59593	0.84	0.76514
0.10	0.11246	0.35	0.37938	0.60	0.60385	0.85	0.77066
0.11	0.12362	0.36	0.38932	0.61	0.61168	0.86	0.77610
0.12	0.13475	0.37	0.39920	0.62	0.61941	0.87	0.78143
0.13	0.14586	0.38	0.40900	0.63	0.62704	0.88	0.78668
0.14	0.15694	0.39	0.41873	0.64	0.63458	0.89	0.79184
0.15	0.16799	0.40	0.42839	0.65	0.64202	0.90	0.79690
0.16	0.17901	0.41	0.43796	0.66	0.64937	0.91	0.80188
0.17	0.18999	0.42	0.44746	0.67	0.65662	0.92	0.80676
0.18	0.20093	0.43	0.45688	0.68	0.66378	0.93	0.81156
0.19	0.21183	0.44	0.46622	0.69	0.67084	0.94	0.81627
0.20	0.22270	0.45	0.47548	0.70	0.67780	0.95	0.82089
0.21	0.23352	0.46	0.48465	0.71	0.68466	0.96	0.82542
0.22	0.24429	0.47	0.49374	0.72	0.69143	0.97	0.82987
0.23	0.25502	0.48	0.50274	0.73	0.69810	0.98	0.83423
0.24	0.26570	0.49	0.51166	0.74	0.70467	0.99	0.83850
						1.00	0.84270

or

$$\text{erf}(x) = 0.01.$$

By Table 18.1,

$$x \simeq 0.01 = \frac{1}{2}\left(\frac{8\gamma\tau_0^2}{\tau - \tau_0}\right)^{1/2}.$$

By definitions:

$$T = \frac{L}{a} = \frac{20}{1200} = 0.0167 \text{ sec.}$$

$$\tau_0 = \frac{Tv}{r^2} = \frac{0.0167(0.18 \times 10^{-3})144}{0.015625} = 0.0277.$$

By (18.20):

$$\tau = 0.0277 + \frac{8(1.4)(0.0277)^2}{(0.02)^2} = 21.512,$$

and

$$t = \frac{\tau r^2}{v} = \frac{21.512(0.015625)}{(0.18 \times 10^{-3})144} = 12.97 \text{ sec}$$

to reach the 99% pressure recovery in air.

18.3 Response of Liquid-filled Pressure Systems

When a pressure-measuring system is filled with a liquid, the natural frequency of the system is significantly lower than that of a gas-filled system [8]. The transient response depends primarily on two factors, the resonant frequency F, and the damping ratio ζ. Both of these factors depend heavily on the tube diameter (d) of the pressure-sensing system. That is, the natural frequency varies directly as the first power of d, and the damping ratio varies inversely as the cube of d.

Brown and Nelson [9] present a theoretical solution for the response of liquid-filled systems. Their work compares very favorably with Karam's simplified solution which, for liquids, reduces from (18.20) to

$$\frac{p_t}{p_i} = \operatorname{erfc}\left[\frac{1}{2}\frac{\tau_0}{\sqrt{\tau - \tau_0}}\right]. \tag{18.21}$$

These two solutions are compared in Figure 18.6.

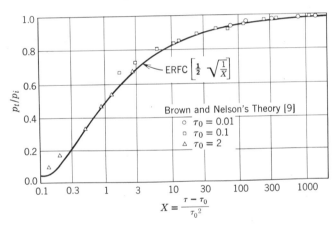

Figure 18.6 Karam's equation (18.20) compared with Brown and Nelson's theoretical liquid solution (after [10]).

Many of these definitions as well as many useful graphs (e.g., Figure 18.3) concerning the dynamic response of pressure transducers are given in the authorative reference [12].

Example 4. If the pressure transmission line of Example 2 was filled with water such that the acoustic velocity was 4900 ft/sec, and the kinematic viscosity was 1.44×10^{-3} in.^{2}/sec, find the 99% pressure recovery time.

By (18.21):
$$\frac{p_t}{p_i} = 1 - \operatorname{erf}(x) = 0.99,$$

or $\operatorname{erf}(x) = 0.01,$

or $x \simeq 0.01$, as in Example 2.

By definitions: $T = \dfrac{L}{a} = \dfrac{20}{4900} = 0.00408$ sec.

$$\tau_0 = \frac{T\nu}{r^2} = \frac{0.00408(1.44 \times 10^{-3})}{0.015625} = 3.76 \times 10^{-4}.$$

Thus, by (18.21): $x = 0.01 = \dfrac{1}{2} \dfrac{\tau_0}{\sqrt{\tau - \tau_0}}$

or $\tau = \tau_0 + \left(\dfrac{\tau_0}{0.02}\right)^2 = 3.76 \times 10^{-4} + 3.5344 \times 10^{-4} = 7.2944 \times 10^{-4}.$

and $t_{99\%} = \dfrac{\pi r^2}{\nu} = \dfrac{7.29 \times 10^{-4}(0.015625)}{1.44 \times 10^{-3}} = 7.9 \times 10^{-3}$ sec.

We conclude that the water-filled system is faster than the air-filled system by a factor of

$$\frac{12.97}{7.9 \times 10^{-3}} = 1642.$$

18.4   References

[1] A. R. Sinclair and A. W. Robbins, "A Method for the Determination of the Time Lag in Pressure Measuring Systems Incorporating Capillaries," NACA TN 2793, September 1952.

[2] R. P. Benedict, "The Response of a Pressure-Sensing System," *Trans. ASME, J. Basic Eng.*, **82**, June 1960, 482.

[3] R. C. Bauer, "A Method of Calculating the Response Time of Pressure Measuring Systems," Arnold Engrg. Dev. Center TR-56-7, November 1956 (available from O.T.S. as PB-131321).

[4] G. J. Delio, G. V. Schwent, and R. S. Cesaro, "Transient Behavior of Lumped-Constant Systems for Sensing Gas Pressures," NACA TN 1988, December 1949.

[5] R. C. Johnson, "Averaging of Periodic Pressure Pulsations by a Total-Pressure Probe," NACA TN 3568, October 1955.

[6] H. L. Weissberg, "The Response Time of Small Pitot Tubes," Union Carbide and Carbon Co. Report KLI-2616, October 1953 (available from O.T.S. as PB-AECU-3242).

[7] J. E. Funk and T. R. Robe, "Transients in Pneumatic Transmission Lines Subjected to Large Pressure Changes," ASME Paper 69-FLCS-42, June 1969.

[8] G. Wad, "Pressure Transducers in Liquid Filled Pressure Systems," DISA Information Bulletin No. 7, p. 19.

[9] F. T. Brown and S. E. Nelson, "Step Responses of Lines with Frequency-Dependent Effects on Viscosity," *Trans. ASME, J. Basic Eng.*, Vol. 87, No. 3, September 1965, pp. 504–510.

[10] J. T. Karam, Jr., "A Simple but Complete Solution for the Step Response of a Semi-Infinite, Circular Fluid Transmission Line," *Trans. ASME, J. Basic Eng.*, June 1972, pp. 455–456.

[11] J. T. Karam, Jr. and R. G. Leonard, "A Simple Yet Theoretically Based Time Domain Model for Fluid Transmission Line Systems," *Trans. ASME, J. Fluids Eng.*, December 1973, p. 498.
[12] "A Guide for the Dynamic Calibration of Pressure Transducers," ASME ANSI B88.1–1972.
[13] *Handbook of Mathematical Functions*, Edited by M. Abramowitz and I. A. Stegun, National Bureau of Standards Applied Mathematics Series 55, June 1964, p. 310.

Nomenclature

Roman

a acoustic velocity
A area
C constant
d diameter; also exact differential
e base of natural logarithm
f a function of
K lumped single order time constant
L length
p absolute pressure
r pressure tubing radius
R specific gas constant
t time
T absolute temperature, isentropic delay time
V directed velocity
$\mathbf{V}$ volume
w specific weight
x axial length

Greek

μ dynamic viscosity
ρ density
ζ damping factor
ω_0 system natural frequency
τ dimensionless time
τ_0 dimensionless delay time
ν kinematic viscosity

Subscripts

e equivalent
f final

i initial

n general subscript

t a function of time

Problems

1. Find the linear time to reach 95% of a small step change in air pressure at a level of 15 psia and a dynamic viscosity of 4×10^{-7} lbf sec/ft^2, using the lumped system model. The geometry is such that $L_1 = 0.5$ in., $L_2 = 10$ ft, $d_1 = 0.03$ in., $d_2 = 0.120$ in., and $\mathbf{V}_\infty = 10$ in.^{3}.

   Ans. $t_{95\%} = 0.2707$ sec.

2. Find the nonlinear time to reach 99% pressure recovery for a 50% step change in air pressure for the system of Example 1.

   Ans. $t_{99\%} = 0.288$ sec.

3. Find the time to reach 95% of a step change in air pressure at a level of 15 psia and a kinematic viscosity of 0.18×10^{-3} ft^2/sec, using the distributed resistance-capacitance (transmission line) approach. The geometry of the pressure line is specified by $L = 10$ ft and $d = 0.120$ in. The air is characterized by an acoustic velocity $= 1160$ ft/sec, and $\gamma = 1.4$.

   Ans. $t_{95\%} = 0.7707$ sec.

4. A 100 ft line of 1/8 in. tubing is filled with water at a kinematic viscosity of 1×10^{-5} ft^2/sec. For an acoustic velocity of 5000 ft/sec, find the 95% pressure recovery time.

   Ans. $t_{95\%} = 0.0388$ sec.

Part III
Flow and Its
Measurement

"... the several aqueducts reach the city at different elevations. Whence it comes that some deliver water on higher grounds, while others cannot elevate themselves to the higher summits..."

Sextus Julius Frontinus (97 A.D.)

INTRODUCTION TO PART III

We undertake in this section a rational explanation of flow and its measurement. Definitions of flow rate for constant density and compressible fluids are stated and their development is traced historically. The standard discharge coefficients for head meters used in open and closed channel flows are given. Factors to correct for the transverse expansions of compressible fluids after they pass through orifices are discussed. Metering of flow when the pressure drop across the meter exceeds the "critical" drop is considered. Fluid properties, such as those required in flow-rate calculations, are provided.

Chapter 19

THE CONCEPT OF FLOW RATE

". . . a river in each part of its length in an equal time gives passage to an equal quantity of water, whatever the width, the depth, the slope, the roughness, the tortuosity . . ."

Leonardo da Vinci (1502)

19.1 The Concepts

From time immemorial the measurement of fluid flow has been a mark of civilization. The Egyptians depended strongly on the indications of their Nile meters. Would the year be favorable or not? Year after year the flow of the Nile was predicted, the rise in spring indicating the coming fertility of their land. Roman engineers were famous not only for their roads, but also for their aqueducts and piping systems leading water to their homes, baths, and fountains. The rate of flow was established then as now by orifices.

One of the basic premises of fluid mechanics is that mass is a conserved quantity. In the absence of sources and sinks (i.e., when mass is neither introduced nor removed spontaneously), this conservation principle requires that the local rate of increase of mass within a given system plus the net rate of efflux of mass across the boundaries of the system must equal zero. In the steady state, the mass entering a system is equal to the mass leaving the system when both are measured over the same time interval (although they are independent of time itself). Thus, with $M/\Delta t$

simulated by $\overset{\circ}{m}$, we can write

$$\overset{\circ}{m}_{in} = \overset{\circ}{m}_{out}. \qquad (19.1)$$

If ρV (density times volume) is substituted for mass, (19.1) becomes

$$\left(\frac{\rho V}{\Delta t}\right)_{in} = \left(\frac{\rho V}{\Delta t}\right)_{out}. \qquad (19.2)$$

The volume passing a given plane in a given time interval is (see Figure 19.1)

$$\frac{V}{\Delta t} = \int \frac{\Delta x \, dA}{\Delta t} = \int V_p \, dA, \qquad (19.3)$$

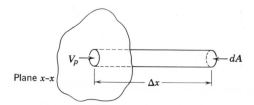

Figure 19.1 Volume passing a given plane in a given time interval is used to define volumetric rate of flow.

where V_p is the fluid velocity at a point. It is often convenient to define an average velocity as

$$V \equiv \frac{\int V_p \, dA}{A}. \qquad (19.4)$$

The use of the average velocity given by (19.4) is not restricted to systems with one-dimensional flow; for example, in a two-dimensional flow, the average velocity is obtained as follows. From Figure 19.2, $dA = b \, dy$,

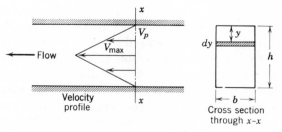

Figure 19.2 Average velocity in a two-dimensional flow system.

$A = bh$, and $V_p = V_{max}(y/\frac{1}{2}h)$; thus the average velocity is

$$V = \frac{2\int_0^{h/2} V_{max}\left(\frac{y}{h/2}\right)b\,dy}{bh} = \frac{4V_{max}}{h^2}\int_0^{h/2} y\,dy = \frac{V_{max}}{2}. \qquad (19.5)$$

In terms of the average velocity, (19.2) is

$$\mathring{m} = \rho A V = \text{Constant}, \qquad (19.6)$$

which is usually called the continuity equation. The acceleration of gravity can be considered constant for systems of reasonable extent. Since the weight and specific weight are, respectively, $W = KMg$ and $w = K\rho g$, and with $W/\Delta t$ simulated by $\mathring{w}$, we can write

$$\mathring{w} = wA V = \text{Constant}. \qquad (19.7)$$

Equation 19.7 expresses the weight flow rate of the fluid in terms of the specific weight, the area perpendicular to the flow, and the average velocity. It further establishes the fact that weight is also conserved in earthbound flow systems.

The simplest method for determining the rate of flow of a fluid is to catch it in a suitable container and weigh it over a given interval of time. Such a procedure avoids, for example, consideration of how specific weight varies with temperature; it obviates the need for point-velocity measurements and an averaging technique; and it makes superfluous a measurement of the area involved in the flow. In ascertaining the flow rate of liquids in closed conduits, this "catching-weighing-timing" approach is the most common and most reliable method in use. The technique is used to determine the flow rates of the huge steam turbines that drive electrical generating equipment. In such cases, after the steam expends most of its energy in the turbine, it is condensed, and the water is weighed over given time periods to establish the flow rate.

It is not always possible, however, to determine a flow rate by weighing the fluid. It would be impossible, for instance, to measure the flow rate of a river in this way. Consider also the flow of a gas that cannot be condensed into a liquid under ordinary conditions and that would escape easily from any open container of the type ordinarily used for catching liquids. There are cases also in which it is impossible to break into a system (such as a closed-loop chemical process) to measure its flow rate. Thus it is clear that the basic weight per unit time method for determining quantity of flow must be augmented by other means better suited for a variety of practical situations.

There is an almost endless variety of instrument types used in the measurement of flow rates: variable-area or float types, electromagnetic types, force or fixed-vane types, turbine or rotating-vane types, thermal types, acoustic types, and also others. Whatever instrument is used, the inherent difficulties of accurate flow measurement make the process a complex one, and it is necessary to take into consideration many of the natural characteristics of the fluid being measured, the fluid surroundings, and the man-made characteristics of the barriers, pipes, channels, etc., used to control the flow. Over the years, many of these characteristics have been collected in charts and tables to simplify the measurement process, but their use requires thorough knowledge of the conditions and basic formulas to which they apply. Such conditions and such equations are discussed in detail in the following chapters [1].

19.2 Historical Resume

As in Parts I and II on temperature and pressure, respectively, a short historical review of the basic concepts involved in flow measurement is presented here.* As is always true in the history of science, the material is instructive and the names are significant.

The principle of continuity (wherein mass is conserved) expressed by (19.1), can certainly be credited to Leonardo da Vinci at about 1502. For it was he who stated "... that a river in each part of its length in an equal time gives passage to an equal quantity of water, whatever the width, the depth, the slope, the roughness, the tortuosity · · · a river of uniform depth will have a more rapid flow at the narrower section than at the wider, to the extent that the greater width surpasses the lesser...."

Isaac Newton's second law of motion, first proposed in 1687, states that "... the net external (unbalanced) force acting on a material body is directly and linearly proportional to, and in the same direction as, its acceleration...." This law, along with the choice (arbitrary) of a standard of mass, fixes the unit of force (hence of weight) and, on a unit volume basis, sets the relation between specific weight and density [given above as (19.7)]. It was Newton, too, who first noted that the computed rate of efflux through an orifice was about $\sqrt{2}$ times as great as that actually measured and who first attributed this to the contraction of the jet.

Christian Huygens and Gottfried Wilhelm Leibniz in about 1700 first discussed the conservation of energy during an elastic impact. In the words of Huygens, "... the sum of the product of their masses times the square of their velocities will remain the same before and after the

---

* Much of the history pertaining to flow measurement may be found in reference [2]. Several of the quotations that appear in this section were obtained from reference [3].

impact . . . ." Daniel Bernoulli, in 1748, proceeded toward the principle of the conservation of the sum of the kinetic and potential energies, stating ". . . we must finally justify the principles which we have so often mentioned. The first is that of the conservation of live forces, or as I have stated it, the equality between the actual descent and the potential ascent . . . ." The effect, however, of a change in pressure head (p/w) on the general energy accounting was still not clearly understood.

It remained for Leonhard Euler, friend and contemporary of D. Bernoulli, to generalize the continuity principle in its current form as

$$\frac{\partial \rho}{\partial t} + \frac{\partial}{\partial x}(\rho u) + \frac{\partial}{\partial y}(\rho v) + \frac{\partial}{\partial z}(\rho w) = 0 \qquad (19.8)$$

(where u, v, w *only* in this equation represent velocity components), and to present the so-called Bernoulli equation for a constant density steady flow in the absence of losses as

$$\frac{p}{w} + \frac{V^2}{2g} + z = \text{Constant.} \qquad (19.9)$$

Jean Charles Borda, in 1766, introduced the concept of elementary stream tubes and showed that not only did the contraction of a jet influence the flow rate through a small opening (relative to the flow passage) but also that a loss of (available) energy accompanied such a flow.

In the 1840s the principle of conservation of energy was extended to include the energies of heat and work by such men as Julius Robert Mayer (". . . energies are indestructible, convertible entities . . .") and James Prescott Joule (". . . my course was to publish only such theories as I had established by experiments calculated to commend them to the scientific public . . ."). In this same period, Julius Weisbach did important work concerning flow losses, contraction coefficients, and approach factors. He also greatly popularized the use of the Bernoulli equation in flow analyses.

In 1895, Clemens Herschel invented the Venturi meter for measuring the discharge of a liquid. More recently, in the twentieth century, Edgar Buckingham and Howard S. Bean of the National Bureau of Standards, and Samuel R. Beitler of the Ohio State University (each also of the ASME Research Committee on Fluid Meters) advanced the art and science of flow measurement by their careful analyses and experiments, primarily on the determination of discharge coefficients of nozzles and orifices.

19.3   References

[1] R. P. Benedict, "Fluid Flow Measurement," *Electro-Technol.*, January 1966, p. 55.
[2] H. Rouse and S. Ince, *History of Hydraulics*, Dover, New York, 1963.
[3] G. Holton, *Introduction to Concepts and Theories in Physical Science*, Addison-Wesley, Cambridge, Mass., 1952.

Nomenclature

Roman

A area
b base
d exact differential
h height
$\mathring{m}$ mass flow rate
M mass
p absolute pressure
t time
V directed velocity
$\mathbf{V}$ volume
w specific weight
W weight
$\mathring{w}$ weight flow rate
x axial length
y vertical length
Z elevation

Greek

Δ finite difference
ρ density
∂ partial differential

Subscripts

p at a point
max   maximum

Chapter 20

OPEN-CHANNEL FLOW

"... when one has a flow of water to convey, either to provide some at a place where there is none, or to drain where there is too much, one is, almost everywhere, obliged to make the most water flow with the least possible slope ..."

Antoine Chezy (1770)

The flow in rivers, canals, and pipes that are not flowing full, that is, where one surface of the liquid is free of solid boundaries, is called open-channel flow. This type of flow is often measured by placing an obstruction across the flow path and metering some characteristic variable resulting from the flow over or under the obstruction. Such dams erected for the purpose of metering the flow of liquids are called weirs or sluice gates, and the measurable quantity of liquid is called the head (see Figure 20.1). These topics are discussed in detail in the following sections.

20.1 General Relations

The energy of a liquid flowing between two stations (x and y) can be accounted for by the generalized Bernoulli relation [1], [2], [3]

$$\left(\frac{p_x}{w} + \frac{V_x^2}{2g} + Z_x\right) + \mathbf{W}_{\text{net}} - h_{\text{loss}} = \left(\frac{p_y}{w} + \frac{V_y^2}{2g} + Z_y\right), \qquad (20.1)$$

where p = effective static pressure;
$\quad w$ = uniform specific weight of liquid;
$\quad V$ = average liquid velocity of continuity;
$\quad Z$ = effective vertical distance from a consistent horizontal datum;

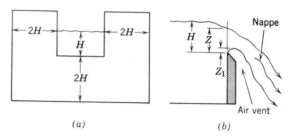

(a)   (b)

Figure 20.1 (a) Ideal dimensions for fully contracted ($n = 2$) rectangular weir; (b) vertical contraction of stream over sharply crested weir.

$\mathbf{W}_{net}$ = net mechanical energy addition between stations per pound of flowing liquid;

h_{loss} = total energy dissipated between stations per pound of flowing fluid.

If the liquid flows in a horizontal bottomed open-channel such that its upper surface is freely exposed to a uniform ambient pressure, and in addition the flow is in the absence of any mechanical energy addition between stations, then (20.1) can be expressed more simply as

$$E_x = E_y + h_{loss}, \tag{20.2}$$

where E denotes the specific energy of the liquid, and is defined as

$$E \equiv D + \frac{V^2}{2g}, \tag{20.3}$$

where D is the depth of the liquid at a station measured from the channel bottom to the free surface (where $p_x = p_y = p_{ambient}$).

For a given flow rate per unit channel width q, where, by continuity,

$$q = DV, \tag{20.4}$$

it follows that the specific energy will reach a minimum value at the special depth

$$D_c{}^3 = \frac{q^2}{g}, \tag{20.5}$$

as indicated by differentiating E with respect to D at constant q. The critical depth D_c will be seen later in this development to divide the flow into two distinct regions. Some useful relations at this critical depth, which follow directly from (20.3) and (20.5) are:

$$V_c = (gD_c)^{1/2} \tag{20.6}$$

and

$$E_c = \tfrac{3}{2}D_c. \tag{20.7}$$

Since a gravity wave is known to propagate in shallow water at a velocity

$$G = (gD)^{1/2},\tag{20.8}$$

another useful quantity, the Froude number (Fr), can be introduced as

$$\text{Fr} \equiv \frac{V}{G} = \frac{V}{(gD)^{1/2}} = \frac{q}{(gD^3)^{1/2}}.\tag{20.9}$$

By combining (20.6) and (20.9) we see that the Froude number equals one at the point of minimum specific energy, further delineating the flow.

These quantities can be pictured as in Figure 20.2, where it is evident that flow in two regimes is possible for a given specific energy, with the dividing criterion being the critical depth. If the actual depth exceeds D_c, the flow is said to be tranquil. Conversely, when the actual depth is less than D_c, the flow is said to be rapid. In general, it is not possible to go from one regime to the other without outside influence. However, the flow may change abruptly with attendant large losses from the rapid to the tranquil regime through the mechanism of a hydraulic jump.

A hydrostatic force balance (per unit channel width) across an abrupt jump in depth indicates that there is a maximum depth attainable in the tranquil regime from a given depth in the rapid regime for a given flow

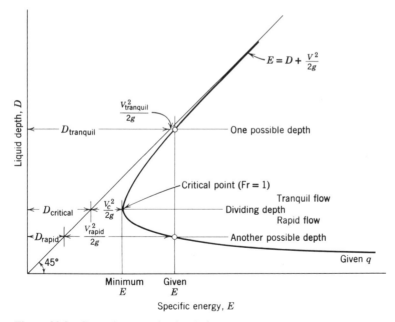

Figure 20.2   General energy-depth relations.

rate. In terms of an initial state (2) and a conjugate state (4), this force balance can be given as

$$F_2 - F_4 = Ma = \left(\frac{wq}{g}\right)(V_4 - V_2), \tag{20.10}$$

which reduces to $\qquad D_2 D_4 (D_2 + D_4) = \dfrac{2q^2}{g}. \tag{20.11}$

This quadratic in D_4 has the real solution

$$D_4 = \frac{D_2}{2}[-1 + (1 + 8\,\mathrm{Fr}_2{}^2)^{1/2}], \tag{20.12}$$

which can be given in the more explicit form as

$$D_4 = \frac{D_2}{2}\left[-1 + \left(\frac{16 E_2}{D_2} - 15\right)^{1/2}\right]. \tag{20.13}$$

The minimum head loss across the jump is given as

$$h_{\substack{\text{loss} \\ \text{jump}}} = E_2 - E_4 = \frac{(D_4 - D_2)^3}{4 D_2 D_4}. \tag{20.14}$$

Some of these quantities are pictured in Figure 20.3.

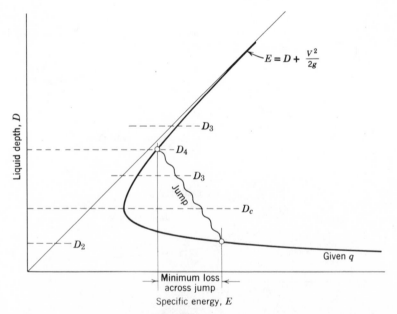

Figure 20.3 Specific energy-depth relations in terms of a hydraulic jump. Notation: D_c, critical depth; D_2, initial depth; D_3, two possible tail water depths; D_4, maximum jump depth.

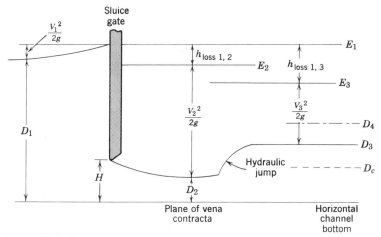

Figure 20.4 Free efflux sluice gate.

If the actual downstream liquid level (tail water depth, D_3) is maintained at less than the conjugate depth D_4, rapid flow at D_2 is initially possible. If the tail water depth is also greater than D_c, there will be an abrupt jump to D_3. Conversely, if D_3 exceeds D_4, initially rapid flow at D_2 is not possible. In terms of a sluice gate, this tail water depth-conjugate depth relationship determines two distinct modes of operation of the gate. If D_3 is less than D_4, the gate will operate with a free efflux (see Figure 20.4); whereas if D_3 exceeds D_4, the gate will operate with a submerged efflux (see Figure 20.5). In the following sections flow under a sluice gate operating in these situations is discussed, as is the flow over a weir.

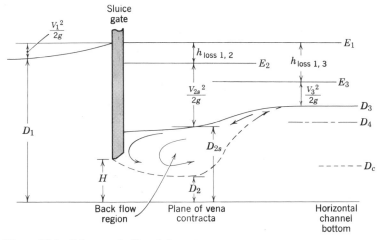

Figure 20.5 Submerged efflux sluice gate.

20.2 Sluice Gate with Free Efflux

An energy relation betwen the various terms pertaining to a vertical sluice gate with free efflux can be given as

$$E_1 = D_1 + \frac{V_1^2}{2g} = D_2 + \frac{V_2^2}{2g} + h_{\text{loss}\,1,2} = D_3 + \frac{V_3^2}{2g} + h_{\text{loss}\,1,3}. \quad (20.15)$$

Continuity, for this same flow condition, is expressed as

$$q = D_1 V_1 = D_2 V_2 = D_3 V_3. \quad (20.16)$$

These quantities are pictured in Figure 20.4.

An ideal flow rate q' must now be defined so that a flow discharge coefficient C can be particularized by the relation

$$C \equiv \frac{q}{q'}. \quad (20.17)$$

For example, on an analytical basis one could define the ideal flow rate that is implied by the measurable head difference $(D_1 - H)$; that is, by continuity,

$$q'_F = D_1 V'_1 = H V'_H, \quad (20.18)$$

whereas, by energy, $\quad D_1 + \dfrac{(V'_1)^2}{2g} = H + \dfrac{(V'_H)^2}{2g}, \quad (20.19)$

where the primes signify ideal quantities, and the subscript F stands for *free* efflux. These equations [(20.18) and (20.19)] lead at once to the ideal flow rate

$$q'_F = H\left[\frac{2gD_1}{1+(H/D_1)}\right]^{1/2} = H\left[\frac{2g(D_1-H)}{1-(H/D_1)^2}\right]^{1/2}. \quad (20.20)$$

Now, according to (20.17), (20.20) and an experimentally determined flow rate q, a particular discharge coefficient C_F is defined for the free efflux sluice gate. H. R. Henry [4] provides experimental work on discharge coefficients for sluice gates, and hence provides a basis by which the characteristics of the discharge coefficient defined by (20.17) and (20.20) can be examined. Since Henry uses for an ideal flow rate the arbitrary

$$q'_{FH} = H[2gD_1]^{1/2}, \quad (20.21)$$

where the subscript FH stands for *free Henry*, the relation between C_F and C_{FH} is simply

$$q_F = C_F q'_F = C_{FH} q'_{FH}. \quad (20.22)$$

Thus $\quad C_F = C_{FH}\left(1 + \dfrac{H}{D_1}\right)^{1/2}. \quad (20.23)$

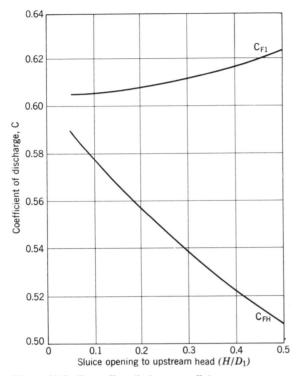

Figure 20.6   Free efflux discharge coefficients.

In Figure 20.6, both discharge coefficients are shown. Both are seen to have acceptable characteristics being easily formed from measurable depths and being only weak functions of the flow. The recommended discharge coefficient for the free efflux sluice gate can be represented quite closely by the parabola

$$C_F = 0.604 + 0.02\left(\frac{H}{D_1}\right) + 0.045\left(\frac{H}{D_1}\right)^2. \tag{20.24}$$

Example 1.   Free Efflux Under a Sluice Gate.   *Problem.* Find the flow rate of water for a free efflux under a sluice gate as in Figure 20.4, where $H = 2$ in., $D_1 = 20$ in., and L the width of the sluice $= 2$ ft. *Solution.* By (20.17), ʻ20.20), and (20.24):

$$C_F = 0.604 + 0.02\left(\frac{2}{20}\right) + 0.045\left(\frac{2}{20}\right)^2 = 0.6064,$$

$$Q'_F = Lq'_F = 2 \times \frac{2}{12}\left[\frac{2 \times 32.2 \times (20/12)}{1 + 2/20}\right]^{1/2} = 3.292 \text{ ft}^3/\text{sec}$$

$$\overset{\circ}{w} = w_{\text{water}}(C_F Q'_F) = 62.4 \times 0.6064 \times 3.292 = 124.2 \text{ lb/sec}.$$

Note that this agrees precisely with the flow rate predicted by Henry's discharge coefficient.

20.3   Sluice Gate with Submerged Efflux

An energy relation between the various terms pertaining to a vertical sluice gate with submerged efflux is

$$E_1 = D_1 + \frac{V_1^2}{2g} = D_{2s} + \frac{V_2^2}{2g} + h_{\text{loss }1,2} = D_3 + \frac{V_3^2}{2g} + h_{\text{loss }1,3}. \quad (20.25)$$

Continuity for this situation is

$$q = D_1 V_1 = D_2 V_2 = D_3 V_3. \quad (20.26)$$

These quantities are pictured in Figure 20.5.

The depth D_2 is used to determine flow rate in the submerged efflux case (rather than D_{2s}) because it represents the only area (per unit channel width) available for through flow. The depth difference $D_{2s} - D_2$ can be looked upon as an additional pressure head at the vena contracta plane, to be included in the energy accounting of (20.25) but not to be considered in the continuity of (20.26). Since $D_3 > D_2$, it follows from (20.26) that $V_2 > V_3$, and hence from (20.25) that $D_{2s} < D_3$, as indicated in Figure 20.5. However, in actual fact D_{2s} may be very close in magnitude to D_3.

Once again the definition of an ideal flow rate, this time for the submerged efflux case, must be agreed on before a flow discharge coefficient can be specified by (20.17). One ideal flow rate that appears in the literature has been defined as

$$q_s' = H[2g(D_1 - D_3)]^{1/2}. \quad (20.27)$$

Bakhmeteff [3] indicates that (20.27) is to be used with the simplified discharge coefficient $C_{sB} = 0.6$ (where the subscript B stands for Bakhmeteff). Rouse [1] notes that it is common practice to use the ideal flow rate of (20.27) with the analytical discharge coefficient

$$C_{sR} = \frac{(C_v C_c)_F [1 - C_{cF}(H/D_1)]^{1/2}}{[1 - (C_v C_c)_F^2 (H/D_1)^2]^{1/2}} \quad (20.28)$$

(where the subscript R stands for Rouse), although Rouse expresses misgivings about the lack of information on the velocity coefficient C_v and the contraction coefficient C_c in the submerged efflux case. In any case, Henry [4], [5], [6] again provides the only recent experimental work in discharge coefficients for sluice gates, and hence provides a basis by which the characteristics of the discharge coefficients for submerged efflux sluice gates can be examined. Henry again employs (20.21) to define his ideal flow rate; thus,

$$q'_{sH} = q'_{FH}, \tag{20.29}$$

and on this basis

$$q_s = C_s q'_s = C_{sH} q'_{sH}, \tag{20.30}$$

so that

$$C_s = \frac{C_{sH}}{(1 - D_3/D_1)^{1/2}}. \tag{20.31}$$

All of these discharge coefficients are compared with Henry's C_{sH} in Figure 20.7, where it can be noted that:

1. Some of the submerged efflux coefficients are more complex than those of the free efflux cases, since they are functions of H/D_3 as well as of H/D_1.

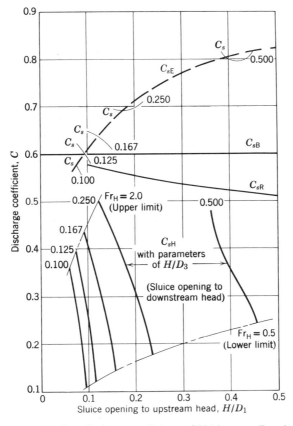

Figure 20.7 Submerged efflux discharge coefficients. (Within same Froude number limits used by Henry.)

2. C_{sB} (of Bakhmeteff) and C_{sR} (of Rouse) do not coincide with C_s of (20.31), although they should since all of these coefficients are to be used with q'_s of (20.27). This indicates either that the empirical C_{sB} is incorrect, that $(C_v C_c)_s \neq (C_v C_c)_F$, as assumed in C_{sR}, or that Henry's work is suspect. (In the absence of newer experimental work, Henry's data is taken here as correct.)

3. C_s of (20.31) varies much less than Henry's C_{sH} for a given H/D_3 between the same Froude number limits.

The locus of any constant Froude number, based on the sluice gate opening, can be defined for Henry's data as

$$C_{sH} = \mathrm{Fr}_H \left[\left(\frac{H}{D_1} \right) \bigg/ 2 \right]^{1/2}. \tag{20.32}$$

Item 3 above suggests that C_s might serve as the basis of an empirical coefficient of discharge, C_{sE}, which will be independent of H/D_3 and yet will yield acceptable flow rates. A simple parabola that closely approximates the C_s plot in Figure 20.7 is

$$C_{sE} = 0.49 + 1.376 \left(\frac{H}{D_1} \right) - 1.43 \left(\frac{H}{D_1} \right)^2. \tag{20.33}$$

This is recommended to define the submerged discharge coefficient when used with (20.27).

Example 2. Submerged Efflux Under a Sluice Gate. *Problem.* Find the flow of water for a submerged efflux under a sluice gate as in Figure 20.5, where $H = 2$ in., $D_1 = 20$ in., $D_3 = 16$ in., and L the width of the sluice $= 2$ ft. *Solution.* By (20.17), (20.27), and (20.33):

$$C_{sE} = 0.49 + 1.376 \left(\frac{2}{20} \right) - 1.43 \left(\frac{2}{20} \right)^2 = 0.6133,$$

$$Q'_s = L q'_s = 2 \times \frac{2}{12} \left[\frac{2 \times 32.2(20 - 16)}{12} \right]^{1/2} = 1.544 \text{ ft}^3/\text{sec},$$

$$\mathring{w} = w_{\text{water}} (C_{sE} Q'_s) = 62.4 \times 0.6133 \times 1.544 = 59.1 \text{ lb/sec}.$$

Note that this is within 3% of the flow rate predicted by Henry's more complex discharge coefficient.

20.4 Weirs

In determining the ideal flow rate over a rectangular weir, the approach velocities are considered to be negligible and the sheet of liquid flowing over the weir (the nappe) is assumed to be surrounded by atmospheric pressure. Hence the nappe is treated as a free falling body. Thus, according to Figure 20.1,

$$V_h = (2gh)^{1/2}, \tag{20.34}$$

and the ideal volumetric flow rate as given by (19.3) and (19.4) is

$$dQ_{ideal} = V_h \, dA. \tag{20.35}$$

Substituting (20.34) and using the dimensions shown in Figure 20.8a, we have

$$dQ_{ideal} = (2gh)^{1/2}(L \, dh). \tag{20.36}$$

After integration between zero and H,

$$Q_{ideal} = \tfrac{2}{3}L(2g)^{1/2}H^{3/2}, \tag{20.37}$$

which shows that the volumetric flow rate should be proportional to the $\tfrac{3}{2}$ power of the head for rectangular weirs. Similarly, the ideal flow rate for a triangular weir (see Figure 20.8b) can be shown to be

$$Q_{ideal} = \frac{4L}{15H}(2g)^{1/2}H^{5/2}. \tag{20.38}$$

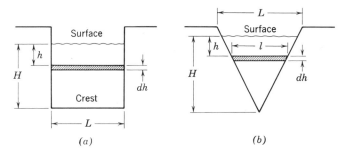

Figure 20.8 Standard weir shapes: (a) rectangular; (b) triangular.

Weight flow rates are related to volumetric flow rates by (19.7), so that

$$\mathring{w} = wQ. \tag{20.39}$$

Empirical equations are always used for actual flow rates; but they are based on the ideal equations given above. For example, the fluid velocity of approach has an effect on flow rates. This could be introduced in the ideal equations, but it has been omitted here for the sake of brevity. The actual installation particulars and the shape of the crest of the weir also are important. One of the most widely accepted empirical equations for rectangular weirs is the Francis formula:

$$Q = 3.33\left(L - \frac{nH}{10}\right)\left[\left(H - \frac{V_0^2}{2g}\right)^{3/2} - \left(\frac{V_0^2}{2g}\right)^{3/2}\right], \qquad (20.40)$$

where n is the number of lateral contractions of the weir and V_0 is the approach velocity. If the weir does not extend over the full width of the approach channel, lateral contractions occur. On the other hand, if all lateral contractions are eliminated, the weir is said to be suppressed ($n = 0$). In this case, if the velocity of approach is negligible, (20.40) becomes

$$Q = 3.33 L H^{3/2}, \qquad (20.41)$$

which agrees well with (20.37), since the actual flow rate must be less than the ideal one because of losses.

Example 3. Water Flow over Rectangular Weir. *Problem.* Find the flow rate of water in lb/sec for a fully contracted weir as in Figure 20.8a, where $n = 2$ for $L = 4$ ft, $H = 3$ ft, and the approach area is 56 ft². *Solution.* For a first try, assume the velocity of approach to be negligible. From (20.40),

$$Q = 3.33\left(4 - \frac{2 \times 3}{10}\right)3^{3/2} = 58.83 \text{ ft}^3/\text{sec}.$$

At this approximate flow rate and with an approach area of 56 ft², the approach velocity is, by (20.35),

$$V_0 = \frac{58.83}{56} = 1.05 \text{ ft/sec},$$

and the approach velocity head is

$$\frac{V_0^2}{2g} = \frac{1.05^2}{64.34} = 0.024 \text{ ft},$$

which certainly can be considered to be negligible compared to the 3-ft head. Hence $Q = 58.83$ ft³/sec. For the flow rate of water in lb/sec by (20.39), $\dot{w} = 62.4 \times 58.83 = 3671$ lb/sec.

20.5   References

[1] H. Rouse, *Fluid Mechanics for Hydraulic Engineers*, Dover, New York, 1961, p. 315.

[2] R. C. Binder, *Fluid Mechanics*, 4th ed., Prentice-Hall, Englewood Cliffs, N.J., 1960.

[3] B. A. Bakhmeteff, *Hydraulics of Open Channels*, McGraw-Hill, New York, 1932, p. 270.

[4] Harold R. Henry, "A Study of Flow From a Submerged Sluice Gate," M.S. thesis State University of Iowa, February 1950.

[5] M. L. Albertson *et al.*, "Diffusion of Submerged Jets," *Trans. ASCE*, **115**, 1950, p. 687. Discussion by H. R. Henry.

[6] V. T. Chow, *Open Channel Hydraulics*, McGraw-Hill, New York, 1959, p. 509.

Nomenclature

Roman

a	acceleration
C	flow discharge coefficient
D	depth
E	specific energy
F	force
Fr	Froude number
g	acceleration of gravity
G	velocity of a gravity wave
h	head
H	vertical opening of sluice gate
M	mass
p	absolute pressure
q	flow rate per unit channel width
V	directed velocity
$\overset{\circ}{w}$	weight rate of flow
w	specific weight

Subscripts

c	critical
s	submerged
x, y	stations in the flow
$1, 2, 3, 4$	stations in the flow

Problems

1. For the flow of water in a rectangular open channel of 3 ft width, 2 in. depth, and a volumetric flow rate of 4 ft³/sec, find q, V, E, D_c, V_c, Fr.

    Ans. $q = 1.33$ ft³/sec, $V = 8$ ft/sec, $E = 1.1613$ ft, $D_c = 4.57$ in, $V_c = 3.5$ ft/sec, Fr $= 3.455$.

2. What is the conjugate depth that goes with the flow of Problem 1?

    Ans. $D_4 = 8.822$ inches.

3. What is the minimum head loss across the hydraulic jump of Problem 2?

    Ans. $h_{\text{loss}} = 0.375$ ft.

4. If a sluice gate is set at $H = 1$ in., and $D_1 = 4$ in., find the free efflux flow rate if $L = 2$ ft (use $w_{H_2O} = 62.4$ lbf/ft^3).

    Ans. $\mathring{w} = 26.358$ lb/sec.

5. Find the flow rate under a submerged sluice gate if $H = 1$ in., $D_1 = 20$ in., $D_3 = 12$ in., and $L = 2$ ft.

    Ans. $\mathring{w} = 38.31$ lb/sec.

Chapter 21

THEORETICAL RATES IN CLOSED-CHANNEL FLOW

"... many a time did I stand beside such a pipe and exert myself to invent how to force these pipes to reveal the secret of their hidden action..."

Clemens Herschel (1898)

21.1 General Remarks

In closed-channel flow (as in pipes, ducts, etc.), the system is full of fluid and, consequently, the fluid is completely bounded. For one-dimensional steady flow, continuity can be expressed as

$$\overset{\circ}{m} = \rho_1 A_1 V_1 = \rho_2 A_2 V_2, \tag{21.1}$$

whence

$$V_1 = \left(\frac{\rho_2}{\rho_1}\right)\left(\frac{A_2}{A_1}\right) V_2. \tag{21.2}$$

Under the same conditions, a general energy accounting for a reversible thermodynamic process (i.e., when mechanical friction and fluid turbulence are negligible) yields, in the absence of mechanical work and elevation change,

$$0 = \frac{dp}{\rho} + \frac{V\,dV}{g_c}. \tag{21.3}$$

Geometric considerations indicate the usefulness of the definitions

$$\beta = \frac{D_2}{D_1} \tag{21.4}$$

and
$$\beta^2 = \left(\frac{A_2}{A_1}\right).$$ (21.5)

It is necessary also to consider the very important differences that exist between the highly compressible gases and the relatively incompressible liquids [1], [2]. They must be considered separately when evaluating densities, velocities, and flow rates. It is conventional in a study of flow rates to examine theoretical relations first. In the interest of simplicity, we also idealize the fluids so that a liquid is taken to exhibit constant density, whereas it is assumed that the equation of state of a gas is given by

$$p = \rho RT.$$ (21.6)

21.2 Constant-Density Fluids

Equation 21.3, when integrated between two arbitrary positions for the constant-density case, yields:

$$\frac{V_2^2 - V_1^2}{2g_c} = \frac{p_1}{\rho} - \frac{p_2}{\rho},$$ (21.7)

which is a form of the Bernoulli equation. Combining (21.2), (21.5), and (21.7) results in:

$$V_2 = \left[\frac{2g_c(p_1 - p_2)}{\rho(1 - \beta^4)}\right]^{1/2}.$$ (21.8)

Thus the theoretical rate of flow of a constant-density fluid in a closed channel is

$$\mathring{m}_{\text{ideal}} = A_2 \left[\frac{2g_c\rho(p_1 - p_2)}{144(1 - \beta^4)}\right]^{1/2}.$$ (21.9)

where $\mathring{m}_{\text{ideal}}$ is in lbm/sec,
  A_2 is in inches^2,
  g_c is in lbm ft/lbf sec^2,
  ρ is in lbm/ft^3,
and Δp is in lbf/in^2.

The flow rate is seen to be directly proportional to the square root of the pressure drop $(p_1 - p_2)$ which, across a constant area section, may be very small. To obtain a measurable pressure drop, the flow is usually *obstructed* in a manner similar to the way in which open-channel flow is obstructed. The obstruction and the required static pressure taps make up the closed-channel fluid meter. The venturi, the nozzle, and the square-edged orifice plate (and their associated pressure taps) are the most common closed-channel fluid meters, although porous plugs or simple restrictions in the walls of a flow tube can suffice to establish a suitable pressure drop (see Figure 21.1).

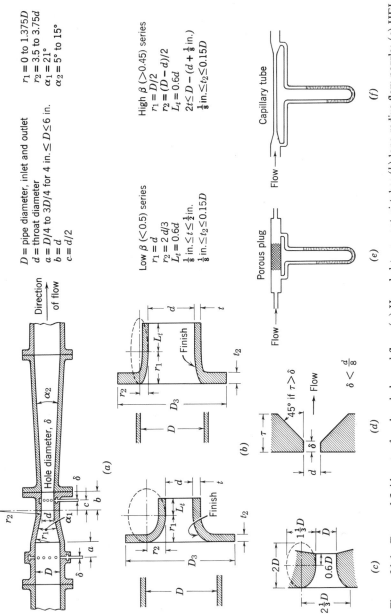

Figure 21.1 Types of fluid meter for closed-channel flow: (*a*) Herschel-type venturi tube; (*b*) long-radius flow nozzle; (*c*) HEI flow nozzle; (*d*) squared-edged orifice: (*e*) porous plug flow meter: (*f*) restrictive-type flow meter. (From ASME Fluid Meters Report, 1971.)

415

21.3 Compressible Fluids

When the thermodynamic process between two arbitrary positions in a system is isentropic (i.e., when there is no heat transfer, no mechanical friction, no fluid turbulence, and no unrestrained expansion), the ideal gas of equation (21.6) also can be characterized by

$$\rho_2 = \rho_1 r^{1/\gamma}, \tag{21.10}$$

where r is the static pressure ratio (p_2/p_1) and γ is the ratio of specific heats, c_p/c_v. General energy (21.3) under these conditions can be integrated to yield

$$\frac{V_2^2 - V_1^2}{2g_c} = \left(\frac{\gamma}{\gamma-1}\right)\left(\frac{p_1}{\rho_1} - \frac{p_2}{\rho_2}\right), \tag{21.11}$$

which, with (21.2) can be expressed as

$$V_2 = \left[\frac{2\gamma g_c p_1 (1 - r^{(\gamma-1)/\gamma})}{(\gamma-1)\rho_1(1 - r^{2/\gamma}\beta^4)}\right]^{1/2}. \tag{21.12}$$

Thus the theoretical rate of flow of a compressible fluid in a closed channel, according to (21.1), (21.10), and (21.12), is [3]

$$\mathring{m}_{\text{ideal}} = A_2 \left[\frac{2\gamma g_c p_1 \rho_1 (r^{2/\gamma} - r^{(\gamma+1)/\gamma})}{(\gamma-1)(1 - r^{2/\gamma}\beta^4)144}\right]^{1/2}, \tag{21.13}$$

for same units as in (21.9). Equation 21.13 also can be given in the useful form

$$\mathring{m}_{\text{ideal}} = \frac{A_2 p_1}{\sqrt{\dfrac{RT_t}{g_c}}} \left[\left(\frac{2\gamma}{\gamma-1}\right)\frac{(r^{2/\gamma} - r^{(\gamma+1)/\gamma})(1 - r^{(\gamma+1)/\gamma}\beta^4)}{(1 - r^{2/\gamma}\beta^4)^2}\right]^{1/2} \tag{21.14}$$

with the same units as in (21.9) except

$$R \text{ is in lbf ft/lbm°R,}$$

and T_t is in °R.

Alternatively, if general energy (21.3) is integrated between static and total states at a single position, we have the general relation

$$\frac{V^2}{2g_c} = \left(\frac{\gamma}{\gamma-1}\right)\left(\frac{p_t}{\rho_t} - \frac{p}{\rho}\right), \tag{21.15}$$

which can be expressed as

$$V = \left[\frac{2\gamma g_c p_t}{(\gamma-1)\rho_t}\left(1 - \left(\frac{p}{p_t}\right)^{(\gamma-1)/\gamma}\right)\right]^{1/2}. \tag{21.16}$$

Table 21.1 Compressible Flow Function for Various Pressure Ratios and Isentropic Exponents

Pressure Ratio, p/p_t	Generalized Compressible Flow Function, Γ		
	$\gamma = 1.4$	$\gamma = 1.3$	$\gamma = 1.115$
0.99	0.20543	0.21072	0.22244
0.98	0.28894	0.29626	0.31244
0.97	0.35193	0.36069	0.38002
0.96	0.40412	0.41401	0.43575
0.95	0.44929	0.46008	0.48376
0.94	0.48938	0.50091	0.52616
0.93	0.52555	0.53770	0.56423
0.92	0.55858	0.57123	0.59879
0.91	0.58898	0.60205	0.63044
0.90	0.61715	0.63056	0.65959
0.89	0.64338	0.65705	0.68657
0.88	0.66790	0.68177	0.71163
0.87	0.69088	0.70491	0.73497
0.86	0.71249	0.72660	0.75675
0.85	0.73284	0.74700	0.77711
0.84	0.75203	0.76618	0.79617
0.83	0.77016	0.78426	0.81402
0.82	0.78729	0.80130	0.83074
0.81	0.80348	0.81737	0.84640
0.80	0.81880	0.83253	0.86106
0.79	0.83330	0.84682	0.87479
0.78	0.84701	0.86030	0.88762
0.77	0.85997	0.87299	0.89961
0.76	0.87222	0.88495	0.91078
0.75	0.88378	0.89619	0.92117
0.74	0.89469	0.90675	0.93082
0.73	0.90497	0.91664	0.93975
0.72	0.91464	0.92590	0.94798
0.71	0.92371	0.93455	0.95554
0.70	0.93222	0.94259	0.96246
0.69	0.94016	0.95006	0.96873
0.68	0.94756	0.95695	0.97439
0.67	0.95444	0.96330	0.97946
0.66	0.96079	0.96911	0.98393
0.65	0.96664	0.97439	0.98783
0.64	0.97199	0.97915	0.99118
0.63	0.97685	0.98340	0.99397
0.62	0.98123	0.98715	0.99622
0.61	0.98514	0.99041	0.99794
0.60	0.98859	0.99313	0.99914
0.59	0.99157	0.99548	0.99982
0.58	0.99409	0.99730	
0.57	0.99616	0.99865	
0.56	0.99778	0.99953	
0.55	0.99896	0.99996	
0.54	0.99970		
0.53	0.99999		

(*continued overleaf*)

Table 21.1 **(contd)**

		Critical Values		
0.58155				1.00000
0.54573			1.00000	
0.52828		1.00000		

Thus the theoretical rate of flow of a compressible fluid in a closed channel is, according to (21.1), (21.10), and (21.16),

$$\mathring{m}_{\text{ideal}} = \frac{A_2 p_t}{T_t^{1/2}} \left[\left(\frac{p_2}{p_t} \right)^{2/\gamma} \left(1 - \left(\frac{p_2}{p_t} \right)^{(\gamma-1)/\gamma} \right) \right]^{1/2} \left[\left(\frac{g_c}{R} \right) \left(\frac{2\gamma}{\gamma-1} \right) \right]^{1/2}.$$

(21.17)

In terms of the generalized compressible flow function Γ, which has been defined [4] as

$$\Gamma = \left\{ \frac{\left(\frac{p}{p_t} \right)^{2/\gamma} \left[1 - \left(\frac{p}{p_t} \right)^{(\gamma-1)/\gamma} \right]^{1/2}}{\left(\frac{2}{\gamma+1} \right)^{2/(\gamma-1)} \left(\frac{\gamma-1}{\gamma+1} \right)} \right\},$$

(21.18)

(21.17) also can be given in the simplified form

$$\mathring{m}_{\text{ideal}} = \left(\frac{A_2 p_t}{T_t^{1/2}} \right) \Gamma_2 K_\Gamma,$$

(21.19)

The constant in (21.19) is simply

$$K_\Gamma = \left[\left(\frac{g_c}{R} \right) \left(\frac{2\gamma}{\gamma-1} \right) \left(\frac{2}{\gamma+1} \right)^{2/(\gamma-1)} \left(\frac{\gamma-1}{\gamma+1} \right) \right]^{1/2},$$

(21.20)

which takes values at standard gravity conditions of

$$\left. \begin{array}{l} K_{\Gamma\text{air}} = 0.531748 \\ K_{\Gamma\text{steam}} = 0.408650 \end{array} \right\}, \qquad \frac{\text{lbm}^\circ R^{1/2}}{\text{lbf sec}}. \qquad \begin{array}{l} (21.21) \\ (21.22) \end{array}$$

For brief tabulations of the Γ function see Table 21.1. For more complete tabulations see [4]. Note that p_t in (21.17) and (21.19) is the isentropic total pressure in the fluid meter, defined in general as [5], [6]

$$p_t = p_1 \left[\frac{1 - r^{(\gamma+1)/\gamma} \beta^4 C_c^2}{1 - r^{2/\gamma} \beta^4 C_c^2} \right]^{\gamma/(\gamma-1)}.$$

(21.23)

In the ideal case, C_c is usually set equal to unity.

21.4   Critical Flow Relations

The flow rate of a compressible fluid was seen (21.13) to be dependent in general on the ratio of the downstream static pressure p_2 to the upstream static pressure p_1. The variation in flow rate with changes in the static pressure ratio is important in studying the critical flow of gases through nozzles. First, the square of the isentropic flow rate (21.13) is differentiated with respect to r to obtain:

$$\frac{d(\mathring{m}^2_{ideal})}{dr} = \left[A_2{}^2 \left(\frac{2\gamma}{\gamma - 1} \right) p_1 \rho_1 g_c \right]$$

$$\times \left[\frac{\left(\dfrac{2}{\gamma} \beta^4 r^{(2-\gamma)/\gamma} \right) (r^{2/\gamma} - r^{(\gamma+1)/\gamma})}{(1 - r^{2/\gamma} \beta^4)^2} \right.$$

$$\left. + \frac{\left(\dfrac{2}{\gamma} r^{(2-\gamma)/\gamma} - \dfrac{\gamma+1}{\gamma} r^{1/\gamma} \right)}{(1 - r^{2/\gamma} \beta^4)} \right]. \tag{21.24}$$

The critical static-pressure ratio (the one that yields the maximum isentropic flow rate for given fluid conditions at inlet and for a given geometry) is obtained by setting (21.24) equal to zero.

The result is $\qquad r_*^{(1-\gamma)/\gamma} + \left(\dfrac{\gamma-1}{2} \right) \beta^4 r_*^{2/\gamma} = \dfrac{\gamma+1}{2},$ $\qquad$ (21.25)

where the asterisk signifies the condition of maximum flow rate. Note that if the geometry is such that $\beta \to 0$ then $p_1 \to p_t$ and (21.25) leads to the familiar critical point function of thermodynamics

$$\left(\frac{p_2}{p_t} \right)_* = \left(\frac{2}{\gamma + 1} \right)^{\gamma/(\gamma-1)}. \tag{21.26}$$

Thus theory reveals and experiment agrees that the flow rate of a convergent nozzle (where $C_c = 1$) attains constancy and is maximized at the critical pressure ratio (21.25). At this critical pressure ratio, the fluid velocity equals the local velocity of sound, and the flow no longer responds to changes in the down-stream pressure.

Although in the case of a flow nozzle the *throat* static pressure is called for in (21.13) to (21.23), it is customary (and usually preferred, see, for example, [7]) to measure the lower pressure in the larger-diameter discharge pipe. This is usually called the back pressure p_b. If the flow is subsonic, p_2 can be taken as the back pressure. On the other hand, if the nozzle is choked (i.e., if for a given inlet pressure the flow is maximum and also independent of the back pressure), the throat static pressure must be greater than the back pressure. In fact, whenever the measured static pressure ratio (p_b/p_1) is less than or equal to r_* of (21.25), the nozzle is choked and the pressure ratio to use in (21.13) to (21.23) is r_* of (21.25). On the other hand, if p_b/p_1 exceeds r_*, subsonic flow exists.

Venturis also are operated as critical flow meters with certain advantages noted in the literature [8]. To verify that critical flow conditions exist in the venturi, it is only necessary to show that throat conditions are independent of the overall pressure ratio across the venturi.

Contrary to the behavior of the convergent and convergent-divergent passages of nozzles and venturis, the square-edged orifice meter does not exhibit a maximum flow rate. For example, Perry [9] and Cunningham [10] both indicate that the flow rate (for constant upstream conditions) continues to increase at all pressure ratios between the critical ratio of (21.25) and zero. This range is thus defined as the "supercritical" range of pressure ratios.

The study of critical flowmeters for compressible flow measurements is a complex and rapidly changing subject for which a rapidly growing literature is developing [5], [8]–[12].

21.5   References

[1] R. P. Benedict, "Fluid Flow Measurement," *Electro-Technol.*, January 1966, p. 55.

[2] R. P. Benedict, "Some Comparisons Between Compressible and Incompressible Treatments of Compressible Fluids," *Trans. ASME, J. Basic Eng.*, September 1964, p. 527.

[3] *Fluid Meters—Their Theory and Application*, Report of ASME Research Committee on Fluid Meters, 6th ed., 1971.

[4] R. P. Benedict and W. G. Steltz, *Handbook of Generalized Gas Dynamics*, Plenum Press, New York, 1966.

[5] R. P. Benedict, "Generalized Contraction Coefficient of an Orifice for Subsonic and Supercritical Flows," *Trans, ASME, J. Basic Eng.*, June 1971, p. 99.

[6] R. P. Benedict and R. D. Schulte, "A Note on the Critical Pressure Ratio Across a Fluid Meter," *Trans. ASME, J. Fluids Eng.*, September 1973, p. 337.

[7] J. S. Wyler and R. P. Benedict, "Comparisons Between Throat and Pipe Wall Tap Nozzles," *Trans. ASME, J. Eng. Power*, October 1975, p. 569.

[8] R. E. Smith, Jr., and R. J. Matz, "A Theoretical Method of Determining Discharge Coefficients for Venturis Operating at Critical Flow Conditions," *Trans. ASME, J. Basic Eng.*, December 1962, p. 434.

[9] J. A. Perry, "Critical Flow Through Sharp-Edged Orifices," *Trans. ASME*, October 1949, p. 757.

[10] R. G. Cunningham, "Orifice Meters with Supercritical Compressible Flow," *Trans. ASME*, July 1951, p. 625.

[11] A. Weir, Jr., J. L. York, and R. B. Morrison, "Two- and Three-Dimensional Flow of Air Through Square-Edged Sonic Orifices," *Trans. ASME*, April 1956, p. 481.
[12] B. T. Arnberg, "Review of Critical Flowmeters for Gas Flow Measurements," *Trans. ASME, J. Basic Eng.*, December 1962, p. 447.

Nomenclature

Roman

A area
c_p specific heat capacity at constant pressure
c_v specific heat capacity at constant volume
d exact differential
D diameter
g acceleration of gravity
K a constant
$\overset{\circ}{m}$ mass flow rate
p absolute pressure
r static pressure ratio
R specific gas constant
t time
T absolute temperature
V directed velocity

Greek

β diameter ratio
γ ratio of specific heats
Γ generalized compressible flow function
Δ finite difference
ρ density

Subscripts

t total
1, 2   stations in the flow
$*$ signifies conditions at the maximum flow rate

Problems

1. Find the ideal constant density flow rate if an orifice of 2 in. bore, installed in a 4 in. pipe, shows a pressure drop of 5 psi when the density of the fluid is 60 lbm/ft^3.

   Ans. $\overset{\circ}{m}_{ideal} = 37.57$ lbm/sec.

2. Find the ideal compressible flow rate of an orifice of 2 in. bore, installed in a 4 in. pipe, if the static pressure ratio across the fluid meter is 0.9, at an inlet pressure of 20 psia and an inlet air density of 0.076 lbm/ft^3. Use $\gamma = 1.4$.

   Ans. $\mathring{m}_{ideal} = 0.795$ lbm/sec.

3. For problem 2, find the total temperature of the air if $R = 53.35$ ft lbf/°R lbm. Hint: use (21.14).

   Ans. $T_t = 711.5$°R.

4. Steam, of $\gamma = 1.3$, flows through a 4 in. throat nozzle at a pressure ratio $p_2/p_t = 0.9$, a total temperature of 540°R, and at a total pressure of 50 psia. Find the ideal flow rate by means of the Γ function.

   Ans. $\mathring{m}_{ideal} = 6.967$ lbm/sec.

5. Find the critical static pressure ratio for air ($\gamma = 1.4$) for a nozzle of $\beta = 0.4$.

   Ans. $r_* = 0.5314$.

Chapter 22

THE DISCHARGE COEFFICIENT

". . . if modern formulae are empirical with scarcely an exception, and are not homogeneous, or even dimensional, then it is obvious that the truth of any such equation must altogether depend on that of the observations themselves, and it cannot in strictness be applied to a single case outside them . . ."

Robert Manning (1890)

22.1 General Remarks

We can now establish theoretical flow rates for compressible and constant density fluids, but we cannot yet determine the actual rates, since no real flow process is entirely reversible. The intentional addition of dams in the flow path only aggravates the fluid turbulence level. Then, in fluid meters with abrupt area changes (such as the orifice meter), the channel is no longer entirely full. This means that the geometric area used in the flow rate equations of Chapter 21 may not be the same as the actual flow area (this is the so-called vena contracta effect in which the fluid jet attains its minimum area). Finally, the static pressure downstream from a fluid meter can vary widely. Hence the flow rate, which is a function of the pressure drop, is strongly dependent on the location of the downstream static taps (see Figure 22.1).

For example, consider the following tap arrangements that are used in orifice metering [1]: *flange taps* are pressure taps located 1 in. from the upstream and downstream faces of the orifice plate; *vena contracta taps* are located a distance of 1 pipe diameter upstream from the orifice plate and at the plane of minimum pressure, which is assumed to be at the vena contracta; *pipe taps* are located $2\frac{1}{2}$ pipe diameters upstream and 8 pipe

424

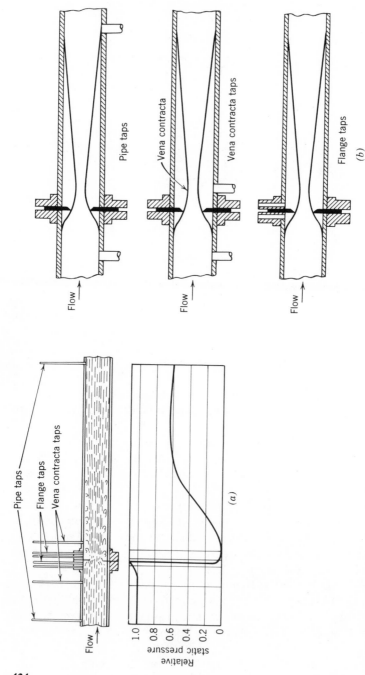

Figure 22.1 (*a*) Fluid flow through an orifice showing tap locations and relative static-pressure distribution. (From ASME Fluid Meters Report, 6th ed., 1971) (*b*) Conventional orifice installations. (From *Thermodynamics of Fluid Flow*, N. A. Hall.) Prentice-Hall, 1951.

diameters downstream from the orifice plate; *one and one-half D taps* are located 1 pipe diameter upstream and $\frac{1}{2}$ pipe diameter downstream of the orifice plate (both measured from the upstream face of the orifice plate). Note that it is the inside diameter and not the nominal or outside diameter called for in the above.

Or again consider the location of pressure taps in a nozzle installation. R. G. Folsom [2] indicates that the nozzle flow characteristics are almost independent of the clearance in the annulus between the pipe wall and the nozzle external diameter at all pressure tap locations between the holding flange and the end of the nozzle. But at pressure taps downstream from the end, appreciable changes in flow characteristics with clearance exist with larger differences corresponding to smaller diameter ratios. It is evident that we need more information to establish an actual flow rate on a firm basis.

Whenever an experiment is possible, some of the desired information can be obtained by a calibration process based on the "catching-weighing-timing" scheme. Under controlled laboratory conditions, a series of flow rates over the range of interest is obtained. At the same time, the static pressure readings across the flow meter are recorded so that the theoretical flow rate can be established. Construction and location of the static taps are specified, so that they may be reproduced in the application. A discharge coefficient for the closed-channel flow can be defined as

$$C_D \equiv \frac{\mathring{m}_{\text{actual}}}{\mathring{m}_{\text{ideal}}}. \tag{22.1}$$

The discharge coefficient can be presented in terms of the ideal flow rate per unit area as shown in Figure 22.2. Then, given the static pressure drop across the meter, C_D can be obtained directly for a given installation handling a given fluid, since the ideal flow rate is a function only of p_2/p_1. But note that this is far from a general presentation of the data.

A second, and far more conventional, method presents C_D in terms of the Reynolds number (Re), defined as

$$\text{Re} = \frac{\rho V D}{\mu} = \frac{48 \mathring{m}_{\text{actual}}}{\pi D \mu}, \tag{22.2}$$

where in the latter part of (22.2) $\mathring{m}$ is in lbm/sec, D in inches, and μ is the dynamic viscosity of the fluid in lbm/ft-sec. Use of the Reynolds number broadens the application of C_D and $\mathring{m}$ to include both constant density and compressible fluids, while β parameters can account for installation differences. If the pipe is not circular, an equivalent diameter is often used in place of D, as indicated in Figure 22.3. Since the discharge

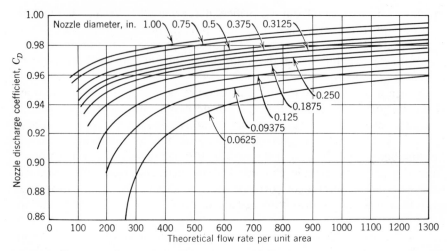

Figure 22.2 Nozzle discharge coefficients for theoretical flow rates of air at 70°F (from Heat Exchange Institute). For temperatures between 0 and 140°F curves give nozzle coefficient with maximum error of ± 0.005. Nozzle coefficients less than 0.93 are extrapolated data.

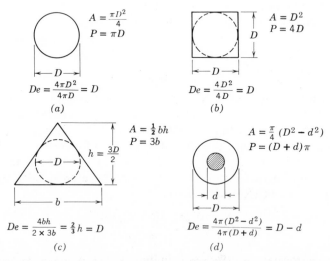

Figure 22.3 Equivalent diameter is an arbitrary definition of a value calculated so that the ratio of pressure forces acting over the flow area to the frictional forces acting along the wetted perimeter of the duct is the same for circular and noncircular pipes: $De = 4A/P$.

coefficient is a function of the flow rate to be determined, an iterative (trial and error) procedure must be used. Fortunately, convergence is generally rapid, and it is usually possible to obtain the correct C_D and the correct flow rate with one or two guesses. (This is demonstrated in examples given later in this chapter.)

Aside from the purely experimental approach, discharge coefficients can be approximated by several methods that are based primarily on established theories of fluid mechanics. Although these rational coefficients cannot be given the weight of experimental coefficients, they are of great interest, suggesting as they do that a physical significance can be attached to the experimental data.

Now having an appreciation of the problem before us, we consider in the next sections both the empirical (curve and equation) fittings to experimental determinations of discharge coefficients and the various theoretical determinations of discharge coefficients. Whenever practical, comparisons are made.

22.2 ASME Flow Nozzles

Rational Methods

The rational methods for approximating discharge coefficients can be subdivided into two groups; that is, those emphasizing a loss factor (f or K), and those founded on a boundary-layer thickness (δ or δ^*). Arranged chronologically, we have the following.

PARDOE METHOD (1943). The oldest method [3] considered here indicates, by a general analysis for a constant-density fluid, that the discharge coefficient of a flow nozzle can be given as

$$C_D = \frac{(1-\beta^4)^{1/2}}{(\alpha_2 - \alpha_1\beta^4 + K_2)^{1/2}}, \tag{22.3}$$

where in the latter part of (22.2) $\dot{m}$ is in lbm/sec, D in inches, and μ is the dynamic viscosity of the fluid in lbm/ft-sec. Use of the Reynolds number the uniform velocity head at the nozzle throat. However, no reliable values for α_1, α_2, and K_2 had been given, so (22.3) was of purely academic interest at that time.

SHAPIRO–SMITH–RIVAS METHOD (1948–1952). From an elementary energy balance, presuming *plenum flow* at the inlet pipe wall tap, and with losses and changes in momentum flux in the nozzle given in terms of an integrated apparent friction factor, $\bar{f}_A$, they obtained [4], [5] for the case of a developing *laminar* boundary layer in the nozzle

$$C_{DL} = \frac{1}{(1 + \bar{f}_{AL} L'/d)^{1/2}}. \tag{22.4}$$

Although (22.4) follows directly from (22.3) when $\beta = 0$, and when one-dimensional flow is considered there is this important difference. Equation 22.4 delineates the loss coefficient of (22.3) in such a manner that C_D can be calculated. Specifically, the mean apparent friction factor was obtained from

$$\bar{f}_{AL} = \frac{13.74}{R_d^{0.5} (L'/d)^{0.5}} \tag{22.5}$$

which is a mathematical fit of a curve based on experimental data [6], [7]. The symbol L' signifies an effective length of the nozzle defined as $L' = L + L_{eq}$, where L is the length of the straight cylindrical throat section, and L_{eq} is the equivalent length of the contraction portion of the nozzle in terms of the throat section. For example, the length to the throat tap of an ASME long radius nozzle of the low β series (see Figure 21.1) is specified [1] as $L = 0.5d$, whereas the equivalent length (L_{eq}) for this same nozzle has been given [4] as about $0.4d$ or again [5] as varying between $0.25d$ and $0.6d$. Note that (22.5) has no application in the turbulent boundary layer region.

Under the foregoing conditions (22.4) should be valid for throat Reynolds number up to 5×10^5. This is the usual boundary layer transition Reynolds number (based on length) of flat-plate theory, and has been confirmed experimentally for tube flow.

SIMMONS METHOD (1955). Another class of rational approximations based on boundary layer thickness was first reported in 1955. Simmons [8], using a two-pronged approach, makes use of the alternate definitions:

$$C_D = \frac{1}{A_d} \int \frac{u}{V} \, dA \tag{22.6}$$

(which ratios the actual to ideal flow rates at the nozzle exit in terms of the boundary layer and free stream velocities), and

$$C_D = 1 - 2 \left(\frac{\delta^*}{r} \right) \tag{22.7}$$

(which ratios the effective to actual flow areas at the nozzle exit in terms of the boundary layer displacement thickness), and obtains

$$C_D = 4 \ln 2 \left[\frac{(2/(1 - \ln 2)(L/d + \frac{1}{4}(L''/d))}{R_d} \right]^{1/2} \tag{22.8}$$

as his basic rational equation, where L is the length of the straight cylindrical portion of the nozzle and L'' is the axial length of the convergent entry of the nozzle. R_d is the Reynolds number based on the throat diameter.

Equation 22.8 can be particularized for this study by noting that for the ASME nozzle $L \simeq 0.6d$ and $L'' = d$. Thus

$$C_D = 1 - (6.525)R_d^{-0.5}, \tag{22.9}$$

said to be valid over the range $10^4 \le R_d \le 10^6$.

HALL METHOD (1959).   Hall [9] idealizes the actual flow nozzle by a simplified straight cylindrical model such that the boundary layer thickness from the actual nozzle is conserved. The cylindrical boundaries of the model are then, in concept, displaced inwardly by an amount equal to the boundary layer displacement thickness such that the potential flow through this hypothetical nozzle is identical to the flow through the actual nozzle in the presence of a boundary layer. By taking the ratio of the throat area of the hypothetical nozzle to that of the actual nozzle, Hall obtains the (22.7) definition of C_D, in agreement with Simmons.

However, Hall greatly simplifies the development by declaring that flat-plate displacement thicknesses are adequately valid approximations for use in nozzle flows. Thus, for laminar boundary layers, Hall suggests the Blasius relation

$$\delta_L^* = 1.73L'R_L^{-0.5}, \tag{22.10}$$

whereas for turbulent boundary layers, the corresponding relation is

$$\delta_T^* = 0.046L'R_L^{-0.2}. \tag{22.11}$$

Before proceeding, one must assume some relation between the length of the hypothetical flow nozzle and the throat diameter of the actual flow nozzle. Hall indicates that the ratio, L'/d, is approximately unity for the ASME nozzle. With this added simplification, (22.10) and (22.11) are inserted into (22.7) to obtain

$$C_{DL} = 1 - (6.92)R_d^{-0.5} \tag{22.12}$$

for the range $10^3 \simeq R_d \le 2 \times 10^5$, and

$$C_{DT} = 1 - (0.184)R_d^{-0.2} \tag{22.13}$$

for the range $10^6 \le R_d \le 10^7$. Hall notes that while initially C_D should follow the relatively steep slope of laminar equation (22.12), there is a transition region (where the boundary layer is neither laminar nor turbulent) in which a hump in the $C_D - R_d$ relation may occur, and that eventually C_D should settle down to the relatively flatter slope of turbulent equation (22.13).

LEUTHEUSSER METHOD (1964). Applying a boundary layer solution specifically to the ASME nozzle profile, Leutheusser [10] obtains by way of (22.7) the particularized rational approximation

$$C_D = 1 - (6.528)R_d^{-0.5} \qquad (22.14)$$

said to be valid over the range $10^3 \leq R_d \leq 10^6$. By comparing (22.14) with (22.9) of Simmons, however, we cannot fail to note their essential coincidence, indicating that the effective axial length of the nozzle (rather than its profile) is the controlling consideration in developing rational expressions for nozzle-discharge coefficients.

BENEDICT–WYLER METHOD (1974). Because all of the above rational formulations ignore the diameter ratio effect and all overlook losses and velocity profile effects upstream of the fluid meter, and because none is used to predict discharge coefficients in nozzles, venturis, or orifices, Pardoe's equation was recalled, generalized, and particularized [11]. That is,

$$C_D = \frac{C_c(1 - \beta^4)^{1/2}}{(K_{1,2} + \alpha_2 - \alpha_1 C_c^2 \beta^4)^{1/2}} \qquad (22.15)$$

was proposed for any differential pressure type fluid meter, where C_c $K_{1,2}$, α_2 and α_1 were specified according to the best empirical information available. In terms of these particular values, (22.15) can also be given as

$$C_D = \left[\frac{1 - \beta^4}{A - \beta^4 + B + C - 0.4505\beta^{3.8}R_d^{-0.2}} \right]^{1/2} \qquad (22.16)$$

where A, B, and C are given in Table 22.1 for the various fluid meters shown in Figure 22.4.

Semirational Approximations

Nozzle discharge coefficients also can be approximated by methods based in part on boundary layer theory and in part on the experimental characteristics of ASME flow nozzles. These semirational approaches, while subscribing to the basic dependence of C_D on δ, are attempts to free the discharge coefficient from some of the unlikely conditions imposed by the rationals. For example, whereas all rational approximations are based on zero longitudinal pressure gradient, uniform free stream

Table 22.1 Summary of Rational Discharge Coefficients for Differential Pressure-type Fluid Meters (after [11]).

Generalized discharge coefficient	$C_D = \left[\dfrac{1-\beta^4}{A - \beta^4 + B + C - 0.4505\,\beta^{3.8}R_d^{-0.2}}\right]^{1/2}$	
Fluid meter type	Boundary layer type	
	Laminar	Turbulent
Throat tap meter	$A = 1$ $B = 0$ $C = 9.7156\,R_d^{-0.5}$	$A = 1$ $B = 0$ $C = 0.17\,R_d^{-0.2}$
pipe wall tap nozzle	$A = 1$ $B = 0$ $C = 13.74\,R_d^{-0.5}$	$A = 1$ $B = 0$ $C = 0.296\,R_d^{-0.2}$
Vena contracta & one-and one half D tap orifice	$A = 1/C_c^2$ $B = 0.26 - 1.311(\beta - 0.35)^2$ $C = -15\,R_d^{-0.5}$	

velocity, zero approach velocity, and known equivalent frictional length of the nozzle, none of these conditions is assured (indeed, some are absurd) in the ASME flow nozzle. Thus the emergence of the semirationals.

MURDOCK METHOD. A graphical method reported recently [12] plots the average ASME data of Figure 22.5 in terms of C_D versus $(10^6/R_d)^{0.5}$ in the laminar region (taken as extending to $R_d = 10^6$), and in terms of C_D versus $(10^6/R_d)^{0.2}$ in the turbulent region (taken as extending above $R_d = 10^6$). The resulting quasilinear plot was found to have a graphical C_D-intercept of 0.9975, and thus the Murdock semirational approximations can be given as

$$C_{DL} = 0.9975 - (6.53)R_d^{-0.5} \tag{22.17}$$

for the range $3 \times 10^4 \le R_d \le 10^6$, and

$$C_{DT} = 0.9975 - 0.1035\,R_d^{-0.2} \tag{22.18}$$

for the range $10^6 \le R_d \le 10^7$.

AUTHOR'S METHOD. A least-squares computer method applied to the data in Figure 22.5 while maintaining $R_d^{-0.5}$ in the laminar region and $R_d^{-0.2}$ in the turbulent region to retain connection with boundary layer theory yields the equations

$$C_{DL} = 0.99822 - (6.59298)R_d^{-0.5} \tag{22.19}$$

and
$$C_{DT} = 0.99822 - (0.10449)R_d^{-0.2} \tag{22.20}$$

over the same ranges given for the Murdock equations.

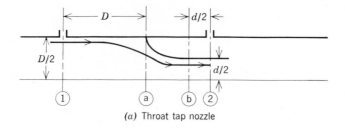

(a) Throat tap nozzle

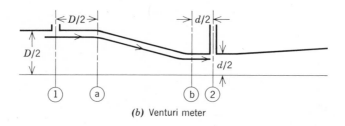

(b) Venturi meter

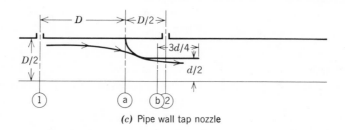

(c) Pipe wall tap nozzle

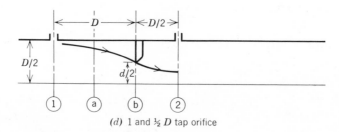

(d) 1 and ½ D tap orifice

Figure 22.4 Various fluid meter types.

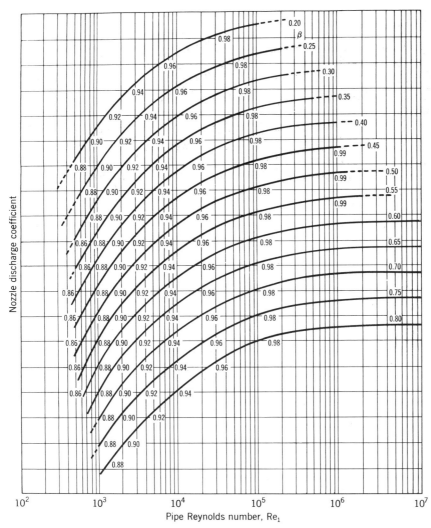

Figure 22.5   Discharge coefficients of ASME long-radius nozzle with pressure taps at $1\,D$ upstream and $D/2$ downstream. (From ASME Fluid Meters Report, 5th ed., 1959.)

Empirical Approximations

Flow-nozzle discharge coefficients also can be approximated by arbitrary mathematical fits of the experimental characteristics of ASME flow nozzles shown in Figure 22.5. The 13 curves of the ASME can be generalized, since a plot of C_D versus R_D/β ($= R_d$) essentially eliminates the diameter-ratio parameter as shown in Figure 22.6. These empirical

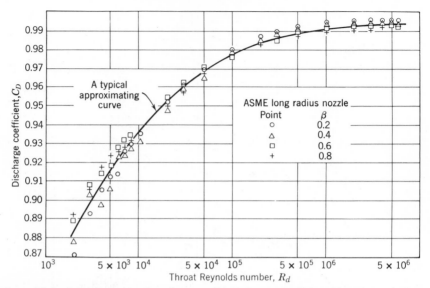

Figure 22.6   ASME long radius flow nozzle discharge coefficients in terms of throat Reynolds number.

approaches are entirely oblivious to the theories of fluid mechanics; they have as their only objective the useful mathematical representation of the available data; and they usually are based on some statistical procedure for maximizing the validity of the results.

BEITLER EQUATIONS.   Beitler [13] has presented several empirical fits of the ASME flow nozzle data of Figure 22.5 in which the rigid dependence of C_D on specified powers of R_d, as indicated by flat-plate boundary layer theory, was relaxed. The suggested expressions are

$$C_{DL} = 1 - (3.598) R_d^{-0.44} \tag{22.21}$$

for the range $2.5 \times 10^3 \le R_d \le 10^6$ and

$$C_{DT} = 0.9975 - (0.0649) R_d^{-0.176} \tag{22.22}$$

for the range $10^6 \le R_d \le 10^7$.

AUTHOR'S EQUATION.   The best empirical fit developed by the author [14] for the ASME flow nozzle data of Figure 22.5 can be expressed by a single polynomial (a cubic) in terms of the natural logarithm of R_d, namely,

$$C_D = 0.19436 + 0.152884(\ln R_d) - 0.0097785(\ln R_d)^2 + 0.00020903(\ln R_d)^3 \tag{22.23}$$

which is offered for the entire range of ASME data. It is based on the data of Table 22.2, which is taken from Figure 22.5.

Table 22.2 Discharge Coefficients, Average Discharge Coefficients, and rms Values for ASME Flow Nozzles in Terms of Throat Reynolds Number

| $R_d \times 10^{-3}$ | BETA | | | | | | | $\bar{C}$ | $\overline{\text{rms}} \times 10^3$ |
	0.2	0.3	0.4	0.5	0.6	0.7	0.8		
6000				0.9943	0.9938	0.9932	0.9916	0.99322	1.016
5000				0.9942	0.9935	0.9932	0.9916	0.99312	0.952
4000				0.9944	0.9938	0.9932	0.9916	0.99325	1.043
3000			0.9942	0.9936	0.9933	0.9928	0.9911	0.99300	1.053
2000		0.9940	0.9930	0.9926	0.9925	0.9922	0.9906	0.99248	1.017
1000	0.9939	0.9919	0.9910	0.9910	0.9910	0.9908	0.9897	0.99133	1.207
900	0.9932	0.9916	0.9906	0.9905	0.9907	0.9905	0.9894	0.99093	1.100
800	0.9928	0.9912	0.9904	0.9901	0.9904	0.9902	0.9890	0.99059	1.086
700	0.9924	0.9906	0.9898	0.9898	0.9901	0.9899	0.9886	0.99017	1.067
600	0.9917	0.9900	0.9892	0.9894	0.9898	0.9896	0.9882	0.98970	0.978
500	0.9910	0.9892	0.9886	0.9885	0.9891	0.9890	0.9876	0.98900	0.958
400	0.9900	0.9881	0.9870	0.9875	0.9882	0.9880	0.9870	0.98797	0.948
300	0.9885	0.9865	0.9858	0.9862	0.9864	0.9868	0.9856	0.98654	0.885
200	0.9858	0.9840	0.9832	0.9831	0.9838	0.9843	0.9828	0.98386	0.935
100	0.9788	0.9767	0.9770	0.9776	0.9781	0.9780	0.9770	0.97760	0.695
90	0.9776	0.9753	0.9760	0.9766	0.9772	0.9771	0.9754	0.97646	0.842
80	0.9762	0.9738	0.9744	0.9752	0.9758	0.9761	0.9741	0.97509	0.917
70	0.9745	0.9718	0.9724	0.9736	0.9741	0.9748	0.9726	0.97340	1.062
60	0.9725	0.9695	0.9702	0.9712	0.9715	0.9730	0.9698	0.97110	1.242
50	0.9694	0.9669	0.9678	0.9684	0.9694	0.9704	0.9672	0.96850	1.194
40	0.9662	0.9659	0.9641	0.9651	0.9660	0.9666	0.9636	0.96536	1.049
30	0.9608	0.9576	0.9592	0.9600	0.9614	0.9615	0.9584	0.95984	1.396
20	0.9524	0.9483	0.9502	0.9523	0.9544	0.9539	0.9510	0.95179	1.972
10	0.9355	0.9303	0.9334	0.9370	0.9400	0.9392	0.9360	0.93599	2.959
9	0.9328	0.9272	0.9305	0.9344	0.9374	0.9374	0.9334	0.93330	3.384
8	0.9296	0.9236	0.9272	0.9314	0.9346	0.9346	0.9306	0.93023	3.650
7	0.9260	0.9194	0.9244	0.9280	0.9313	0.9316	0.9273	0.92686	3.894
6	0.9218	0.9140	0.9190	0.9238	0.9272	0.9280	0.9233	0.92244	4.466
5	0.9134	0.9064	0.9128	0.9184	0.9216	0.9234	0.9182	0.91631	5.414
4	0.9046	0.8984	0.9045	0.9110	0.9148	0.9171	0.9120	0.90891	6.143
3	0.8927	0.8884	0.8938	0.8996	0.9048	0.9082	0.9042	0.89881	6.809

Comparisons

With this host of rational, semirational, and empirical approximations to choose from, it is small wonder that no definitive relation for the ASME flow nozzle discharge coefficients has been given. There is also no dearth of graphical forms to choose from when seeking the best coordinate system on which to present these comparisons for visual digestion. For convenience many of the equations considered in this study are tabulated in Table 22.3 Graphical comparisons are presented in Figures 22.7–22.11. The Hall plot of Figure 22.8 appears best for studying and portraying the rational approximations Note the sharp change in slope evident between $R_d = 10^5$ and 10^6, clearly delineating the laminar and turbulent boundary-layer regimes, Figures 22.7 and 22.8 decisively indicate the advantage of the empirical cubic of (22.23) in best representing the ASME flow nozzle data. Thus this equation has been solved at several R_ds to form a table of most probable discharge coefficients (see Table 22.2).

Table 22.3   Summary of Some Equations Used to Approximate ASME Flow Nozzle Data

Source	Boundary Layer Region	
	Laminar	Turbulent
	Rationals	
Shapiro-Smith (22.4)	$C_D = \dfrac{1}{(1 + f_A L'/d)^{1/2}}$	—
Simmons (22.9) Leutheusser (22.14)	$C_D = 1 - (6.525)R_d^{-0.5}$	—
Hall $\begin{cases}(22.12)\\(22.13)\end{cases}$	$C_D = 1 - (6.92)R_d^{-0.5}$	$C_D = 1 - (0.184)R_d^{-0.2}$
	Semirationals	
Murdock $\begin{cases}(22.17)\\(22.18)\end{cases}$	$C_D = 0.9975 - (6.53)R_d^{-0.5}$	$C_D = 0.9975 - (0.1035)R_d^{-0.2}$
Author $\begin{cases}(22.19)\\(22.20)\end{cases}$	$C_D = 0.99822 - (6.59298)R_d^{-0.5}$	$C_D = 0.99822 - (0.10449)R_d^{-0.2}$
	Empiricals	
Beitler $\begin{cases}(22.21)\\(22.22)\end{cases}$	$C_D = 1 - (3.598)R_d^{-0.44}$	$C_D = 0.9975 - 0.0649R_d^{-0.176}$
Author (22.23)	$C_D = 0.19436 + 0.152884\,(\ln R_d)$ $- 0.0097785\,(\ln R_d)^2 + 0.00020903\,(\ln R_d)^3$	

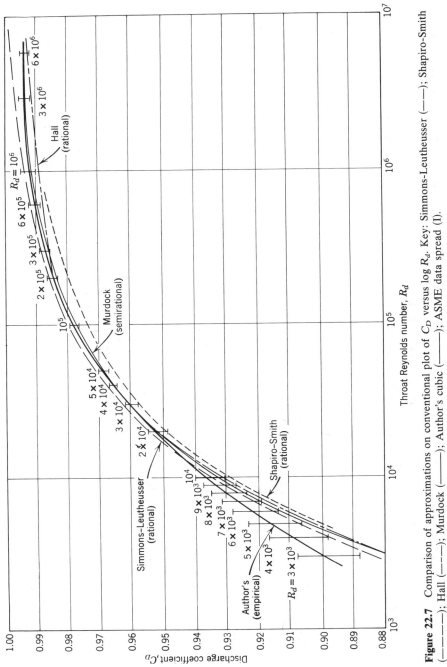

Figure 22.7 Comparison of approximations on conventional plot of C_D versus log R_d. Key: Simmons-Leutheusser (——); Shapiro-Smith (———); Hall (——); Murdock (———); Author's cubic (——); ASME data spread (I).

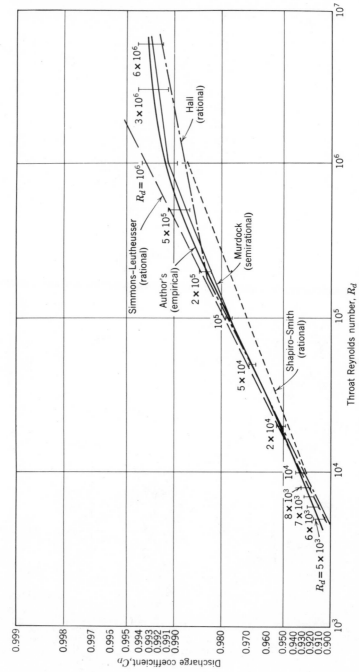

Figure 22.8 Comparison of approximations on Hall plot of $\log(1 - C_D)$ versus $\log R_d$. Key: Simmons-Leutheusser (———); Shapiro-Smith (———); Hall (———); Murdock (———); Author's cubic (———); ASME data spread (I).

438

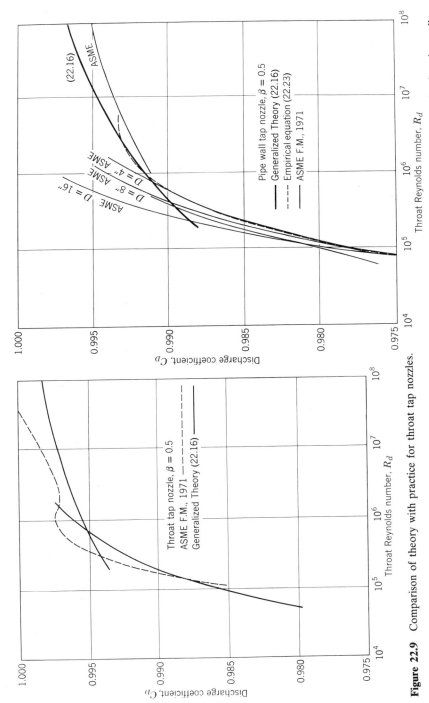

Figure 22.9 Comparison of theory with practice for throat tap nozzles.

Figure 22.10 Comparison of theory with ASME practice for pipe wall tap nozzles.

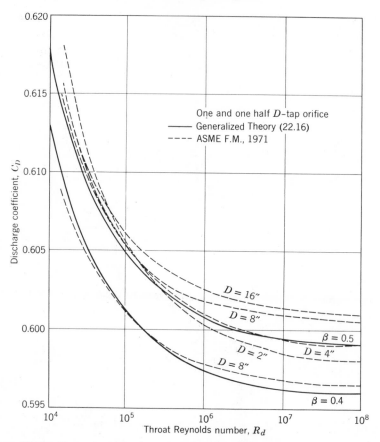

Figure 22.11   Comparison of theory with ASME practice for $1\,D$ and $\frac{1}{2}\,D$ tap orifices.

Example 1.   Subsonic Air Flow Through ASME Nozzle.   *Problem.* Find the flow rate of air in lbm/sec when $p_1 = 14.64417$ psia, $p_2 = 11.34677$ psia, $D = 10.136$ in., $d = 4.054$ in., $T_1 = 80°F$, and $\mu = 0.1237 \times 10^{-4}$ lbm/ft-sec. *Solution.* For $\gamma = 1.4$, $\beta = 4.054/10.136 = 0.4$ and $\beta^4 = 0.0256$, we have

$$A_2 = 0.7853975(4.054)^2 = 12.9079 \text{ in.}^2$$

$$\frac{p_2}{p_1} = \frac{11.34677}{14.65517} = 0.77425$$

From (21.23)

$$p_t = 14.65517\left[\frac{1-(0.77425)^{2.4/1.4} \times 0.0256}{1-(0.77425)^{2/1.4} \times 0.0256}\right]^{1.4/0.4} = 14.72063 \text{ psia,}$$

and

$$\frac{p_2}{p_t} = \frac{11.34677}{14.72063} = 0.77081.$$

From Table 21.1 or from (21.18), $\Gamma = 0.85896$. Then from (20.19)

$$\overset{\circ}{m}_{\text{ideal}} = 12.9079 \left[\frac{0.531748 \times 14.72063 \times 0.85896}{540^{1/2}} \right] = 3.738 \text{ lbm/sec.}$$

The discharge coefficient can now be obtained by iteration. First it can be assumed arbitrarily that the pipe Reynolds number is 10^5. The C_D from Figure 22.5 or from (22.23) is 0.984. The actual flow rate (22.1) is

$$\overset{\circ}{m}_{\text{actual}} = 0.984 \times 3.738 = 3.678 \text{ lbm/sec.}$$

The corresponding Reynolds number (22.2) is

$$R_D = \frac{48 \times 3.678}{10.136\pi \times 0.1237 \times 10^{-4}} = 4.48 \times 10^5.$$

Since this value of R_D differs from the assumed value, a second trial is necessary. Using 4.48×10^5 as a second guess, $C_D = 0.991$, $\overset{\circ}{m} = 3.704$ lbm/sec, and $R_D = 4.51 \times 10^5$. Since this value of R_D does not differ significantly from the assumed value, the best value of flow rate has been obtained.

22.3 HEI Flow Nozzles

These nozzles, sponsored by the Heat Exchange Institure, are accepted as flow metering standards by some. The characteristics of this type of

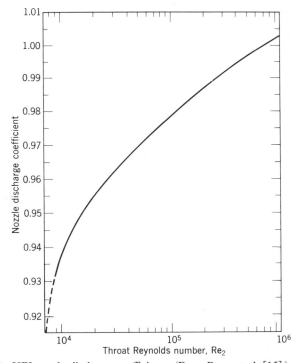

Figure 22.12 HEI nozzle discharge coefficients. (From Bean et al. [15].)

nozzle are based on the calibration work of H. S. Bean, R. M. Johnson, and T. R. Blakeslee [15]. Their information is thus in the form of graphical representations of experimental data (see Figure 22.12). They plot discharge coefficient versus a throat Reynolds number, where throat and pipe Reynolds numbers are related by

$$R_d = (R_D/\beta)(\mu_1/\mu_2). \tag{22.24}$$

Thus the plot is similar the R_D/β plot of Figure 22.6 of the ASME data so that the β parameter is unnecessary. The curve in Figure 22.12 has been represented by the least-squares cubic

$$C_D = -0.413 + 0.319104 \ln R_d - 0.0248987(\ln R_d)^2$$
$$+ 6.67923 \times 10^{-4} (\ln R_d)^3, \tag{22.25}$$

which fits the curve within ± 0.002 on the discharge coefficient.

Example 2.   Critical Air Flow Through HEI Nozzle.   *Problem.* Find the flow rate of air in lbm/sec when $p_1 = 33.35162$ psia, $p_b = 14.78162$ psia, $D = 0.942$ in., $d = 0.375$ in., $T_1 = 78°F$, and $\mu = 0.1234 \times 10^{-4}$ lbm/ft-sec. *Solution.* For $\gamma = 1.4$, $\beta = 0.375/0.942 = 0.3981$, we have

$$A_2 = 0.7853975 \times 0.375^2 = 0.110446 \text{ in.}^2$$

and

$$\frac{P_b}{P_1} = \frac{14.78162}{33.35162} = 0.44321.$$

From (21.25), $(p_2/p_1)_* = 0.53143$. Since $p_b/p_1 < (p_2/p_1)_*$, the nozzle is choked and

$$p_2 = 33.35162 \times 0.53143 = = 17.72402 \text{ psia.}$$

From (21.26), $p_2/p_t = 0.52828$, so that $p_t = 33.55042$. From Table 21.1 or from (21.18), $\Gamma = 1$. Then according to (21.19).

$$\dot{m}_{ideal} = 0.110446 \left[\frac{0.531748 \times 33.55042 \times 1.0}{538^{1/2}} \right] = 0.0850 \text{ lbm/sec.}$$

For a first try, assume that the pipe Reynolds number is 10^5. The discharge coefficient for the HEI nozzle, however, is a function of the throat Reynolds number (22.24).

$$R_d \approx \frac{10^5}{0.3981} = 2.5 \times 10^5.$$

From (22.25) the coefficient for this assumed throat Reynolds number is $C_D = 0.990$. The actual flow rate is given by (22.1):

$$\dot{m} = 0.990 \times 0.0850 = 0.0842 \text{ lbm/sec.}$$

The corresponding pipe Reynolds number (22.2) is

$$R_D = \frac{48 \times 0.0842}{0.942 \pi \times 0.1234 \times 10^{-4}} = 1.107 \times 10^5,$$

and the throat Reynolds number is

$$R_d \approx \frac{1.107 \times 10^5}{0.3981} = 2.8 \times 10^5.$$

It would be possible to stop here with a flow rate of 0.0842 lbm/sec or try another step in the iteration. One additional step leads to values of $C_D = 0.9915$, $R_D = 1.108 \times 10^5$, $R_d = 2.8 \times 10^5$, and $\dot{m} = 0.0843$ lbm/sec. The figures indicate that stopping at the second step is justified.

22.4 ASME Venturi Meters

Characteristic curves for venturi meters are shown in Figure 22.13. These are for a diameter ratio of 0.5. No general empirical equations are available for this type of flow meter [16].

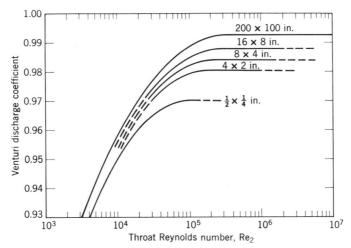

Figure 22.13 ASME venturi discharge coefficients. (After Giles [16].)

22.5 ASME Square-Edged Orifices

The ASME currently offers [1] for each set of pressure tap locations (as mentioned in Section 22.1) the flow coefficient

$$K = \frac{C_D}{(1 - \beta^4)^{1/2}} \tag{22.26}$$

as a function of the pipe Reynolds number ($R_D = \mathrm{Re}_1$), the pipe diameter (D), and the diameter ratio ($\beta = d/D$, where d represents the orifice

Table 22.4 Table of Most Probable Discharge Coefficients for ASME Long Radius Flow Nozzles[a]

$R_d \times 10^{-3}$	1	2	3	4	5	6	7	8	9
C_D	—	—	0.89886	0.90898	0.91630	0.92195	0.92650	0.93028	0.93350
$R_d \times 10^{-4}$	1	2	3	4	5	6	7	8	9
C_D	0.93628	0.95242	0.96024	0.96512	0.96855	0.97113	0.97316	0.97482	0.97620
$R_d \times 10^{-5}$	1	2	3	4	5	6	7	8	9
C_D	0.97737	0.98373	0.98647	0.98804	0.98907	0.98979	0.99032	0.99074	0.99106
$R_d \times 10^{-6}$	1	2	3	4	5	6	7	8	9
C_D	0.99132	0.99251	0.99288	0.99305	0.99315	0.99322	—	—	—

[a] Based on Equation 22.23 (for pipe wall taps).

diameter); that is

$$K = f(R_D, D, \beta). \tag{22.27}$$

As early as 1935, S. R. Beitler [17] found that K could be expressed as a linear function of $1/R_D$ or $1/(R_D)^{1/2}$. Empirical equations were developed on this basis, primarily by Buckingham [1], by which tables of flow coefficients could be generated (see sample in Table 22.5). These are summarized in the ASME Fluid Meters Report as follows:

For Flange Taps

$$A = d\left(830 - 5000\beta + 9000\beta^2 - 4200\beta^3 + \frac{530}{D^{1/2}}\right) \tag{22.28}$$

$$K_e = 0.5993 + \frac{0.007}{D} + \left(0.364 + \frac{0.076}{D^{1/2}}\right)\beta^4$$

$$+ 0.4\left(1.6 - \frac{1}{D}\right)^5\left(0.07 + \frac{0.5}{D} - \beta\right)^{5/2}$$

$$- \left(0.009 + \frac{0.034}{D}\right)(0.5 - \beta)^{3/2} + \left(\frac{65}{D^2} + 3\right)(\beta - 0.7)^{5/2} \tag{22.29}$$

$$K_0 = K_e\frac{(10^6 d)}{(10^6 d + 15A)} \tag{22.30}$$

For Pipe Taps.

$$A = d\left(905 - 5000\beta + 9000\beta^2 - 4200\beta^3 + \frac{875}{D}\right) \tag{22.31}$$

$$K_e = 0.5925 + \frac{0.0182}{D} + \left(0.44 - \frac{0.06}{D}\right)\beta^2 + \left(0.935 + \frac{0.225}{D}\right)\beta^5$$

$$+ 1.35\beta^{14} + \left(\frac{1.43}{D^{1/2}}\right)(0.25 - \beta)^{5/2} \tag{22.32}$$

$$K_0 = K_e\frac{(10^6 d)}{(10^6 d + 15A)} \tag{22.33}$$

where K_e is a particular value of K when $R_D = 10^6 d\beta/15$.

For One and One-Half D Taps

$$b = 0.0002 + \frac{0.0011}{D} + \left(0.0038 + \frac{0.0004}{D}\right)[\beta^2 + (16.5 + 5D)\beta^{16}]$$

(22.34)

$$K_0 = 0.6014 - 0.01352D^{-0.25} + (0.376 + 0.07257D^{-0.25})$$
$$\times \left(\frac{0.00025}{(D^2\beta^2 + 0.0025\,D)} + \beta^4 + 1.5\beta^{16}\right) \quad (22.35)$$

Table 22.5 Flange Taps [*values of the discharge coefficient* C_D *as a function of the throat Reynolds number* R_d *and diameter ratio* β; *for 2 in., pipe,* $D = 2.067$ *in.*]

β \ R_d	30,000	40,000	50,000	75,000	100,000	500,000	1,000,000
0.1500	0.6030	0.6020	0.6014	0.6006	0.6002	0.5993	0.5992
0.2000	0.6012	0.6001	0.5995	0.5986	0.5982	0.5972	0.5970
0.2500	0.6012	0.6001	0.5994	0.5985	0.5980	0.5969	0.5968
0.3000	0.6027	0.6015	0.6007	0.5998	0.5993	0.5981	0.5979
0.3500	0.6049	0.6036	0.6028	0.6017	0.6012	0.5999	0.5997
0.4000	0.6075	0.6059	0.6050	0.6038	0.6032	0.6017	0.6015
0.4500	0.6104	0.6086	0.6075	0.6061	0.6053	0.6036	0.6034
0.5000	0.6138	0.6116	0.6102	0.6085	0.6076	0.6055	0.6052
0.5500	0.6174	0.6147	0.6130	0.6109	0.6098	0.6072	0.6068
0.5750	0.6193	0.6163	0.6145	0.6121	0.6109	0.6080	0.6076
0.6000	0.6212	0.6179	0.6158	0.6132	0.6118	0.6086	0.6082
0.6250	0.6230	0.6193	0.6171	0.6141	0.6126	0.6091	0.6087
0.6500	0.6245	0.6205	0.6180	0.6148	0.6131	0.6092	0.6087
0.6750	0.6256	0.6212	0.6185	0.6149	0.6131	0.6088	0.6083
0.7000	0.6261	0.6212	0.6183	0.6144	0.6124	0.6077	0.6071
0.7250	0.6271	0.6218	0.6186	0.6143	0.6122	0.6071	0.6065
0.7500	0.6323	0.6265	0.6230	0.6183	0.6160	0.6104	0.6097

(From ASME Fluid Meters Report, 6th ed., 1971.)

Thus on a linear basis the flow coefficient for orifices in both flange and pipe tap installations is given by

$$K = K_0\left(1 + \frac{\beta A}{R_D}\right),$$

(22.36)

whereas for one and one-half D tap orifices the flow coefficient is

$$K = K_0 + \frac{1000b}{(R_D)^{1/2}}$$

(22.37)

in agreement with the findings of Beitler, where K_0 is the limiting value of K when R_D increases indefinitely. It should be noted, in (22.29) and (22.32), one or more terms may go "imaginary," and in such cases, the whole term is to be dropped.

These formulations of the flow coefficient are complicated mathematical expressions and are presented here for possible use in computer applications only. For hand calculation, on the other hand, the elementary linear equation

$$C_D = C_0 + \Delta C \left(\frac{10^4}{R_d} \right)^a \tag{22.38}$$

has recently been presented by J. W. Murdock [18]. The values of C_0 (the discharge coefficient at $R_d \to \infty$) and of ΔC (representing the increase in the discharge coefficient for the arbitray Reynolds number change $10^4 < R_d < \infty$) are tabulated for given D, β, and tap setups (see sample in Table 22.6). The exponent a in (22.38) takes the value of one for flange and pipe taps and the value of $\frac{1}{2}$ for all others. Murdock prefers an equation of the form of (22.38) rather than (22.36) and (22.37) because C_D shows less variation with β than does K for any Reynolds number, and C_D shows less variation with β when expressed as a function of R_d rather than R_D.

Example 3. *Water Flow Through Orifice with Flange Taps (Using ASME Tables).* *Problem.* Find the flow rate of water in lbm/sec when $p_1 = 17.74159$ psia, $p_2 = 14.63401$ psia, $D = 2$ in., $d = 1.209$ in., and $T_1 = 100°F$. *Solution.* $\beta = d/D = 1.209/2 = 0.6045$, $(1-\beta^4)^{1/2} = 0.93084$, $\mu = 0.4584 \times 10^{-3}$ lbm/ft-sec, and $\rho = 62$ lbm/ft^3.

$$A_2 = \frac{\pi D_2{}^2}{4} = 0.7853975 \times 1.209^2 = 1.148 \text{ in.}^2,$$

$$p_1 - p_2 = 17.74159 - 14.63401 = 3.10758 \text{ lbf/in.}^2$$

According to (21.9), the theoretical rate of flow is

$$\mathring{m}_{\text{ideal}} = \frac{1.148}{0.93084} \left(\frac{64.34 \times 62 \times 3.10758}{144} \right)^{1/2} = 11.4428 \text{ lbm/sec.}$$

Iteration is used to obtain the discharge coefficient. First assume that the throat Reynolds number is 10^5. The discharge coefficient is then obtained by referring to the ASME tables for the 2 in. pipe with flange taps (see Table 22.5) or by solving (22.31) through (22.36). The portion of the ASME table [1], by linear interpolation at $R_d = 10^5$ and $\beta = 0.6045$, yields

$$C_D = 0.61194.$$

The actual flow rate for the first try is calculated according to (22.1):

$$\mathring{m} = 0.61194 \times 11.4428 = 7.00231 \text{ lbm/sec.}$$

Table 22.6 Values for use in Equation 22.38 to Obtain Discharge Coefficients for Orifices with Flange Taps[a]

	D = 1.5 in.		D = 2 in.		D = 3 in.		D = 4 in.		D = 6 in.	
β	C_0	ΔC	C_0	ΔC	C_0	ΔC	C_0	ΔC	C_0	ΔC
0.10	0.60265	0.00767	0.60370	0.00955	0.59902	0.01297	0.59554	0.01622	0.59273	0.02248
0.11	0.60210	0.00811	0.60265	0.01005	0.59796	0.01360	0.59480	0.01697	0.59252	0.02345
0.12	0.60160	0.00850	0.60169	0.01050	0.59706	0.01415	0.59423	0.01761	0.59249	0.02426
0.13	0.60113	0.00886	0.60083	0.01090	0.59631	0.01462	0.59384	0.01815	0.59260	0.02493
0.14	0.60070	0.00918	0.60006	0.01125	0.59570	0.01503	0.59360	0.01861	0.59284	0.02546
0.15	0.60031	0.00947	0.59937	0.01156	0.59523	0.01538	0.59350	0.01899	0.59317	0.02588
0.16	0.59996	0.00973	0.59877	0.01184	0.59489	0.01568	0.59353	0.01930	0.59353	0.02619
0.17	0.59965*	0.00997	0.59825	0.01209	0.59468	0.01594	0.59368	0.01956	0.59388	0.02640
0.18	0.59938	0.01019	0.59781	0.01231	0.59458	0.01615	0.59392	0.01976	0.59423	0.02654
0.19	0.59915	0.01040	0.59745	0.01251	0.59458	0.01634	0.59423	0.01992	0.59456	0.02660
0.20	0.59895	0.01059	0.59717	0.01269	0.59468	0.01650	0.59458	0.02004	0.59488	0.02661
0.21	0.59879	0.01077	0.59696	0.01286	0.59486	0.01664	0.59492	0.02014	0.59520	0.02658
0.22	0.59867	0.01094	0.59682	0.01302	0.59511	0.01677	0.59525	0.02021	0.59550	0.02651
0.23	0.59859	0.01111	0.59675	0.01317	0.59542	0.01689	0.59557	0.02027	0.59580	0.02642
0.24	0.59854	0.01127	0.59674	0.01332	0.59575	0.01700	0.59589	0.02033	0.59609	0.02633
0.25	0.59852	0.01144	0.59680	0.01348	0.59609	0.01712	0.59620	0.02038	0.59638	0.02623
0.26	0.59855	0.01161	0.59691	0.01364	0.59642	0.01724	0.59651	0.02045	0.59666	0.02615
0.27	0.59860	0.01179	0.59707	0.01381	0.59674	0.01737	0.59681	0.02052	0.59693	0.02609
0.28	0.59869	0.01197	0.59729	0.01399	0.59706	0.01752	0.59710	0.02062	0.59720	0.02606
0.29	0.59882	0.01217	0.59755	0.01418	0.59738	0.01769	0.59739	0.02075	0.59747	0.02608
0.30	0.59897	0.01238	0.59784	0.01440	0.59769	0.01789	0.59768	0.02091	0.59773	0.02614
0.31	0.59916	0.01261	0.59817	0.01463	0.59800	0.01812	0.59797	0.02112	0.59799	0.02627
0.32	0.59938	0.01285	0.59852	0.01489	0.59831	0.01838	0.59825	0.02136	0.59825	0.02646
0.33	0.59963	0.01312	0.59888	0.01518	0.59862	0.01868	0.59853	0.02166	0.59851	0.02673
0.34	0.59990	0.01340	0.59923	0.01549	0.59892	0.01902	0.59882	0.02201	0.59876	0.02708
0.35	0.60020	0.01371	0.59959	0.01583	0.59923	0.01941	0.59910	0.02243	0.59902	0.02752
0.36	0.60053	0.01405	0.59995	0.01621	0.59954	0.01984	0.59938	0.02291	0.59927	0.02807
0.37	0.60088	0.01441	0.60030	0.01662	0.59985	0.02033	0.59966	0.02346	0.59953	0.02871
0.38	0.60125	0.01480	0.60066	0.01707	0.60016	0.02088	0.59995	0.02409	0.59978	0.02947
0.39	0.60164	0.01522	0.60102	0.01756	0.60047	0.02148	0.60023	0.02479	0.60004	0.03035
0.40	0.60205	0.01568	0.60139	0.01809	0.60079	0.02215	0.60052	0.02558	0.60030	0.03135
0.41	0.60246	0.01616	0.60175	0.01866	0.60110	0.02288	0.60081	0.02645	0.60056	0.03247
0.42	0.60287	0.01668	0.60212	0.01928	0.60142	0.02368	0.60110	0.02741	0.60082	0.03373
0.43	0.60328	0.01723	0.60249	0.01995	0.60174	0.02454	0.60139	0.02846	0.60108	0.03513

0.44	0.60370	0.01783	0.60286	0.02066	0.60206	0.02548	0.60168	0.02961	0.60134	0.03667
0.45	0.60412	0.01845	0.60323	0.02142	0.60238	0.02649	0.60198	0.03086	0.60160	0.03836
0.46	0.60453	0.01912	0.60360	0.02223	0.60270	0.02758	0.60227	0.03220	0.60186	0.04019
0.47	0.60494	0.01982	0.60397	0.02310	0.60302	0.02874	0.60256	0.03365	0.60211	0.04218
0.48	0.60535	0.02057	0.60434	0.02401	0.60334	0.02998	0.60285	0.03520	0.60237	0.04432
0.49	0.60575	0.02135	0.60470	0.02498	0.60365	0.03130	0.60313	0.03686	0.60262	0.04662
0.50	0.60614	0.02217	0.60504	0.02600	0.60394	0.03270	0.60340	0.03862	0.60285	0.04908
0.51	0.60649	0.02304	0.60536	0.02708	0.60422	0.03418	0.60365	0.04049	0.60307	0.05170
0.52	0.60685	0.02394	0.60568	0.02821	0.60450	0.03574	0.60390	0.04246	0.60329	0.05448
0.53	0.60722	0.02489	0.60600	0.02939	0.60477	0.03739	0.60414	0.04455	0.60350	0.05742
0.54	0.60758	0.02588	0.60632	0.03063	0.60504	0.03911	0.60438	0.04674	0.60370	0.06052
0.55	0.60794	0.02691	0.60664	0.03193	0.60530	0.04092	0.60461	0.04905	0.60389	0.06379
0.56	0.60829	0.02798	0.60694	0.03328	0.60555	0.04281	0.60482	0.05146	0.60406	0.06722
0.57	0.60863	0.02909	0.60724	0.03469	0.60579	0.04478	0.60502	0.05398	0.60422	0.07081
0.58	0.60896	0.03024	0.60751	0.03614	0.60600	0.04683	0.60520	0.05660	0.60436	0.07456
0.59	0.60926	0.03144	0.60777	0.03766	0.60619	0.04896	0.60535	0.05933	0.60446	0.07846
0.60	0.60954	0.03267	0.60799	0.03922	0.60635	0.05117	0.60547	0.06216	0.60453	0.08252
0.61	0.60979	0.03394	0.60818	0.04083	0.60647	0.05345	0.60555	0.06510	0.60456	0.08673
0.62	0.60999	0.03525	0.60833	0.04250	0.60655	0.05580	0.60559	0.06813	0.60454	0.09108
0.63	0.61015	0.03659	0.60843	0.04421	0.60657	0.05823	0.60557	0.07125	0.60447	0.09557
0.64	0.61025	0.03797	0.60846	0.04596	0.60653	0.06072	0.60548	0.07446	0.60433	0.10020
0.65	0.61029	0.03938	0.60843	0.04776	0.60642	0.06327	0.60532	0.07776	0.60412	0.10495
0.66	0.61024	0.04082	0.60832	0.04960	0.60623	0.06589	0.60508	0.08114	0.60381	0.10982
0.67	0.61010	0.04229	0.60811	0.05147	0.60594	0.06856	0.60474	0.08458	0.60341	0.11480
0.68	0.60985	0.04378	0.60779	0.05338	0.60554	0.07127	0.60429	0.08810	0.60290	0.11988
0.69	0.60949	0.04529	0.60735	0.05531	0.60501	0.07403	0.60371	0.09167	0.60226	0.12505
0.70	0.60898	0.04682	0.60677	0.05727	0.60434	0.07683	0.60298	0.09529	0.60147	0.13029
0.71	0.60858	0.04838	0.60619	0.05926	0.60359	0.07967	0.60215	0.09896	0.60055	0.13560
0.72	0.60898	0.05003	0.60602	0.06132	0.60297	0.08257	0.60136	0.10269	0.59960	0.14100
0.73	0.61055	0.05181	0.60648	0.06349	0.60259	0.08554	0.60066	0.10650	0.59864	0.14647
0.74	0.61354	0.05375	0.60769	0.06577	0.60250	0.08861	0.60009	0.11039	0.59770	0.15202
0.75	0.61811	0.05586	0.60976	0.06818	0.60274	0.09176	0.59966	0.11435	0.59677	0.15764
0.76	0.62436	0.05818	0.61272	0.07073	0.60332	0.09501	0.59937	0.11838	0.59584	0.16331
0.77	0.63237	0.06072	0.61662	0.07344	0.60423	0.09835	0.59922	0.12248	0.59489	0.16904
0.78	0.64216	0.06350	0.62143	0.07632	0.60545	0.10179	0.59916	0.12665	0.59389	0.17479
0.79	0.65369	0.06653	0.62714	0.07936	0.60696	0.10532	0.59917	0.13086	0.59280	0.18054
0.80	0.66692	0.06981	0.63368	0.08257	0.60868	0.10892	0.59918	0.13510	0.59156	0.18626

a After J. W. Murdock [18].

A check of the Reynolds number (22.2) results in

$$R_d = \frac{48 \times 7.00231}{1.209\pi \times 0.4584 \times 10^{-3}} = 1.9305 \times 10^5.$$

Since this calculated value of R_d differs from the assumed value, a second trial is necessary using the calculated value of the Reynolds number. Thus $C_D = 0.61118$, and the flow rate is

$$\mathring{m} = 0.61118 \times 11.4428 = 6.99361 \text{ lbm/sec.}$$

The new value of the Reynolds number is 1.9281×10^5. Since this does not differ significantly from the previous value, it can be concluded that a sufficiently close value of the flow rate has been obtained.

Example 4. Water Flow Through Orifice with One and One-Half D Taps (Using Murdock Tables). *Problem.* Find the flow rate of water in lbm/sec. when all variables are as in the previous problem. *Solution.* The theoretical rate of flow is thus

$$\mathring{m}_{\text{ideal}} = 11.4428 \text{ lbm/sec.}$$

Iteration is used to obtain the discharge coefficient by first referring to the Murdock tables for the 2-in. pipe with one and one-half D taps. The following portion of the Murdock tables [18] gives C_0 and ΔC as functions of β only:

$$D = 2 \text{ in.}$$

β	C_0	ΔC
0.60	0.60355	0.02674
0.61	0.60412	0.02706

Linear interpolation of the table at $\beta = 0.6045$ yields $C_0 = 0.60381$ and $\Delta C = 0.02688$. Then (22.38) is applied, taking as a first guess, $R_d = 10^5$. Thus

$$C_D = 0.60381 + 0.02688\left(\frac{10^4}{10^5}\right)^{1/2} = 0.61231.$$

The actual flow rate for the first try is calculated according to (22.1):

$$\mathring{m} = 0.61231 \times 11.4428 = 7.0065 \text{ lbm/sec.}$$

A check of the Reynolds number (22.2) results in

$$R_d = \frac{48 \times 7.0065}{1.209\pi \times 0.4584 \times 10^{-3}} = 1.932 \times 10^5.$$

Since this calculated value of R_d differs from the assumed value, a second trial is necessary using the calculated value of the Reynolds number in (22.38). Thus we obtain $C_D = 0.60993$, and the revised flow rate is

$$\mathring{m} = 0.60993 \times 11.4428 = 6.9793 \text{ lbm/sec.}$$

The new value of the Reynolds number is 1.924×10^5. Since this does not differ significantly from the previous value, the flow rate is as given.

22.6 References

[1] *Fluid Meters—Their Theory and Application*, Report of ASME Research Committee on Fluid Meters, 6th ed., 1971.

[2] R. G. Folsom, "Determination of ASME Nozzle Coefficients for Variable Nozzle External Dimensions," *Trans. ASME*, **72**, July 1950 p. 651.

[3] W. S. Pardoe, "The Effect of Installation on the Coefficient of Venturi Meters," *Trans. ASME*, **65**, 1943, p. 337.

[4] A. H. Shapiro and R. D. Smith, "Friction Coefficients in the Inlet Length of Smooth, Round Tubes," NACA TN 1785, November 1948.

[5] M. A. Rivas, Jr., and A. H. Shapiro, "On the Theory of Discharge Coefficients for Rounded-Entrance Flow Meters and Venturis," *Trans. ASME*, **78**, April 1956, p. 489.

[6] S. J. Kline and A. H. Shapiro, "Experimental Investigation of the Effects of Cooling on Friction and Boundary Layer Transition for Low-Speed Gas Flow at the Entry of a Tube," NACA TN 3048, November, 1953.

[7] A. H. Shapiro, R. Siegel, and S. J. Kline, "Friction Factor in the Laminar Entry Region of a Smooth Tube," *Proc. Second U.S. National Congress of Applied Mech.*, June 1954, p. 733.

[8] F. S. Simmons, "Analytic Determination of the Discharge Coefficients of Flow Nozzles," NACA TN 3447, February 1955.

[9] G. W. Hall, "Application of Boundary-Layer Theory to Explain Some Nozzle and Venturi Flow Peculiarities," *Proc. Inst. Mech. Eng.*, London, **173**, No. 36, 1959 p. 837.

[10] H. J. Leutheusser, "Flow Nozzles With Zero Beta Ratio," *Trans. ASME, J. Basic Eng.*, September 1964, p. 538.

[11] R. P. Benedict and J. S. Wyler, "A Generalized Discharge Coefficient for Differential Type Fluid Meters," *Trans. ASME, J. Eng. Power*, October 1974, p. 440

[12] J. W. Murdock, "A Rational Equation for ASME Coefficients for Long-Radius Flow Nozzles Employing Pipe Wall Taps at $1D$ and $\frac{1}{2}D$," ASME Paper 64-WA/FM-7.

[13] S. R. Beitler, discussion of reference [10].

[14] R. P. Benedict, "Most Probable Discharge Coefficients for ASME Flow Nozzles," *Trans. ASME, J. Basic Eng.*, December 1966, p. 734.

[15] H. S. Bean, R. M. Johnson, and T. R. Blakeslee, "Small Nozzles and Low Values of Diameter Ratio," *Trans. ASME*, **76**, August 1954, p. 863.

[16] R. V. Giles, *Fluid Mechanics and Hydraulics*, Schaum, New York, 1962.

[17] S. R. Beitler, "The Flow of Water Through Orifices," Ohio State University Engineering Experimental Station Bulletin 89, 1935.

[18] J. W. Murdock, "Tables for the Interpolation and Extrapolation of ASME Coefficients for Square-Edged Concentric Orifices," *ASME Paper 64-WA/FM-6*.

[19] R. P. Benedict, "Fluid Flow Measurement," *Electro-Technol.*, January 1966, p. 55.

Nomenclature

a exponent

A area

C_D discharge coefficient
d throat diameter
D pipe diameter
f friction factor
K Loss coefficient; also flow coefficient
$\dot{m}$ mass flow rate
N number of values
r radius
R_d Reynolds number, throat
R_D Reynolds number, pipe
Re Reynolds number, general
t time
u boundary layer velocity
v directed velocity

Greek

α kinetic energy coefficient
β diameter ratio
δ boundary layer thickness
Δ finite difference
μ dynamic viscosity
ρ density
Σ summation

Subscripts

A approximate
L laminar
t total
T turbulent
1,2 stations in the flow

Problems

1. Show numerically that both forms of (22.2) yield identical results
   for the Reynolds number, given: $\rho = 62.4 \, \text{lbm/ft}^3$, $V = 10 \, \text{ft/sec}$,
   $D = 6$ in. and $\mu = 60.6653 \times 10^{-5} \, \text{lbm/ft sec}$.

   Ans. Re $= 5.143 \times 10^5$.

2. Find the discharge coefficient by the generalized equation (22.16) for
   a pipe wall tap nozzle installation of $\beta = 0.5$ at a throat Reynolds
   number of 10^6 (i.e., for a turbulent boundary layer).

   Ans. $C_D = 0.99124$.

3. Find C_D for Problem 2 by the empirical equation (22.23) and check by Table 22.4.

   Ans. $C_D = 0.99132$, for a difference of 0.01%.

4. For a flange tap orifice of bore = 1 in. installed in a 2 in. pipe, find the flow rate of water if: $\rho = 62.4$ lbm/ft^3, $\mu = 0.60665 \times 10^{-3}$ lbm/ft sec, and the pressure drop across the orifice is 10 psi.

   Ans. $\dot{m} = 8.216$ lbm/sec.

Chapter 23

THE EXPANSION FACTOR

"... a jet of liquid from a sharp-edged orifice contracts more than a jet of gas because the gas expands after leaving the orifice ..."

Edgar Buckingham (1931)

23.1 General Remarks

For purposes of uniqueness in determining a discharge coefficient, an ideal flow rate can be defined as

$$\mathring{m}_i' \equiv A_2 \left[\frac{2 g_c \rho_1 (p_1 - p_2)}{1 - \beta^4} \right]^{1/2}, \tag{23.1}$$

where the prime indicates an ideal formulation and the subscript i indicates an *incompressible* quantity. This was chosen to coincide with (21.9) and hence represents the theoretical flow rate resulting from a one-dimensional workless energy analysis of a reversible steady flow of a constant density fluid through a horizontal fluid meter.

Now the general discharge coefficient of (22.1) can be particularized in terms of (23.1) as

$$C_{Di} \equiv \frac{\mathring{m}_i}{\mathring{m}_i'} \tag{23.2}$$

This discharge coefficient can be firmly established for any fluid meter in terms of a flowing liquid by obtaining $\mathring{m}_{\text{actual}}$ via a "weighing-catching-timing" experiment while at the same time forming $\mathring{m}_i'$ of (23.1) from a measured Δp across the fluid meter.

454

The discharge coefficient so established by means of *liquid* tests is actually *always* used for all fluids, compressible or not. We must carefully note, moreover, that the validity of this choice is not subject to experimental confirmation; For example, for compressible fluids, an ideal flow rate has been defined as

$$\overset{\circ}{m}{}_c' \equiv A_2 \left[\frac{2 \gamma g_c p_1 \rho_1 (r^{2/\gamma} - r^{(\gamma+1)/\gamma})}{(\gamma - 1)(1 - r^{2/\gamma} \beta^4)} \right]^{1/2}, \tag{23.3}$$

where the subscript c indicates a *compressible* quantity. Since this coincides with (21.13), it represents the theoretical flow rate resulting from a one-dimensional workless analysis of a reversible, adiabatic steady flow of a perfect gas through a horizontal fluid meter. Thus if the fluid is condensible (as steam), and if the solid boundaries of the fluid meter constrain the fluid to expand in the axial direction only from inlet to throat (as in a nozzle and venturi), a reasonable discharge coefficient could be established in terms of flowing steam by obtaining $\overset{\circ}{m}_{\text{actual}}$ as before while at the same time forming $\overset{\circ}{m}{}_c'$ via (23.3) from a measured Δp across the meter.

However, and this is the salient point, failure of $C_{D\text{-steam}}$ to coincide with $C_{D\text{-liquid}}$ at the same Reynolds number may only indicate failure of (23.3) to represent the theoretical compressible flow, failure of (23.1) to represent the theoretical constant density flow, or both. Actually, $C_{D\text{-steam}}$ so determined agrees remarkably well with $C_{D\text{-liquid}}$. But in general, it is far more convenient to settle on the use of a single C_D-Re relationship for all fluids (namely, the one established by liquid tests), and to introduce an expansion factor Y to account for deviations between compressible fluids and the arbitrary hydraulic fluid of (23.1). That is

$$Y \equiv \frac{\overset{\circ}{m}_{\text{actual}}}{\overset{\circ}{m}_i}. \tag{23.4}$$

Thus for *any* fluid, through *any* fluid meter, the actual flow rate can be determined from

$$\overset{\circ}{m}_{\text{actual}} = Y(C_{Di} \overset{\circ}{m}{}_i') \tag{23.5}$$

where $Y_{\text{liquid}} = 1$, by (23.4).

In general, Y is to be determined empirically from experiment, although it must reflect and be strongly influenced by theoretical considerations; for example, with C_{Di} established by liquid tests, the product $(C_{Di} Y)$ could be obtained for steam via equation (23.5). A plot of $(C_{Di} Y)$ versus $(\Delta p / p_1 \gamma)$, this latter group based on theoretical reasoning, would yield in general the polynomial

$$(C_{Di}Y) = (C_{Di}Y)_{\Delta p=0} + A\left(\frac{\Delta p}{p_1\gamma}\right) + B\left(\frac{\Delta p}{p_1\gamma}\right)^2 + \cdots \qquad (23.6)$$

Noting that the intercept $(C_{Di}Y)_{\Delta p=0}$ reduces to C_{Di}, since $Y_{\Delta p=0}=1$, (23.6) also can be written as

$$Y = 1 + \left(\frac{A}{C_{Di}}\right)\left(\frac{\Delta p}{p_1\gamma}\right) + \left(\frac{B}{C_{Di}}\right)\left(\frac{\Delta p}{p_1\gamma}\right)^2 + \cdots \qquad (23.7)$$

23.2 Experimental Determination of Y

To use the method of (23.5) to determine the compressible flow rate it is clear, for gases and vapors, that Y must be available. Actually, Y is determined experimentally by means of a version of (23.5), namely,

$$Y = \frac{(\mathring{m}_{\text{actual}}/\mathring{m}_i')}{C_{Di}} = \frac{(C_{Di}Y)}{C_{Di}} \qquad (23.8)$$

where (a) $\mathring{m}_{\text{actual}}$ is determined by a catching-weighing-timing procedure for condensible vapors; by catching-timing-and thermodynamic deduction for the permanent gases; or by installing a standard fluid meter in series with the test meter whose expansion factor is to be determined.

(b) $\mathring{m}_i'$ is determined quite simply from (23.1).

(c) C_{Di} is obtained either from ASME tables or equations, if they are established for the given test meter; or by extrapolation of the experimental data, as shown typically in Figure 23.1. Note that C_{Di} from ASME shows very close agreement with that from extrapolated data, for established fluid meters.

23.3 Analytical Determination of Y

It is always useful to predict a factor such as Y analytically: first, to prove our understanding of the phenomena involved; then, to aid our reasoning when unusual fluid meters are involved; and finally to help in determining the flow rate through fluid meters whose characteristics are not available from experiment.

It has been shown by Marxman and Burlage [1] and by the author [2] that (23.4) leads to a generalized theoretical expansion factor

$$Y_{\text{general}} = \frac{C_{vc}C_{cc}}{C_{vi}C_{ci}}\left[\left(\frac{\gamma}{\gamma-1}\right)\frac{r^{2/\gamma}(1-r^{(\gamma-1)/\gamma})(1-C_{ci}^2\beta^4)}{(1-r)(1-r^{2/\gamma}C_c^2\beta^4)}\right]^{1/2}. \qquad (23.9)$$

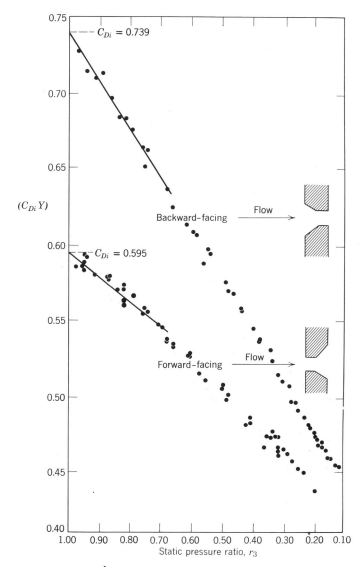

Figure 23.1 Experimental determination of C_{Di} and Y. Sharp-edged orifice $\beta = 0.2405$ — line for intercept purpose only, ... experimental points.

This equation displays the factors that should influence the expansion factor. Specifically these are: the velocity coefficients, C_{vc} and C_{vi}; the contraction coefficients, C_{cc} and C_{ci}; the ratio of specific heats, γ; the static pressure ratio across the fluid meter, r; and the diameter ratio across the fluid meter, β. Note that all these factors are dimensionless ratios.

Equation 23.9 has been extended [2] to the supercritical regions, where the actual pressure ratio (r) across the fluid meter is less than the critical pressure ratio (r_*), defined by

$$r_*^{(1-\gamma)/\gamma} + \left(\frac{\gamma-1}{2}\right) r_*^{2/\gamma} C_{c*}^2 \beta^4 = \frac{\gamma+1}{2}. \qquad (23.10)$$

This critical pressure ratio was first defined by Cunningham [3], and later expanded [4]. The generalized expansion factor for supercritical flow can be given in terms of r_* as

$$Y_{\text{general}}^* = \frac{C_{v*} C_{c*}}{C_{vi} C_{ci}} \left[\left(\frac{\gamma}{\gamma-1}\right) \frac{r_*^{2/\gamma}(1 - r_*^{(\gamma-1)/\gamma})(1 - C_{ci}^2 \beta^4)}{(1-r)(1 - r_*^{2/\gamma} C_{c*}^2 \beta^4)} \right]^{1/2}. \qquad (23.11)$$

Values for the generalized expansion factor, subsonic and supercritical, for various contractions and various betas are given in Tables 23.1–23.6. The contraction coefficients used in defining (23.9) and (23.11) are obtained from [5], and samples are given in Tables 23.7 and 23.8. The velocity coefficients are never specified. Instead, C_{vc} is set equal to C_{vi} for the generation of these tables, with little loss in generality.

23.4   Comparisons of Experimental and Analytical Ys

Typical comparisons of Y determined experimentally, as in Figure 23.1, and analytically, as in Tables 23.1–23.6, are given in Figure 23.2. The agreement is satisfactory.

A word should be said to clarify the use of the symbols r and R in these figures and tables. The symbol r is consistently the static-to-static pressure ratio across the fluid meter. The symbol R represents the downstream static pressure ratioed to the upstream total pressure. When flow is from an inlet plenum where $p_1 = p_t$, it follows that $r = R$.

23.5   Expansion Factors for Nozzles and Venturis

For nozzles and venturis, compressible fluid expansions (in the fluid meter) are limited to the axial direction. Thus the restrictions that

Table 23.1 Generalized expansion factors. ($\gamma=1.4$, $\beta=0$)

r	C_{ci} = 0.6	0.7	0.8	0.9	1.0
.98	.994982	.993538	.992109	.990668	.989228
.96	.989898	.987030	.984146	.981240	.978337
.94	.984785	.980453	.976103	.971721	.967325
.92	.979598	.973808	.967984	.962105	.956187
.90	.974370	.967112	.959790	.952395	.944920
.88	.969085	.960350	.951523	.942584	.933518
.86	.963744	.953524	.943161	.932663	.921979
.84	.958345	.946630	.934738	.922636	.910296
.82	.952883	.939654	.926218	.912502	.898466
.80	.947355	.932618	.917621	.902262	.886483
.78	.941767	.925515	.908932	.891915	.874341
.76	.936098	.918326	.900158	.881450	.862035
.74	.930365	.911078	.891292	.870872	.849558
.72	.924569	.903736	.882336	.860178	.836903
.70	.918680	.896311	.873302	.849381	.824064
.68	.912723	.888811	.864166	.838455	.811033
.66	.906688	.881228	.854934	.827417	.797801
.64	.900551	.873543	.845619	.816267	.784361
.62	.894335	.865773	.836202	.805007	.770702
.60	.888021	.857914	.826687	.793637	.756814
.58	.881607	.849953	.817086	.782165	.742687
.56	.875088	.843516	.809069	.772217	.729789
.54	.868711	.833967	.797817	.759190	.713882
.52	.861707	.825433	.787684	.747243	.698851
.50	.854853	.817061	.777735	.735531	.684731
.48	.847943	.808666	.767810	.723970	.671434
.46	.841009	.800279	.757971	.712641	.658883
.44	.834083	.791950	.748276	.701604	.647010
.42	.827162	.783707	.738742	.690899	.635757
.40	.820302	.775562	.729411	.680543	.625071
.38	.813491	.767536	.720294	.670536	.614907
.36	.806742	.759641	.711394	.660876	.605223
.34	.800076	.751891	.702723	.651561	.595982
.32	.793497	.744283	.694278	.642586	.587152
.30	.787008	.736830	.686068	.633925	.578704
.28	.780622	.729524	.678077	.625572	.570610
.26	.774326	.722360	.670314	.617517	.562846
.24	.768133	.715391	.662768	.609745	.555391
.22	.762040	.708556	.655429	.602236	.548224
.20	.756054	.701870	.648301	.594985	.541328
.18	.750164	.695335	.641375	.587976	.534685
.16	.744384	.688947	.634638	.581198	.528282
.14	.738697	.682709	.628083	.574638	.522103
.12	.733119	.676610	.621711	.568288	.516136
.10	.727641	.670639	.615511	.562143	.510369
.08	.722263	.664809	.609484	.556178	.504791
.06	.716977	.659111	.603609	.550396	.499392
.04	.711791	.653537	.597893	.544792	.494162
.02	.706700	.648091	.592323	.539346	.489094
0.00	.701695	.642760	.586903	.534059	.484178

($\gamma = 1.4$, $\beta = 0$)

Table 23.2 Generalized expansion factors. ($\gamma=1.4$, $\beta=0.4$)

	For Incompressible Contraction Coefficients, C_{ci}, of				
r	C_{ci} = 0.6	0.7	0.8	0.9	1.0
.98	.994903	.993412	.991921	.990399	.988858
.96	.989740	.986781	.983774	.980710	.977610
.94	.984551	.980082	.975552	.970937	.966252
.92	.979287	.973318	.967257	.961075	.954780
.90	.973983	.966505	.958893	.951126	.943190
.88	.968625	.959629	.950460	.941084	.931478
.86	.963211	.952691	.941935	.930940	.919640
.84	.957739	.945688	.933355	.920696	.907671
.82	.952207	.938604	.924682	.910354	.895567
.80	.946608	.931463	.915936	.899913	.883321
.78	.940952	.924258	.907104	.889374	.870929
.76	.935214	.916968	.898191	.878723	.858385
.74	.929414	.909623	.889191	.867968	.845683
.72	.923553	.902186	.880105	.857105	.832815
.70	.917600	.894669	.870945	.846107	.819776
.68	.911579	.887079	.861689	.835067	.806557
.66	.905482	.879409	.852342	.823884	.793151
.64	.899283	.871638	.842916	.812597	.779548
.62	.893008	.863787	.833393	.801208	.765740
.60	.886635	.855847	.823777	.789718	.751717
.58	.880164	.847809	.814080	.778134	.737467
.56	.873588	.841212	.805759	.767835	.724217
.54	.867085	.831593	.794539	.754856	.708353
.52	.860158	.823148	.784504	.743017	.693439
.50	.853376	.814857	.774644	.731401	.679426
.48	.846534	.806536	.764799	.719927	.666227
.46	.839661	.798217	.755031	.708674	.653765
.44	.832792	.789948	.745400	.697706	.641974
.42	.825924	.781760	.735923	.687063	.630796
.40	.819112	.773664	.726644	.676764	.620180
.38	.812345	.765682	.717572	.666807	.610081
.36	.805635	.757828	.708713	.657194	.600456
.34	.799006	.750113	.700078	.647921	.591272
.32	.792461	.742537	.691666	.638985	.582495
.30	.786002	.735113	.683485	.630360	.574095
.28	.779644	.727833	.675521	.622040	.566048
.26	.773372	.720712	.667783	.614016	.558328
.24	.767202	.713744	.660259	.606273	.550915
.22	.761131	.706928	.652941	.598792	.543789
.20	.755165	.700260	.645832	.591568	.536932
.18	.749292	.693742	.638925	.584585	.530328
.16	.743530	.687369	.632206	.577832	.523961
.14	.737859	.681146	.625669	.571297	.517819
.12	.732296	.675061	.619314	.564973	.511887
.10	.726833	.669105	.613132	.558852	.506156
.08	.721470	.663290	.607123	.552913	.500614
.06	.716199	.657607	.601266	.547157	.495251
.04	.711029	.652048	.595570	.541580	.490058
.02	.705954	.646620	.590021	.536162	.485027
0.00	.700966	.641308	.584624	.530905	.480150

(γ = 1.4, β = 0.4)

Table 23.3 Generalized expansion factors. ($\gamma=1.4$, $\beta=0.6$)

r	For Incompressible Contraction Coefficients, C_{ci}, of				
	C_{ci} = 0.6	0.7	0.8	0.9	1.0
.98	.994565	.992866	.991088	.989183	.987139
.96	.989070	.985701	.982131	.978316	.974237
.94	.983552	.978478	.973119	.967404	.961289
.92	.977965	.971201	.964057	.956441	.948292
.90	.972343	.963886	.954946	.945431	.935240
.88	.966670	.956520	.945789	.934365	.922130
.86	.960947	.949102	.936562	.923235	.908955
.84	.955171	.941631	.927301	.912043	.895713
.82	.949339	.934090	.917968	.900791	.882396
.80	.943445	.926503	.908585	.889478	.868999
.78	.937499	.918863	.899137	.878104	.855518
.76	.931477	.911149	.889630	.866656	.841945
.74	.925397	.903392	.880056	.855142	.828274
.72	.919261	.895554	.870420	.843558	.814499
.70	.913038	.887647	.860732	.831916	.800612
.68	.906752	.879680	.850968	.820190	.786606
.66	.900395	.871644	.841136	.808399	.772472
.64	.893942	.863518	.831247	.796540	.758202
.62	.887418	.855324	.821282	.784617	.743787
.60	.880801	.847053	.811247	.772631	.729216
.58	.874091	.838696	.801153	.760587	.714480
.56	.868299	.830996	.791511	.748874	.699827
.54	.860373	.821704	.780734	.736390	.684497
.52	.853770	.813631	.771114	.725004	.670078
.50	.847264	.805640	.761569	.713719	.656436
.48	.840645	.797543	.751938	.702459	.643495
.46	.833943	.789371	.742290	.691314	.631193
.44	.827190	.781177	.732689	.680360	.619476
.42	.820387	.772994	.723161	.669646	.608294
.40	.813591	.764838	.713756	.659202	.597608
.38	.806792	.756737	.704493	.649037	.587380
.36	.800006	.748708	.695384	.639163	.577580
.34	.793260	.740770	.686450	.629584	.568179
.32	.786559	.732927	.677698	.620306	.559154
.30	.779912	.725201	.669142	.611312	.550483
.28	.773336	.717568	.660778	.602605	.542147
.26	.766823	.710111	.652622	.594183	.534130
.24	.760392	.702770	.644672	.586039	.526417
.22	.754044	.695572	.636924	.578161	.518994
.20	.747793	.688519	.629392	.570552	.511852
.18	.741631	.681619	.622073	.563201	.504978
.16	.735581	.674876	.614963	.556106	.498366
.14	.729631	.668300	.608063	.549261	.492007
.12	.723802	.661887	.601380	.542665	.485894
.10	.718092	.655636	.594911	.536318	.480022
.08	.712509	.649565	.588664	.530205	.474386
.06	.707051	.643673	.582628	.524334	.468983
.04	.701735	.637963	.576818	.518709	.463812
.02	.696565	.632451	.571234	.513323	.458874
0.00	.691554	.627146	.565903	.508201	.454183

$(\gamma = 1.4, \beta = 0.6)$

461

Table 23.4 Generalized expansion factors. ($\gamma=1.3$, $\beta=0$)

r	$C_{ci} = 0.6$	0.7	0.8	0.9	1.0
.98	.994631	.993177	.991864	.991161	.988404
.96	.989194	.986277	.983614	.982140	.976692
.94	.983706	.979305	.975280	.972921	.964859
.92	.978171	.972274	.966828	.963502	.952903
.90	.972572	.965160	.958284	.953887	.940820
.88	.966925	.957988	.949626	.944081	.928605
.86	.961211	.950731	.940867	.934060	.916254
.84	.955428	.943400	.932006	.923847	.903763
.82	.949586	.936002	.923041	.913436	.891128
.80	.943662	.928521	.913960	.902819	.878343
.78	.937693	.920962	.904779	.892009	.865403
.76	.931638	.913315	.895493	.880997	.852303
.74	.925499	.905596	.886106	.869801	.839036
.72	.919303	.897790	.876609	.858410	.825597
.70	.913017	.889895	.867011	.846840	.811979
.68	.906647	.881915	.857310	.835095	.798175
.66	.900183	.873842	.847499	.823185	.784177
.64	.893635	.865674	.837590	.811128	.769977
.62	.886997	.857416	.827581	.798910	.755566
.60	.880243	.849048	.817465	.786577	.740935
.58	.873395	.840586	.807258	.774124	.726074
.56	.866438	.832020	.796955	.761585	.710972

($\gamma = 1.3$, $\beta = 0$)

Table 23.5 Generalized expansion factors. ($\gamma=1.3$, $\beta=0.4$)

r	$C_{ci} = 0.6$	0.7	0.8	0.9	1.0
.98	.994546	.993043	.991667	.990897	.988007
.96	.989025	.986012	.983225	.981616	.975912
.94	.983455	.978911	.974703	.972144	.963711
.92	.977839	.971754	.966068	.962476	.951400
.90	.972160	.964517	.957345	.952618	.938976
.88	.966434	.957224	.948513	.942576	.926434
.86	.960643	.949849	.939585	.932324	.913772
.84	.954783	.942402	.930560	.921888	.900983
.82	.948866	.934893	.921436	.911260	.888064
.80	.942868	.927302	.912200	.900434	.875009
.78	.936827	.919635	.902870	.889423	.861814
.76	.930701	.911885	.893440	.878216	.848473
.74	.924492	.904064	.883914	.866835	.834979
.72	.918227	.896160	.874283	.855266	.821328
.70	.911874	.888169	.864557	.843528	.807512
.68	.905439	.880097	.854732	.831622	.793524
.66	.898910	.871935	.844804	.819562	.779356
.64	.892299	.863680	.834783	.807363	.765001
.62	.885600	.855339	.824667	.795013	.750450
.60	.878786	.846891	.814450	.782557	.735694
.58	.871879	.838352	.804148	.769991	.720722
.56	.864865	.829711	.793755	.757349	.705523

($\gamma = 1.3$, $\beta = 0.4$)

Table 23.6 Generalized expansion factors. ($\gamma=1.3$, $\beta=0.6$)

| r | For Incompressible Contraction Coefficients, C_{ci}, of | | | | |
	$C_{ci} = 0.6$	0.7	0.8	0.9	1.0
.98	.994185	.992463	.990798	.989700	.986160
.96	.988307	.984864	.981508	.979252	.972296
.94	.982387	.977207	.972161	.968639	.958402
.92	.976425	.969508	.962722	.957861	.944475
.90	.970407	.961740	.953220	.946923	.930511
.88	.964348	.953931	.943631	.935835	.916503
.86	.958230	.946051	.933969	.924568	.902447
.84	.952048	.938112	.924234	.913151	.888337
.82	.945815	.930123	.914423	.901578	.874169
.80	.939507	.922066	.904524	.889843	.859937
.78	.933162	.913947	.894555	.877959	.845634
.76	.926737	.905756	.884510	.865918	.831255
.74	.920236	.897509	.874394	.853742	.816794
.72	.913685	.889191	.864197	.841419	.802242
.70	.907052	.880799	.853929	.828966	.787594
.68	.900343	.872340	.843588	.816388	.772842
.66	.893546	.863804	.833167	.803696	.757977
.64	.886674	.855189	.822679	.790909	.742991
.62	.879720	.846501	.812120	.778014	.727876
.60	.872657	.837718	.801484	.765056	.712621
.58	.865508	.828859	.790790	.752032	.697217
.56	.858258	.819913	.780028	.738974	.681652

($\gamma = 1.3$, $\beta = 0.6$)

accompanied (23.3) should be fairly well met. Specifically, there should be little loss ($C_v \to 1$), and little contraction ($C_c \to 1$). Hence (23.4) reduces to

$$Y' = \frac{\mathring{m}'_c}{\mathring{m}'_i} \qquad (23.12)$$

where the prime consistently indicates the ideal situation. From either (23.12) or from (23.9) there results

$$Y' = \left[\left(\frac{\gamma}{\gamma - 1} \right) \frac{r^{2/\gamma}(1 - r^{(\gamma - 1)/\gamma})(1 - \beta^4)}{(1 - r)(1 - r^{2/\gamma}\beta^4)} \right]^{1/2} \qquad (23.13)$$

Equation 23.13 is known as the adiabatic expansion factor [6], and this has been shown to be a very close approximation to the facts (as also borne out by the close agreement between $C_{D\text{-liquid}}$ and $C_{D\text{-steam}}$ for nozzles and venturis). Hence for nozzles and venturis there is no need to search for the empirical expansion factor of the form of (23.7). We simply use (23.13). That is,

$$\mathring{m}_{\text{nozzle-}\atop\text{venturi}} = C_{Di} \times Y' \times \mathring{m}'_i. \qquad (23.14)$$

Note that the adiabatic expansion factor of (23.13) appears in Tables 23.1–23.6 in the last column where $C_{ci} = 1$.

Table 23.7 Compressible contraction coefficients. ($\gamma=1.4$, $\beta=0$)

r	For Incompressible Contraction Coefficients, C_{ci}, of			
	$C_{ci} = 0.6$	0.7	0.8	0.9
.98	.603490	.703050	.802330	.901310
.96	.607090	.706220	.804750	.902670
.94	.610830	.709500	.807260	.904090
.92	.614690	.712900	.809870	.905570
.90	.618700	.716440	.812590	.907120
.88	.622860	.720120	.815430	.908740
.86	.627180	.723950	.818380	.910430
.84	.631670	.727940	.821480	.912200
.82	.636340	.732090	.824710	.914060
.80	.641200	.736430	.828100	.916020
.78	.646270	.740970	.831650	.918090
.76	.651550	.745710	.835380	.920270
.74	.657070	.750690	.839300	.922580
.72	.662850	.755900	.843430	.925030
.70	.668890	.761370	.847800	.927650
.68	.675230	.767130	.852410	.930430
.66	.681890	.773200	.857290	.933410
.64	.688880	.779590	.862480	.936610
.62	.696250	.786350	.867990	.940060
.60	.704020	.793510	.873860	.943790
.58	.712230	.801100	.880140	.947840
.56	.720920	.809160	.886850	.952270
.54	.730130	.817750	.894060	.957120
.52	.739820	.826790	.901690	.962320
.50	.749070	.835280	.908660	.966770
.48	.757730	.843070	.914830	.970420
.46	.765850	.850220	.920310	.973430
.44	.773480	.856810	.925210	.975940
.42	.780640	.862900	.929590	.978060
.40	.787400	.868530	.933540	.979870
.38	.793770	.873750	.937110	.981420
.36	.799780	.878600	.940340	.982760
.34	.805470	.883120	.943280	.983930
.32	.810860	.887330	.945960	.984970
.30	.815970	.891270	.948420	.985880
.28	.820830	.894950	.950670	.986690
.26	.825440	.898410	.952750	.987420
.24	.829830	.901660	.954670	.988080
.22	.834010	.904720	.956440	.988670
.20	.838000	.907600	.958090	.989210
.18	.841800	.910320	.959630	.989700
.16	.845440	.912890	.961060	.990150
.14	.848910	.915330	.962390	.990560
.12	.852240	.917640	.963640	.990940
.10	.855430	.919820	.964810	.991300
.08	.858490	.921900	.965920	.991620
.06	.861420	.923880	.966950	.991920
.04	.864240	.925760	.967930	.992210
.02	.866950	.927560	.968650	.992470
0.00	.869550	.929270	.969730	.992720

($\gamma = 1.4$, $\beta = 0$)

Table 23.8 Compressible contraction coefficients. ($\gamma=1.3$, $\beta=0$)

For Incompressible Contraction Coefficients, C_{ci}, of				
r	$C_{ci} = 0.6$	0.7	0.8	0.9
.98	.603780	.703380	.802800	.902510
.96	.607680	.706870	.805670	.905020
.94	.611720	.710480	.808640	.907520
.92	.615910	.714230	.811690	.910010
.90	.620250	.718110	.814850	.912500
.88	.624760	.722150	.818110	.915000
.86	.629440	.726340	.821490	.917490
.84	.634300	.730700	.825000	.920000
.82	.639360	.735250	.828650	.922530
.80	.644620	.739990	.832440	.925080
.78	.650120	.744940	.836400	.927670
.76	.655850	.750110	.840540	.930300
.74	.661830	.755530	.844880	.933000
.72	.668100	.761210	.849430	.935770
.70	.674660	.767170	.854220	.938640
.68	.681540	.773440	.859270	.941630
.66	.688760	.780040	.864600	.944770
.64	.696360	.787000	.870250	.948100
.62	.704370	.794360	.876250	.951630
.60	.712810	.802140	.882630	.955440
.58	.721740	.810400	.889450	.959560
.56	.731200	.819180	.896750	.964070
($\gamma = 1.3$, $\beta = 0$)				

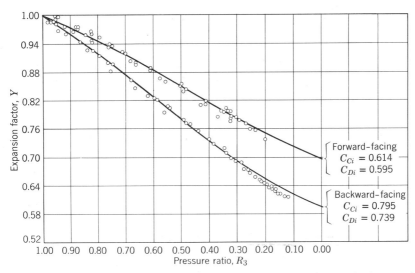

Figure 23.2 Comparison between experimental results and the generalized expansion factor. Orifice, $\beta = 0.2405$. —— Y generalized, ∘∘∘∘∘ Y experimental.

465

23.6   Expansion Factors for Orifices

For square-edged orifices, on the other hand, it is clear that radial as well as axial expansions are to be expected as a compressible fluid passes from the throat of the orifice towards the vena contracts. This two-dimensional expansion means that Y' of (23.13) is no longer a valid approximation to the facts, and one must either seek a new theoretical expression for $Y_{2\text{-}D}$ or resort to an empirical-experimental expression of the form of (23.7).

Buckingham [7], [8] laid the groundwork for a theoretical expansion factor and found by correlation (based primarily on the experimental work of Bean) that the relation $Y = f(r)$ was very nearly linear for

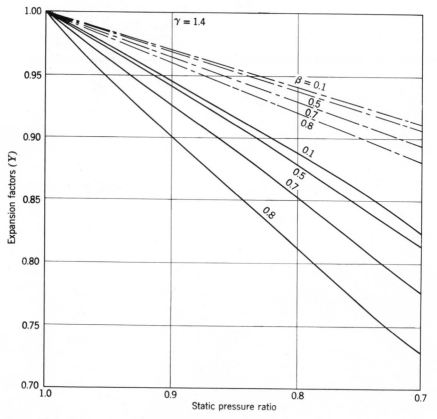

Figure 23.3 Comparison of expansion factors for nozzles, venturis, and orifices. Key: (—––—) flange taps; (———) venturis and nozzles.

orifices. His empirical results were given as

$$Y_{2\text{-}D} = 1 - \frac{(0.41 + 0.35\beta^4)(1-r)}{\gamma}.$$ (23.15)

This equation was adopted by the ASME [6] for flange and 1 and $\frac{1}{2}D$ taps.

Note that this equation is in the form of (23.7). Figure 23.3 summarizes the two expansion factors given.

Cunningham [3] extended both by theoretical reasoning and by experiment, the present range of the ASME orifice expansion factors to include "supercritical" flow.

Murdock and Foltz [9] experimentally confirmed that the ASME linear expansion factors could be applied with equal precision to the flow of steam through orifices, for subsonic flows.

Thus, for flange and $1\,D$ and $\frac{1}{2}D$ tap orifices, we use the empirical (23.15). That is,

$$\mathring{m}_{\text{orifice}} = C_{Di} \times Y_{2\text{-}D} \times \mathring{m}'_i.$$ (23.16)

23.7 A Problem in Fluid Metering

It is clear from an examination of Tables 23.1–23.3 that the Y of (23.9) and (23.11) does not always equal the Y' of (12.12) and (23.13). That is, in general

$$Y \geq Y'.$$ (23.17)

If a compressible discharge coefficient is defined, patterned after (23.2), that is,

$$C_{Dc} = \frac{\mathring{m}_c}{\mathring{m}'_c}$$ (23.18)

it follows from (23.4) that

$$Y = \left(\frac{C_{Dc}}{C_{Di}}\right)\left(\frac{\mathring{m}'_c}{\mathring{m}'_i}\right) = \frac{C_{Dc}}{C_{Di}} \times Y'.$$ (23.19)

But, in view of (23.17), this requires that $C_{Dc} \geq C_{Di}$. Although this has been pointed out in the literature [10], the fact that in a consistent system C_{Dc} will generally exceed C_{Di} has not received wide acceptance. It remains one of the unresolved problems in fluid metering whether C_{Dc} can be used interchangeably with C_{Di}.

23.8 References

[1] G. A. Marxman and H. Burlage, Jr., "Expansion Coefficients for Orifice Meters in Pipes Less Than One Inch in Diameter," *Trans. ASME, J. Basic Eng.*, June 1961, p. 289.

[2] R. P. Benedict, "Generalized Expansion Factor of an Orifice for Subsonic and Supercritical Flows," *Trans. ASME, J. Basic Eng.*, June 1971, p. 121.

[3] R. G. Cunningham, "Orifice Meters with Supercritical Compressible Flow," *Trans. ASME*, July 1951, p. 625.

[4] R. P. Benedict and R. D. Schulte, "A Note on the Critical Pressure Ratio across a Fluid Meter," *Trans. ASME, J. Fluids Eng.*, September 1973, p. 337.

[5] R. P. Benedict, "Generalized Contraction Coefficient of an Orifice for Subsonic and Supercritical Flows," *Trans. ASME, J. Basic Eng.*, June 1971, p. 99.

[6] *Fluid Meters—Their Theory and Application*, Report of the ASME Research Committee on Fluid Meters, 6th ed., 1971.

[7] E. Buckingham, "Notes on Contraction Coefficients of Jets of Gas," *J. Research, Nat. Bur. Std.*, **6**, 1931, p. 765.

[8] E. Buckingham, "Notes on the Orifice Meter; The Expansion Factor for Gases," *J. Research, Nat. Bur. Std.*, **9**, 1932, p. 61.

[9] J. W. Murdock and C. J. Foltz, "Experimental Evaluation of Expansion Factors for Steam," *Trans. ASME*, July 1953, p. 953.

[10] R. P. Benedict, A. R. Gleed, and R. D. Schulte, "Air and Water Studies on a Diffuser-Modified Flow Nozzle," *Trans. ASME, J. Fluids Eng.*, June 1973, p. 169.

Nomenclature

Roman

A area; also a constant
B a constant
C_c contraction coefficient
C_D discharge coefficient
C_v velocity coefficient
g acceleration of gravity
$\overset{\circ}{m}$ mass rate of flow
p absolute pressure
r static pressure ratio
t time
Y expansion factor

Greek

β diameter ratio
γ ratio of specific heats
Δ finite difference

Subscripts

c compressible
i incompressible

Problems

1. A fluid meter calibration in water yields $C_{Di} = 0.98$ at $R_d \geq 10^6$. If the ideal incompressible flow rate for steam is computed at 4 lbm/sec, and the actual flow rate determined by condensing the steam and weighing it is 3.5 lbm/sec, find the expansion factor.

    Ans. $Y = 0.89286$.

2. A nozzle is used to meter air from a plenum inlet at a pressure ratio $R = 0.9$. Compute the appropriate expansion factor.

    Ans. $Y' = 0.94492$.

3. For the values given in Problem 2, find Y for a flange tap orifice.

    Ans. $Y_{2-D} = 0.97071$.

Chapter 24

INSTALLATIONS AND UNCERTAINTIES

"... but I have nothing more to say ... nor can I see any reason for doubt after what has been said. But I still feel and cannot help feeling uncertain in my own mind..."

Plato (c. 400 B.C.)

24.1 Installation of Flow Meters

The usual condition specified for accurate flow measurements, with respect to installation effects, is that the fluid entering the flow meter exhibit a fully developed turbulent velocity profile, free from swirls or vortices. Since bends, tees, valves, abrupt changes in area, and the like will complicate an otherwise uniform velocity profile by inducing secondary flows and skewed velocity distributions, it is apparent that some requirements as to location of flow meters be established [1].

The ASME indicates that such flow disturbances will be minimized by the use of adequate lengths of straight pipes both preceding and following the flow meter. Minimum lengths of such piping are shown in the eight diagrams of Figure 24.1.

Murdock, Foltz, and Gregory [2] point out that changes of indicated flow rate due to the presence of a globe valve placed six pipe diameters before an orifice will be under 2% for all βs less than 0.75, and that furthermore this result is independent of the fluid and of the pressure tap setup. They also indicate that the effect of nonstandard conditions becomes smaller at the smaller βs, and that from 20 to 40 pipe diameters of straight pipe are required to avoid any effect of such valves.

The use of single pressure taps is especially dangerous when nonstandard flow conditions prevail. For example, Ferron [3] has shown that

470

whereas four ring-manifolded taps indicated at $\frac{1}{2}$% variation in the discharge coefficient for a given change in the nonstandard installation, a single tap under these same conditions showed a ±10% change in C_D, depending on where the tap was located around the pipe.

Straightening vanes have been suggested as a means for reducing the length of straight pipe required by the various nonstandard approaches. These can be very effective provided they are properly designed and fastened. For the tubular type, for example, the ASME [1] recommends that the maximum distance between tube centers should not exceed $D/4$, and the overall length of the tubes should be at least eight times the distance between tube centers.

An approach different from that of the ASME has been suggested by Clark [4], [5] wherein one settles for a given percentage off in flow depending on the spacewise length of straight pipe he can afford in a given installation (see Table 24.1).

24.2 Uncertainties

The ASME indicates in broad empirical terms some tolerances to be expected in their discharge coefficients and expansion factors. Briefly, the tolerance on discharge coefficients of venturis can vary linearly from ±2.25% at $R_D = 5 \times 10^4$ down to ±0.75% at $R_D = 2 \times 10^5$, when installed in pipes above 4 in. in diameter. Flow nozzle discharge coefficients can vary ±0.95% for $R_D \geq 3 \times 10^4 \times D$, whereas orifices in pipes of 2-in. diameter and above are characterized by discharge coefficient tolerances of about ±0.55% for $R_D \geq 5 \times 10^3 \times D$, where D preceding is in inches. ASME [6] suggests a ±0.5% addition to all these tolerances to account for installation variations. On top of these figures, one must also consider a tolerance of up to ±0.5% on the expansion factor, where applicable. It is clear that claims of ±0.1% in flow measurements are exceedingly unlikely, while flow rates said to be determined to ±1.5% are within the reach of careful experimenters.

Of course, there are nonstandard fluid meters or nonstandard installations that require amplified uncertainties over those given by the ASME. Some recent papers on some of these effects concern orifice plate eccentricity [7], fluid meters not manufactured for controlled test programs [8], nozzle pipe wall taps installed on test sites [9], and the effects of edge sharpness on orifice discharge coefficients [10].

Entirely separate from the question of tolerance in discharge coefficient and expansion factor is the matter of the experimental basis of the pressure ratio r required in all the compressible flow equations of Chapter

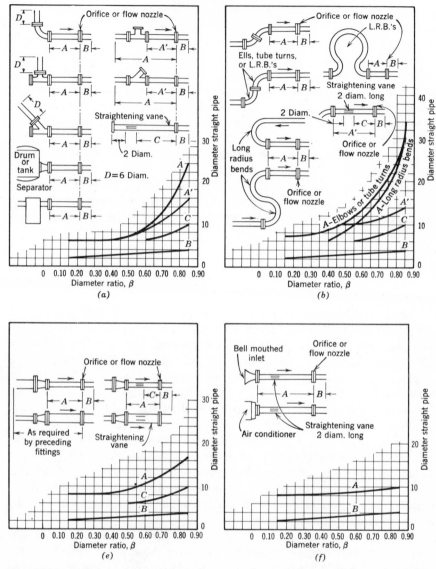

Figure 24.1 Piping requirements for orifices, flow nozzles, and venturi tubes. (From *Flow Meter Computation Handbook*, 1961.) *Note* 1: All control valves must be installed on outlet side of primary element. *Note* 2: In diagram *h* the distances shown are double those at which there seemed to be no effect.

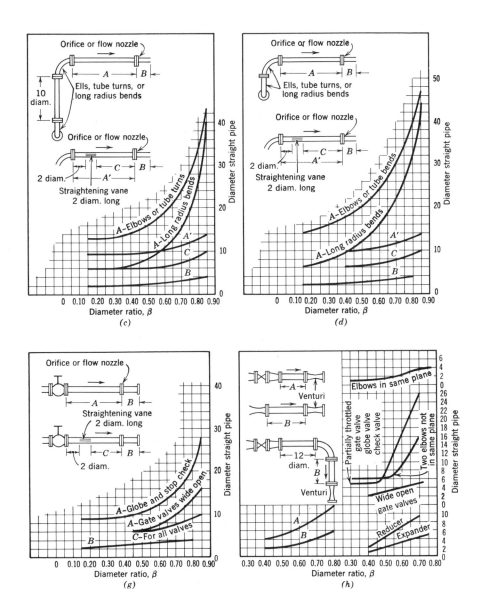

Table 24.1 Recommended Straight Pipe Requirements using Single Tappings to Give Errors in C_D Values not Exceeding $\pm 1\%$, $\pm 1\frac{1}{2}\%$, and $\pm 2\%$, Respectively, After Application of Appropriate Correction Factors[a].

Range of Error	m or β^2	β	A Single, Right-Angle Bend, Two Bends in One Plane, or a Tee	A Globe Valve or a Sluice Valve From Open to $\frac{3}{4}$ Closed	Correction Factor to Basic C_D
$\pm 1\%$	0.5	0.71	8	14	0.99
	0.4	0.63	7	10	
	0.3	0.55	5	8	
	0.2	0.45	4	6	
	0.1	0.32	3	5	
$\pm 1\frac{1}{2}\%$	0.5	0.71	5	9	0.985
	0.4	0.63	4	7	
	0.3	0.55	3	7	
	0.2	0.45	3	6	
	0.1	0.32	3	5	
$\pm 2\%$	0.5	0.71	4	7	0.98
	0.4	0.63	3	6	
	0.3	0.55	3	6	
	0.2	0.45	3	6	
	0.1	0.32	3	5	

Column header spanning the two middle columns: "Upstream Pipe Lengths in Terms of D After"

[a] From Clark [4].

21. This, too, is of prime importance in determining the uncertainty in flow rate and is discussed next.

In forming a ratio, intuitively the difference method (i.e., measuring p_1 and $\Delta p_{1,2}$) seems preferable to the level method (i.e., measuring p_1 and p_2) because, among other reasons, the difference represents a simultaneous sensing of the two levels. However, when possible, let us seek a mathematical basis for our actions [11].

In the level method the pressure ratio is based on the independent measurement of two levels. Thus

$$r = \frac{p_2}{p_1}. \tag{24.1}$$

By logarithmic differentiation,

$$\frac{dr}{r} = \frac{dp_2}{p_2} - \frac{dp_1}{p_1}, \tag{24.2}$$

which expresses the relative variation of the ratio in terms of the relative variation of the levels. To ensure that maximum variations are obtained, (24.2) is rewritten according to the rules of numerical analysis as

$$\left(\frac{\epsilon_r}{r}\right)_L = \pm\left(\frac{\epsilon_1}{p_1} + \frac{\epsilon_2}{p_2}\right), \tag{24.3}$$

where ϵ represents the uncertainty, that is, the possible variation a given quantity might assume, and the subscript L signifies that this relative uncertainty of the ratio is based on level measurements.

The rule used for obtaining (24.3) from (24.1) is the following: "The relative uncertainty of a product or quotient function is not greater than the sum of the products formed by multiplying the relative uncertainty of each independent variable by its exponent." (See Chapter 10 for an extended uncertainty discussion.)

Consider now the difference method. Here the ratio is based on the independent measurement of level one and the difference. Thus

$$r = \frac{p_1 - (p_1 - p_2)}{p_1} = 1 - \frac{\Delta p}{p_1}. \tag{24.4}$$

Again, by differentiating,

$$dr = -d\left(\frac{\Delta p}{p_1}\right), \tag{24.5}$$

which, unlike (24.2), expresses variations in terms of absolute values. Putting (24.5) on a relative basis and applying the same rule used for obtaining (24.3) from (24.1), yields

$$\frac{\epsilon_r}{\Delta p/p_1} = \pm\left(\frac{\epsilon_\Delta}{\Delta p} + \frac{\epsilon_1}{p_1}\right). \tag{24.6}$$

Equation 24.6 can be restated in the form of (24.3) via (24.4) as

$$\left(\frac{\epsilon_r}{r}\right)_\Delta = \pm\frac{(\epsilon_\Delta/\Delta p + \epsilon_1/p_1)}{(p_1/\Delta p - 1)}, \tag{24.7}$$

where the subscript Δ signifies that this relative uncertainty in the ratio involves a difference measurement.

Analysis of (24.3) and (24.7) indicates that as long as $\Delta p < p_1/2$, there is a definite advantage (i.e., a smaller ϵ_r/r) to the difference method of forming pressure ratios. The relative uncertainty in the ratio based on independent level and difference measurements is given very closely by

$$\frac{(\epsilon_r/r)_\Delta}{(\epsilon_1/p_1)} = \pm\left[\frac{(p_1/\Delta p)(\Delta_{max}/L_{max}) + 1}{p_1/\Delta p - 1}\right], \tag{24.8}$$

where Δ_{max} and L_{max} indicate the full scale (max) values of the respective instruments employed (see Figure 24.2).

Having established the fact the pressure ratio always should be based on experimental data obtained by the difference method, we inquire next as to how uncertainties in the variables of (21.14) propagate into the resultant flow rate. Uncertainties in the variables A_2, p_1, and T_t require no special consideration here. These are defined in detail in the numerical examples to follow. But the bracketed terms in (21.14), the so-called fluid meter function,

$$X = \left[\frac{(1 - r^{(\gamma+1)/\gamma}\beta^4)(r^{2/\gamma} - r^{(\gamma+1)/\gamma})}{(1 - r^{2/\gamma}\beta^4)^2} \right]^{1/2} \qquad (24.9)$$

does bear some discussion as to its uncertainty. This is not immediately obvious, since the various functions of r and β in X are not independent of one another. The fluid meter function is given in Tables 24.2 and 24.3.

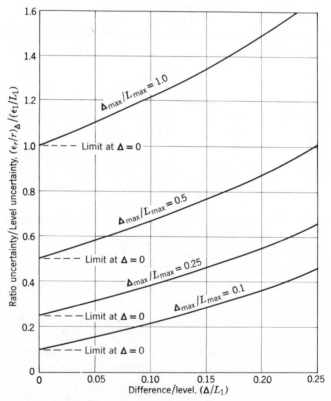

Figure 24.2 Relative uncertainty in ratio based on difference method (and equation 24.8).

Table 24.2 Generalized Fluid Meter Function X

(Isentropic Exponent=1.40) Flow Equation Constant=2.055597

p_2/p_1	D_2/D_1 0.0	0.1	0.2	0.3	0.4	0.5	0.6	0.7	0.8
0.99	0.047761	0.047763	0.047798	0.047953	0.048376	0.049305	0.051143	0.054676	0.061878
0.98	0.067149	0.067152	0.067201	0.067415	0.068002	0.069291	0.071835	0.076714	0.086610
0.97	0.081754	0.081758	0.081817	0.082074	0.082779	0.084326	0.087376	0.093211	0.104984
0.96	0.093838	0.093842	0.093909	0.094200	0.094997	0.096747	0.100194	0.106771	0.119975
0.95	0.104280	0.104285	0.104358	0.104677	0.105551	0.107467	0.111237	0.118414	0.132751
0.94	0.113535	0.113540	0.113619	0.113961	0.114899	0.116955	0.120994	0.128667	0.143916
0.93	0.121874	0.121879	0.121962	0.122325	0.123317	0.125491	0.129758	0.137843	0.153834
0.92	0.129474	0.129480	0.129567	0.129947	0.130986	0.133261	0.137720	0.146151	0.162745
0.91	0.136460	0.136466	0.136556	0.136951	0.138030	0.140391	0.145014	0.153735	0.170817
0.90	0.142921	0.142927	0.143020	0.143427	0.144541	0.146976	0.151739	0.160701	0.178174
0.89	0.148926	0.148933	0.149028	0.149446	0.150589	0.153087	0.157967	0.167131	0.184910
0.88	0.154529	0.154536	0.154634	0.155061	0.156229	0.158780	0.163759	0.173086	0.191099
0.87	0.159772	0.159779	0.159879	0.160314	0.161503	0.164099	0.169159	0.178617	0.196801
0.86	0.164691	0.164697	0.164799	0.165240	0.166447	0.169080	0.174206	0.183767	0.202064
0.85	0.169312	0.169319	0.169422	0.169869	0.171090	0.173753	0.178932	0.188569	0.206930
0.84	0.173662	0.173668	0.173772	0.174223	0.175456	0.178143	0.183361	0.193051	0.211432
0.83	0.177759	0.177765	0.177870	0.178324	0.179566	0.182270	0.187517	0.197239	0.215599
0.82	0.181621	0.181628	0.181733	0.182190	0.183437	0.186154	0.191419	0.201153	0.219456
0.81	0.185263	0.185270	0.185375	0.185834	0.187086	0.189810	0.195083	0.204811	0.223026
0.80	0.188698	0.188705	0.188811	0.189270	0.190524	0.193251	0.198523	0.208230	0.226327
0.79	0.191938	0.191945	0.192051	0.192510	0.193764	0.196489	0.201753	0.211422	0.229376
0.78	0.194993	0.195000	0.195105	0.195564	0.196816	0.199536	0.204783	0.214402	0.232188
0.77	0.197871	0.197878	0.197983	0.198441	0.199689	0.202399	0.207623	0.217179	0.234776
0.76	0.200580	0.200587	0.200692	0.201148	0.202391	0.205089	0.210283	0.219763	0.237153
0.75	0.203128	0.203135	0.203240	0.203693	0.204929	0.207611	0.212679	0.222164	0.239329
0.74	0.205521	0.205528	0.205632	0.206082	0.207311	0.209974	0.215089	0.224389	0.241313
0.73	0.207764	0.207771	0.207874	0.208321	0.209540	0.212182	0.217250	0.226446	0.243115
0.72	0.209863	0.209870	0.209972	0.210416	0.211624	0.214241	0.219258	0.228340	0.244742
0.71	0.211822	0.211829	0.211930	0.212369	0.213566	0.216156	0.221116	0.230079	0.246202
0.70	0.213646	0.213653	0.213753	0.214187	0.215371	0.217932	0.222832	0.231667	0.247502
0.69	0.215338	0.215344	0.215443	0.215873	0.217043	0.219573	0.224407	0.233109	0.248646

Table 24.3 Generalized Fluid Meter Function X

(Isentropic Exponent=1.30   Flow Equation Constant=1.802958)

p_2/p_1	D_2/D_1								
	0.0	0.1	0.2	0.3	0.4	0.5	0.6	0.7	0.8
0.99	0.053165	0.053168	0.053207	0.053379	0.053851	0.054888	0.056937	0.060879	0.068919
0.98	0.074779	0.074782	0.074837	0.075076	0.075732	0.077172	0.080017	0.085474	0.096557
0.97	0.091082	0.091086	0.091152	0.091440	0.092229	0.093962	0.097379	0.103923	0.117152
0.96	0.104589	0.104594	0.104669	0.104995	0.105890	0.107854	0.111725	0.119121	0.134005
0.95	0.116277	0.116283	0.116365	0.116724	0.117707	0.119863	0.124107	0.132199	0.148410
0.94	0.126653	0.126658	0.126747	0.127133	0.128190	0.130508	0.135067	0.143740	0.161037
0.93	0.136015	0.136021	0.136115	0.136525	0.137646	0.140103	0.144929	0.154093	0.172288
0.92	0.144562	0.144568	0.144667	0.145097	0.146273	0.148850	0.153907	0.163488	0.182428
0.91	0.152430	0.152437	0.152540	0.152987	0.154211	0.156892	0.162148	0.172084	0.191641
0.90	0.159721	0.159728	0.159834	0.160297	0.161563	0.164334	0.169760	0.179999	0.200065
0.89	0.166509	0.166516	0.166625	0.167102	0.168404	0.171254	0.176827	0.187323	0.207803
0.88	0.172854	0.172861	0.172973	0.173461	0.174796	0.177713	0.183413	0.194124	0.214936
0.87	0.178804	0.178811	0.178925	0.179423	0.180785	0.183760	0.189568	0.200459	0.221531
0.86	0.184396	0.184403	0.184520	0.185026	0.186411	0.189436	0.195335	0.206373	0.227640
0.85	0.189662	0.189670	0.189788	0.190302	0.191707	0.194773	0.200747	0.211903	0.233309
0.84	0.194630	0.194638	0.194757	0.195277	0.196698	0.199799	0.205834	0.217082	0.238574
0.83	0.199320	0.199328	0.199449	0.199974	0.201409	0.204538	0.210621	0.221935	0.243467
0.82	0.203753	0.203761	0.203883	0.204412	0.205858	0.209009	0.215128	0.226487	0.248016
0.81	0.207945	0.207953	0.208075	0.208608	0.210061	0.213229	0.219374	0.230757	0.252244
0.80	0.211910	0.211918	0.212041	0.212576	0.214035	0.217214	0.223374	0.234762	0.256173
0.79	0.215661	0.215669	0.215792	0.216328	0.217792	0.220976	0.227142	0.238517	0.259822
0.78	0.219209	0.219217	0.219340	0.219877	0.221342	0.224528	0.230690	0.242037	0.263205
0.77	0.222564	0.222572	0.222695	0.223232	0.224696	0.227879	0.234030	0.245334	0.266338
0.76	0.225734	0.225742	0.225865	0.226401	0.227863	0.231040	0.237171	0.248417	0.269235
0.75	0.228727	0.228735	0.228858	0.229392	0.230850	0.234017	0.240122	0.251297	0.271906
0.74	0.231550	0.231559	0.231681	0.232213	0.233665	0.236818	0.242890	0.253983	0.274362
0.73	0.234210	0.234218	0.234340	0.234870	0.236315	0.239450	0.245482	0.256481	0.276613
0.72	0.236712	0.236720	0.236841	0.237368	0.238804	0.241918	0.247905	0.258800	0.278667
0.71	0.239061	0.239069	0.239189	0.239712	0.241138	0.244229	0.250165	0.260946	0.280533
0.70	0.241261	0.241269	0.241389	0.241908	0.243322	0.246387	0.252266	0.262924	0.282218
0.69	0.243317	0.243325	0.243444	0.243958	0.245360	0.248396	0.254213	0.264741	0.283728

By taking derivatives according to the relation

$$dX = \left(\frac{\partial X}{\partial r}\right) dr + \left(\frac{\partial X}{\partial \beta^4}\right) d\beta^4, \tag{24.10}$$

where dr is established precisely by (24.7) or approximately by (24.8) or by Figure 24.2, and noting that $d\beta^4$ can be obtained for a given ϵ_β/β by the relation

$$\epsilon_\beta^4 = 4\beta^4 \left(\frac{\epsilon_\beta}{\beta}\right), \tag{24.11}$$

the required maximum relative uncertainty (dX/X) can be obtained. Representative results that are perfectly general (for the fluids characterized by $\gamma = 1.3$ and $\gamma = 1.4$) are given in Figures 24.3 and 24.4.

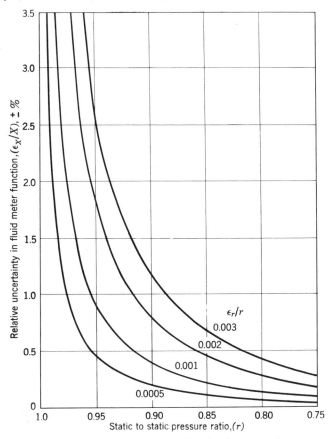

Figure 24.3   Relative uncertainty in fluid meter function in terms of pressure ratio and its relative uncertainty. Parameters: $1.3 \le \gamma \le 1.4$; $0 \le \beta \le 0.2$; $0 \le \epsilon_\beta/\beta \le 0.01$.

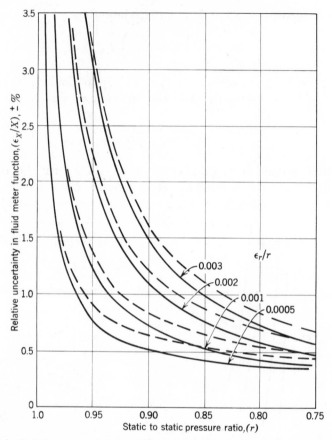

Figure 24.4 Relative uncertainty in fluid meter function in terms of pressure ratio and its relative uncertainty. Parameters: $1.3 \leq \gamma \leq 1.4$; $\beta = 0.6$. Curves: (————), $\epsilon_\beta/\beta = 0.005$; (----), $\epsilon_\beta/\beta = 0.010$.

Now we are in a position to estimate the maximum relative uncertainty in the ideal flow rate of (21.14) by the relation

$$\left(\frac{\epsilon_{\text{flow}}}{\mathring{m}}\right)_{\text{ideal}} = \pm\left[\frac{\epsilon_A}{A_2} + \frac{\epsilon_p}{p_1} + \frac{1}{2}\frac{\epsilon_T}{T_t} + \frac{\epsilon_X}{X}\right], \qquad (24.12)$$

which follows directly from applying the previously given uncertainty rule to (21.14).

The actual flow rate and its maximum relative uncertainty can be determined from (22.1) as

$$\left(\frac{\epsilon_{\text{flow}}}{\mathring{m}}\right)_{\text{max}} = \pm\left[\left(\frac{\epsilon_{\text{flow}}}{\mathring{m}}\right)_{\text{ideal}} + \frac{\epsilon_{CD}}{C_D}\right]. \qquad (24.13)$$

A more realistic combination of the uncertainties of the independent variables might be obtained by recognizing the possibility of some beneficial canceling effects of the various independent uncertainties; that is, all the measurements may not be at their maximum uncertainties at the same time. When the independent variables can be represented by normal error distributions, the uncertainty in a product or quotient function has been shown [12] to be given more realistically by a second power equation (see also Chapter 10). Applied to the actual flow rate, this yields

$$\left(\frac{\epsilon_{\text{flow}}}{\overset{\circ}{m}}\right)_{\text{probable}} = \pm\left[\left(\frac{\epsilon_A}{A_2}\right)^2 + \left(\frac{\epsilon_p}{p_1}\right)^2 + \left(\frac{1}{2}\frac{\epsilon_T}{T_t}\right)^2 + \left(\frac{\epsilon_x}{X}\right)^2 + \left(\frac{\epsilon_{CD}}{C_D}\right)^2\right]^{1/2}.$$
(24.14)

Note that if one of the variables is itself a function of several other variables, as $X = X(\beta, r)$, it seems reasonable to apply the RMS idea of (24.14) to it first to get a probable ϵ_x/X, and *then* to apply it again when combining X with the other variables. Although the uncertainty of (24.14) perhaps is more probable than that of (24.13), one no longer can claim that all experimental determinations of the flow rate will fall within this somewhat reduced uncertainty interval.

The flow rate of a constant density fluid (21.9) involves independent variables only, and hence its uncertainty is simply

$$\left(\frac{\epsilon_{\text{flow}}}{\overset{\circ}{m}}\right)_{\text{max}} = \pm\left(\frac{\epsilon_A}{A_2} + \frac{1}{2}\frac{\epsilon_p}{\rho} + \frac{1}{2}\frac{\epsilon_\Delta}{\Delta p} + \frac{1}{2}\frac{\epsilon_\beta^4}{(1-\beta^4)} + \frac{\epsilon_{CD}}{C_D}\right),$$
(24.15)

with a probable uncertainty patterned after (24.14).

24.3 Examples in Determining Flow Rate Uncertainties

Example 1. An ASME flow nozzle is used, uncalibrated, to meter the flow of air into a piping system from the atmosphere. The throat diameter is 4.054 ± 0.002 in., and the diameter ratio can be taken as zero. Inlet air temperature is $80 \pm 1°F$. Inlet pressure is 14.62 ± 0.01 psia, indicated by a 15-psia gauge. The pressure drop across the meter is 91.9 ± 0.1 in. of water, measured on a 100-in. manometer. Find the actual flow rate and its uncertainty.

Now

$$\Delta p = 91.9 \times 0.036 \text{ psi/in. } H_2O = 3.3084 \text{ psi,}$$

and
$$r = 1 - \frac{3.3084}{14.62} = 0.7736.$$

The fluid meter function at this r and at $\beta = 0$ is, from Table 24.2

$$X = 0.22137.$$

The discharge coefficient, by iteration at $R_d = 1.179 \times 10^6$, is by (22.20),

$$C_D = 0.9915.$$

The actual flow rate by (21.14) is

$$\mathring{m} = 0.9915 \left(\frac{\pi \times 4.054^2}{4} \right) \left(\frac{14.62}{540^{1/2}} \right)(0.22137)(2.055597) = 3.665 \text{ lbm/sec.}$$

The uncertainty in this flow rate is based on the factors of (24.12), where

$$\frac{\epsilon_A}{A_2} = \pm 2 \frac{\epsilon_D}{D_2} = \pm 2 \left(\frac{0.002}{4.054} \right) \approx \pm 0.0010,$$

$$\frac{\epsilon_p}{p_1} = \pm \frac{0.01}{14.62} = \pm 0.00068 \approx \pm 0.0007,$$

$$\frac{1}{2} \frac{\epsilon_T}{T_t} = \pm \frac{1}{2} \left(\frac{1}{540} \right) = \pm 0.00093 \approx 0.0009,$$

$$\frac{\epsilon_X}{X} = f\left(\frac{\epsilon_r}{r}, r \right).$$

By (24.7), the correct

$$\frac{\epsilon_r}{r} = \pm \left(\frac{0.1/91.9 + 0.01/14.62)}{(14.62/3.3084 - 1)} \right) = \pm 0.0005.$$

With $\epsilon_r/r = 0.0005$, one obtains from Figure 24.3

$$\frac{\epsilon_X}{X} = \pm 0.0006.$$

Since the fluid meter installation is uncalibrated, the tolerance of the discharge coefficient is, as mentioned in Section 24.2, about $\pm 1.0\%$. To this is added the tolerance of $\pm 0.5\%$ as suggested to account for installation differences. Thus,

$$\frac{\epsilon_{CD}}{C_D} \sim \pm 0.0150.$$

These independent uncertainties combine according to (24.12) and (24.13) to give the maximum relative uncertainty in flow rate as

$$\left(\frac{\epsilon_{\text{flow}}}{\mathring{m}} \right)_{\text{max}} = \pm [(10 + 7 + 9 + 6 + 150)10^{-4}] = \pm 1.8\%.$$

On the probable basis of (24.14), there results

$$\left(\frac{\epsilon_{\text{flow}}}{\mathring{m}} \right)_{\text{probable}} = \pm [(100 + 49 + 81 + 36 + 22,500)10^{-8}]^{1/2} = \pm 1.5\%$$

Example 2. An ASME venturi is used, uncalibrated, to meter the flow of steam. The throat diameter is 1.209 ± 0.001 in., and the diameter ratio is 0.600 ± 0.006. Inlet total temperature of steam is $300 \pm 3°F$. Inlet pressure is 14.70 ± 0.03 psia. Pressure drop across meter is 2.70 ± 0.03 psi. Both pressure measurements are made on a 30-psi range gauge. Find the actual flow rate and its uncertainty.

Now

$$r = 1 - \frac{2.7}{14.7} = 0.81633.$$

The fluid meter function at this r and at $\beta = 0.6$ is, from Table 24.3,

$$X = 0.19279.$$

The discharge coefficient, at $R_D = 1.7 \times 10^5$, is approximately

$$C_D = 0.983.$$

The actual flow rate is, from (21.14) (with an added factor of 1.003 to account for the thermal expansion of the meter, from [1], page 156):

$$\mathring{m} = 0.983 \left(\frac{\pi \times 1.209^2}{4} \right) \left(\frac{14.7}{760^{1/2}} \right) (1.003)(0.19279)(1.802958) = 0.210 \text{ lbm/sec}.$$

The uncertainty in this flow rate is based on the factors of (24.12), where

$$\frac{\epsilon_A}{A_2} = \pm 2 \left(\frac{0.001}{1.209} \right) \approx \pm 0.0017,$$

$$\frac{\epsilon_P}{P_1} = \pm \frac{0.03}{14.7} = \pm 0.00204 \approx \pm 0.0020,$$

$$\frac{1}{2} \frac{\epsilon_T}{T_t} = \pm \frac{1}{2} \left(\frac{3}{760} \right) = \pm 0.00197 \approx \pm 0.0020,$$

$$\frac{\epsilon_X}{X} = f \left(\frac{\epsilon_r}{r}, \frac{\epsilon_\beta}{\beta}, r, \beta \right),$$

By (24.7) the correct.

$$\frac{\epsilon_r}{r} = \pm \frac{(0.03/2.7 + 0.03/14.7)}{(14.7/2.7 - 1)} = \pm 0.003.$$

With $\epsilon_r / r = \pm 0.003$, one obtains from Figure 24.4:

$$\frac{\epsilon_X}{X} = \pm 0.0070.$$

The tolerance of the discharge coefficient, as indicated in Section 24.2, is at least

$$\frac{\epsilon_{CD}}{C_D} = \pm 0.0125.$$

These independent uncertainties combine according to (24.12) to give the maximum relative uncertainty in flow rate as

$$\left(\frac{\epsilon_{\text{flow}}}{\mathring{m}} \right)_{\text{max}} = \pm [(17 + 20 + 20 + 70 + 125)10^{-4}] = \pm 2.5\%.$$

On the probable basis of (24.14), there results

$$\left(\frac{\epsilon_{\text{flow}}}{\mathring{m}} \right)_{\text{probable}} = \pm [(189 + 400 + 400 + 4900 + 15,625)10^{-8}]^{1/2} = \pm 1.5\%.$$

Example 3. An ASME orifice with pipe taps is used, uncalibrated, to meter the flow of water. The orifice diameter is 1.209 ± 0.001 in., and the diameter ratio is 0.6045 ± 0.006. The water temperature is at 75°F yielding a density of 61.97 ± 0.02 lbm/ft^3. The pressure drop across the meter is 6.86 ± 0.05 in. Hg under water. Find the actual flow rate and its uncertainty.

The discharge coefficient, by iteration at $R_D = 1.48 \times 10^5$ is, by (22.27) to (22.32),

$$C_D = 0.7747.$$

The actual flow rate is, from (21.9)

$$\mathring{m} = 0.7747 \left(\frac{\pi \times 1.209^2}{4} \right) \left[\frac{64.34 \times 61.97 \times 3.107}{(1 - 0.6045^4) \times 144} \right]^{1/2} = 8.864 \text{ lbm/sec.}$$

The uncertainty in this flow rate is based on the factors of (24.15).

$$\frac{\epsilon_A}{A_2} = \pm 2 \frac{\epsilon_D}{D_2} = \pm 2 \left(\frac{0.001}{1.209} \right) \approx \pm 0.0017,$$

$$\frac{1}{2} \frac{\epsilon_\rho}{\rho} = \pm \frac{1}{2} \left(\frac{0.02}{61.97} \right) = \pm 0.00016 \approx \pm 0.0002,$$

$$\frac{1}{2} \frac{\epsilon_\Delta}{\Delta p} = \pm \frac{1}{2} \left(\frac{0.05}{6.86} \right) = \pm 0.00364 \approx \pm 0.0036,$$

$$\frac{1}{2} \frac{\epsilon_\beta^4}{(1 - \beta^4)} = \pm \frac{1}{2} \left(\frac{4 \times 0.13 \times 0.01}{0.87} \right) = \pm \frac{0.0026}{0.87} \approx 0.0030,$$

The tolerance in the discharge coefficient, as indicated in Section 24.2, is at least

$$\frac{\epsilon_{CD}}{C_D} = \pm 0.0105.$$

These independent uncertainties combine according to (24.15) to give the maximum relative uncertainty in flow rate as

$$\left(\frac{\epsilon_{\text{flow}}}{\mathring{m}} \right)_{\text{max}} = \pm [(17 + 2 + 36 + 30 + 105)10^{-4}] = \pm 1.9\%.$$

On the probable basis, patterned after (24.14), there results

$$\left(\frac{\epsilon_{\text{flow}}}{\mathring{m}} \right)_{\text{probable}} = \pm [(189 + 4 + 1296 + 900 + 11{,}025)10^{-8}]^{1/2} = \pm 1.2\%.$$

These uncertainties could be reduced if the fluid meters were calibrated and their individual discharge coefficients were applied. In such situation, the relative uncertainties in the other factors of (24.12) and (24.14) would assume more importance. These uncertainties could be reduced further if repeated measurements were taken of all the variables, assuming that a portion of the uncertainty reported for any individual measurement were still of a random nature (once again check Chapter 10).

24.4 References

[1] *Fluid Meters—Their Theory and Application*, Report of ASME Research Committee on Fluid Meters, 6th ed., 1971.

[2] J. W. Murdock, C. J. Foltz, and C. Gregory, Jr., "Effect of Globe Valve in Approach Piping on Orifice-Meter Accuracy," *Trans. ASME*, **78**, 1956, p. 369.

[3] A. G. Ferron, "Velocity Profile Effects on the Discharge Coefficient of Pressure Differential Meters," *ASME Paper 62-HYD-8.*

[4] W. J. Clark, "Fluid Flow Through Square-Edge Orifice Plates, Some Practical Aspects," *Trans. Soc. Instrument Technology*, **4**, 1952, p. 895.

[5] P. S. Starrett, P. F. Halfpenny, and H. B. Nottage, "Survey of Information Concerning the Effects of Nonstandard Approach Conditions Upon Orifice and Venturi Meters," *ASME Paper 65-WA/FM-5.*

[6] *Flowmeter Computation Handbook*, Report of ASME Research Committee on Fluid Meters, 1961.

[7] R. W. Miller and O. K. Kreisel, "Experimental Study of the Effects of Orifice Plate Eccentricity on Flow Coefficients," *Trans. ASME, J. Basic Eng.* March 1969, p. 121.

[8] R. W. Miller and O. Kreisel, "A Comparison Between Orifice and Flow Nozzle Laboratory Data and Published Coefficients," *Trans. ASME, J. Fluids Eng.*, June 1974, p. 139.

[9] J. S. Wyler and R. P. Benedict, "Comparisons Between Throat and Pipe Wall Tap Nozzles," *Trans. ASME, J. Eng. Power*, October 1975, p. 569.

[10] R. P. Benedict, J. S. Wyler, and G. B. Brandt, "The Effect of Edge Sharpness on the Discharge Coefficient of an Orifice," *Trans. ASME, J. Eng. Power*, October 1975, p. 576.

[11] R. P. Benedict, "Uncertainties in Compressible Flow Measurements," *ASME Paper 66-WA/PTC-7.*

[12] S. J. Kline and F. A. McClintock, "Describing Uncertainties in Single-Sample Experiments," *Mech. Eng.*, **75**, January 1953.

Nomenclature

Roman

A area
C_D discharge coefficient
d exact differential
$\mathring{m}$ mass flow rate
p absolute pressure
r static pressure ratio
R_D Reynolds number, pipe
t time
T_t absolute total temperature
X fluid meter function

Greek

β diameter ratio
Δ finite difference
ϵ uncertainty
ρ density

SOLUTIONS
TO PROBLEMS

1.1 $y = mx + b$
    $y = R$, varies from 0 to 12
    $x = F$, varies from 32 to 212
    $m = \Delta y / \Delta x = \Delta R / \Delta F = 12/180 = 1/15$
    $b = y$ at $x = 0$
    at $R = 0$, $F = 32$ or $0 = 32/15 + b$ and $b = -32/15$
    $\therefore$ $R = 1/15 \ (F - 32)$

1.2 $R = F$
    $F = 1/15 \ (F - 32)$
    $15F - F = -32$
      $14F = -32$
    $\therefore$ $F = -32/14$

1.3 $y = mx + b$
    $y = \text{Cel}$, varies from 100 to 0
    $x = \text{Cen}$, varies from 0 to 100
    $m = \Delta y / \Delta x = \Delta \text{Cel} / \Delta \text{Cen} = -100/100 = -1$
    at $\text{Cel} = 0$, $\text{Cen} = 100$ or $0 = -100 + b$ and $b = 100$
    $\therefore$ $\text{Cel} = -\text{Cen} + 100$

1.4 Scales are equal at 50
    $(\text{Cel} - 50) = -(\text{Cen} - 50)$
    $\therefore$ $\text{Cel} = -(\text{Cen} - 50) + 50$
      $\text{Cen} = -(\text{Cel} - 50) + 50$

1.5 $C = 5/9(F - 32)$ $C = 5/9(F + 40) - 40$
    $C = 5/9(59 - 32)$ $C = 5/9(59 + 40) - 40$
    $C = 15°C$ $C = 55 - 40 = 15°C$

2.1 $MW_{H_2O} = 18.016$, $MW_{air} = 38.96$, $MW_{CO_2} = 44$

$$\therefore \quad R_{H_2O} = \frac{1545}{18.016} = 85.76$$

$$R_{air} = \frac{1545}{28.96} = 53.35$$

$$R_{CO_2} = \frac{1545}{44} = 35.11 \text{ ft lbf/°R lbm}$$

2.2 $\quad T = t - t_0 + 1/\alpha, \ °A$
$\therefore \quad T_{ice} = 0 - 0 + 273 = 273°A$
$\quad T_{steam} = 100 - 0 + 273 = 373°A$

2.3 $\quad T = t - t_0 + 1/\alpha, \ °A$
$\therefore \quad T_{ice} = 0 - 0 + 267 = 267°A$
$\quad T_{steam} = 100 - 0 + 267 = 367°A$

2.4 $\quad pv = \text{constant}$
$p_1 v_1 = p_2 v_2$
$p_2 = (v_1/v_2)p_1$
$\therefore \quad p_2 = (1/2)p_1 = 0.5 \text{ atmosphere}$

2.5 $\quad pv = R(t + 273)$
$R_{air} = 287.1 \ J/kg \cdot K$
$\quad v = 287.1(127 + 273)/1 = 114,840 \ m^3/kg$
$\therefore \quad \mathbf{V} = vM = 114,840 \times 1 \text{ kg} = 114,840 \ m^3$

3.1 $\dfrac{(pv)_x^0}{(pv)_s^0} = 1.5 = \dfrac{T_x}{T_s}; \quad \dfrac{(pv)_s^0}{(pv)_y^0} = 1.5 = \dfrac{T_s}{T_y}; \quad T_s - T_y = 150°A$

$$T_s = \frac{T_x}{1.5} = 1.5 \quad T_y = 150 + T_y$$

$\therefore \quad 0.5 \ T_y = 150, \quad T_y = 300°A$
$\quad T_s = 150 + 300 = 450°A$
$\quad T_x = 1.5 \times 1.5 T_y = 2.25(300) = 675°A$

3.2 $\dfrac{T_{plat}}{T_{steam}} = 5.47; \quad T_{steam} = 373.15K$
$\therefore \quad T_{plat} = 5.47 \times 373.15 = 2041.1305K = 1767.98°C$

3.3 $\quad T_{Hg} = -38.841°C = 234.309K$
$\quad T_{ice} = 273.15K$
$\therefore \quad \dfrac{T_{ice}}{T_{Hg}} = \dfrac{(pv)_{ice}^0}{(pv)_{Hg}^0} = \dfrac{273.15}{234.309} = 1.1658$

4.1 $\Delta t = 0.045 \left(\dfrac{t'}{100}\right)\left(\dfrac{t'}{100} - 1\right)\left(\dfrac{t'}{419.58} - 1\right)\left(\dfrac{t'}{630.74} - 1\right)$
Assume $t' = t_{Callendar} \approx t_{68} = 231.9681°C$
$\Delta t \approx 0.045(2.319681)(1.319681)(-0.447142)(-0.632229)$
$\therefore \quad \Delta t \approx 0.038943°C$ for first try

4.2 Figure 4.3 indicates that at a temperature level of 232°C, Δt is $\approx 0.04°C$

4.3 Table 4.1 indicates
at 30.72 "Hg   b.p. = 213.332°F
at 29.56 "Hg   b.p. = 211.389°F
$\therefore \quad \Delta T_{b.p.} = 1.943°F$

5.1 $t_1 = t_{ind} = 600°F$, $N = 600 - 400 = 200°F$, $t_2 = 150°F$
$C_{s1} = 0.00009 \times 200 \times 450 = 8.1°F$
$C_{s2} = 0.00009 \times 200 \times (608.1 - 150) = 8.2°F$
∴ $t_{bulb} = 608.2°F$

5.2 $t_1 = 250°C$, $N = 250 - 100 = 150°C$, $t_2 = 150°C$
$C_{s1} = 0.00016 \times 150 \times 100 = 2.4°C$
$C_{s2} = 0.00016 \times 150 \times (252.4 - 150) = 2.5°C$
∴ $t_{bulb} = 252.5°C$

5.3 $t_2 = 500°F$, $N = 300°F$, $t_{sp} = 100°F$
$C_{s1} = 0.00009 \times 300 \times (100 - 500) = -10.8°F$
$C_{s2} = 0.00009 \times 300 \times (100 - 489.2) = -10.5°F$
∴ $t_{bulb} = 489.5°F$

5.4 a. Enter Figure 5.3 with $t_1 - t_2 = 450°F$ at $N = 200°F$ read $C_s \approx 8°F$
     b. Enter Figure 5.3 with $t_1 - t_2 = 400°F$ at $N = 300°F$ read $C_s \approx 11°F$

6.1 $a = 0.06\ \Omega$, $b = 8000°R$, $R = 10,000\ \Omega$

$$R = ae^{b/T}, \qquad \frac{b}{T} = \ln R - \ln a$$

$$T = \frac{b}{\ln R - \ln a} = \frac{8000}{9.21034 - (-2.81341)}$$

$$T = 665.35°R$$

∴ $t = 665.35 - 459.67 = 205.68°F$

6.2 $R_{100}/R_0 = 1.3926$, $\delta = 1.495$
$R_0 = 25.550\ \Omega$, $R_t = 35.550\ \Omega$

$$t^2 - \left(\frac{10^4}{\delta} + 10^2\right)t + \frac{10^6}{\delta}\left(\frac{R_t - R_0}{R_{100} - R_0}\right) = 0$$

$$t^2 - 6788.9632t + \frac{10^6}{1.495}\left(\frac{35.550 - 25.550}{35.58093 - 25.550}\right) = 0$$

$$t^2 - 6788.9632t + 666833.804 = 0$$

Form $ax^2 + bx + c = 0$

where $a = 1$, $b = -6788.9632$, $c = 666{,}833.804$

$$x = \frac{-b \pm \sqrt{b^2 - 4ac}}{2a} = -A - \sqrt{A^2 - c}$$

where $\dfrac{b}{2a} = A = -3394.4816$, $A^2 = 11{,}522{,}505.33$

$$t = 3394.4816 - \sqrt{10855671.53}$$

∴ $t = 99.687°C$

6.3 $R_0 = 20\ \Omega$, $R_t = 45\ \Omega$
Table 6.3 indicates $R_{100}/R_0 = 1.663$ *and* $R_{200}/R_0 = 2.501$

$$R_{100} = 1.663 \times 20 = 33.26\ \Omega$$

$$R_{200} = 2.501 \times 20 = 50.02\ \Omega$$

If relation between R and t is linear

$$R_t = R_0(1 + at)$$

$$\frac{R_{100}}{R_0} = 1 + 100a \text{ leads to } a = 0.00663.$$

But using this value of a leads to

$$\frac{R_{200}}{R_0} = 1 + 200(0.00663) = 2.326$$

which does not check Table 6.3. $\therefore$ not linear. If parabolic relation exists between R and t

$$R_t = R_0(1 + at + bt^2)$$
$$1.663 = 1 + 100a + 10^4 b$$
$$2.501 = 1 + 200a + 4 \times 10^4 b$$

Simultaneous solution yields

$$a = 0.005755, \qquad b = 8.75 \times 10^{-6}$$

$\therefore \quad R_t/R_0 = 45/20 = 2.25 = 1 + 0.005755t + 8.75 \times 10^{-6} t^2$
$t^2 + 657.7143t - 142857.143 = 0$

Form $\qquad\qquad\qquad\qquad ax^2 + bx + c = 0$

where $\qquad\qquad a = 1, \qquad b = 657.7143, \qquad c = -142857.143$

$$x = \frac{-b \pm \sqrt{b^2 - 4ac}}{2a} = -A + \sqrt{A^2 - c}$$

where $\qquad\qquad \dfrac{b}{2a} = A = 328.85715, \qquad A^2 = 108147.0251$

$$t = -328.85715 + \sqrt{251004.1681}$$

$\therefore \quad t = 172.146°C$

6.4 $\quad R_0 = 25.5440\ \Omega, \qquad R_{SN} = 48.3434\ \Omega, \qquad R_{ZN} = 65.6091\Omega$
$\quad t_0 = 0°C, \qquad t_{SN} = 231.92916°C, \qquad t_{ZN} = 419.58°C$

a. $\quad c = \dfrac{R_0(t_{ZN} - t_{SN}) - R_{SN}t_{ZN} + R_{ZN}t_{SN}}{t_{SN}t_{ZN}(t_{ZN} - t_{SN})}$

$$c = \frac{25.544(187.65084) - 48.3434(419.58) + 65.6091(231.92916)}{231.92916 \times 419.58 \times 187.65084}$$

$\therefore \quad c = -273.907263/18260835.6 = -1.499971135 \times 10^{-5}$

$$b = \frac{R_{ZN} - R_0 - c \times t_{ZN}^2}{t_{ZN}}$$

$$b = \frac{65.6091 - 25.5440 + 1.499971135 \times 10^{-5} \times (419.58)^2}{419.58}$$

$\therefore \quad b = 42.70575983/419.58 = 0.101782163$

$\quad FI = 100b + c \times 10^4 = 10.0282192$

$\quad C = FI/100R_0 = \alpha = 0.003925861$

$$\delta = \frac{t_{ZN} - 100(R_{ZN} - R_0)/FI}{\left(\dfrac{t_{ZN}}{100} - 1\right)\left(\dfrac{t_{ZN}}{100}\right)}$$

$$\delta = \frac{419.58 - 100(65.6091 - 25.5440)/10.0282192}{\left(\dfrac{419.58}{100} - 1\right)(4.1958)}$$

$$\therefore \quad \delta = 1.49575$$

b. $R_t = 50.0521 \ \Omega$

     $R_t = R_0 + bt + ct^2$

$$t^2 + \left(\frac{b}{c}\right)t + \frac{(R_0 - R_t)}{c} = 0$$

Form

$$Ax^2 + Bx + C = 0$$

where

$$A = 1, \qquad B = \frac{b}{c} = -6785.608111,$$

and

$$C = \frac{R_0 - R_t}{c} = 1633904.775$$

$$x = \frac{-B \pm \sqrt{B^2 - 4AC}}{2A} = -\frac{B}{2} - \sqrt{\left(\frac{B}{2}\right)^2 - C}$$

$$\therefore \quad t = 3392.80406 - 3142.803619 = 250.000°C$$

6.5 $R_{ox} = 6.230 \ \Omega, \qquad R_0 = 25.540 \ \Omega, \qquad R_{100} = 35.590 \ \Omega, \qquad R_{ZN} = 65.650 \ \Omega$
     $t_{ox} = -182.962°C, \qquad t_0 = 0°C, \qquad t_{100} = 100°C, \qquad t_{ZN} = 419.59°C$
     The platinum temperature of zinc is

$$pt_{ZN} = 100(R_{ZN} - R_0)/(R_{100} - R_0) = 399.104478°C$$

$$\therefore \quad \delta = \frac{t_{ZN} - pt_{ZN}}{\left(\dfrac{t_{ZN}}{100} - 1\right)\left(\dfrac{t_{ZN}}{100}\right)} = \frac{419.58 - 399.104478}{3.1958 \times 4.1958} = 1.527005535$$

$$\beta = \frac{t_{ox} - \dfrac{100(R_{ox} - R_0)}{(R_{100} - R_0)} - \delta\left(\dfrac{t_{ox}}{100} - 1\right)\left(\dfrac{t_{ox}}{100}\right)}{\left(\dfrac{t_{ox}}{100} - 1\right)\left(\dfrac{t_{ox}}{100}\right)^3}$$

$$\therefore \quad \beta = 0.0733850237$$

7.1 $T_M = 500°C, \qquad T_R = 0°C$

metal	$a, \ \mu V/°C$	$b, \ \mu V/(°C)^2$
chromel	30.5132	7.6296×10^{-3}
alumel	-9.4332	2.2984×10^{-3}

$$a_{Ch-Al} = a_{Ch} - a_{Al} = 30.5132 - (-9.4332) = 39.9464 \ \mu V/°C$$

$$b_{Ch-Al} = b_{Ch} - b_{Al} = (7.6296 - 2.2984) \times 10^{-3} = 5.3312 \times 10^{-3} \ \mu V(°C)^2$$

$$E_s = at + \tfrac{1}{2}bt^2$$

$$= 39.9464(500) + \tfrac{1}{2}(5.3312 \times 10^{-3})(500)^2 = 20639.6 \ \mu V$$

(Table 7.6 indicates 20640 μV for Type K at 500°C)

$$\alpha = a + bt$$
$$\alpha_{0°} = 39.9464 + (5.3312 \times 10^{-3})(0) = 39.9464 \ \mu V/°C$$
$$\alpha_{500°} = 39.9464 + (5.3312 \times 10^{-3})(500) = 42.612 \ \mu V/°C$$

$\pi = \alpha$ Tabs

$\pi_{0^\circ} = 39.9464(273.15) = 10911.36 \ \mu V$

$\pi_{500^\circ} = 42.612(773.15) = 32945.47 \ \mu V$

$E_T = \frac{1}{2}b(T_{ref}^2 - T^2)$

$= \frac{1}{2}(5.3312 \times 10^{-3})(273.15^2 - 773.15^2) = -1394.51 \ \mu V$

Check:

$$E_s = \pi_2 - \pi_1 + E_T = 32945.47 - 10911.36 - 1394.51$$

$$E_s = 20639.6 \ \mu V$$

7.2

T	JP	JN	TP	TN, EN	KP, EP	KN	Table
400°C	9.7	45.4	16.3	45.5	34.5	7.7	7.2

	E	J	K	T			
400°C	80.0	55.1	42.2	61.8			7.3

From Figure 7.18 or 7.21, it is clear that emf from positive leg must be added to emf of negative leg to obtain the net emf of the thermocouple.

$$J: JP + JN = 9.7 + 45.4 = 55.1 \text{ mV}$$

$$K: KP + KN = 34.5 + 7.7 = 42.2 \text{ mV}$$

$$E: EP + EN = 34.5 + 45.5 = 80.0 \text{ mV}$$

$$T: TP + TN = 16.3 + 45.5 = 61.8 \text{ mV}$$

Conclude that all values in Tables 7.2 and 7.3 are consistent with each other.

7.3 *Given circuit*

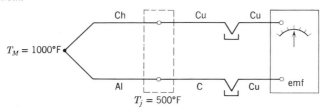

Equivalent circuit

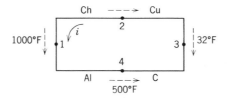

$$E_{net} = E_1 - E_2 - E_3 + E_4$$

$$E_{net} = (Ch-Al)_{1000°F} - (Ch-Cu)_{500°F} - (Cu-C)_{32°f} + (Al-C)_{500°F}$$

$$= (Ch-Al)_{1000°} - (Ch-Al)_{500°} + (Cu-C)_{500°}$$

$$= 22.251 - 10.560 + 12.572 = 24.263 \text{ mV}$$

whereas correct emf should have been 22.251 mV

The net temperature error (on the chromel–alumel scale) = $1085 - 1000 = 85°F$.

7.4   Given circuit

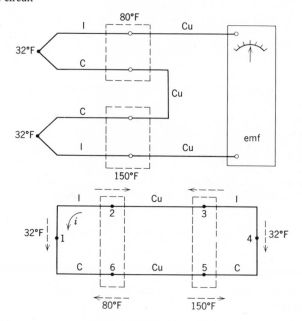

$$E_{net} = E_1 - E_2 + E_3 - E_4 + E_5 - E_6$$
$$= 0 - (ICu)_{80°} + (ICu)_{150°} - 0 + (CuC)_{150°} - (CuC)_{80°}$$
$$= -I_{80°} + Cu_{80°} + I_{150°} + Cu_{150°} - C_{150°} - Cu_{80°} + C_{80°}$$
$$E_{net} = (IC)_{150°} - (IC)_{80°}$$
$$= 3.411 - 1.363 = 2.048 \text{ mV},$$

whereas correct emf should have been 0 mV.

The net temperature error $= 104 - 32 = 72°F$. (Would expect $150 - 80 = 70°F$ for temperature error, but change in slope of E vs. t accounts for apparent discrepancy.)

7.5   Given circuit

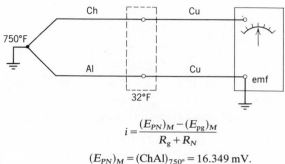

$$i = \frac{(E_{PN})_M - (E_{pg})_M}{R_g + R_N}$$

where $(E_{PN})_M = (ChAl)_{750°} = 16.349 \text{ mV}.$

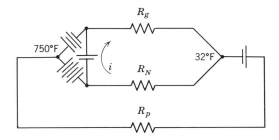

and

$$(E_{pg})_M = (ChCu)_{750°} = (ChC)_{750°} - (CuC)_{750°}$$

$$= 28.854 - 20.801 = 8.053 \text{ mV.}$$

$$i = \frac{16.349 - 8.053}{0.0646 + 4.108} = 1.988 \text{ mA.}$$

$$E_{net} = iRg + (E_{pg})_M = -iR_N + (E_{PN})_M$$

$$= 1.988 \times 0.0646 + 8.053 = -1.988 \times 4.108 + 16.349$$

$$E_{net} = 8.1814 \quad \text{or} \quad = 8.1823$$

Say $\qquad E_{net} = 8.182 \text{ mV.}$

The temperature error is the difference between the temperature of the measuring junction and the temperature equivalent of the net voltage on the chromel-alumel scale, i.e.,

$$\text{Temp. error} = 750 - 394 = 356°F.$$

8.1 $T_B = 2000°F$ for liquid gold of $\epsilon = 0.22$
$T_B = 1093°C = 1366 \text{ K}$
By Table 8.3, $\Delta T = 199 + 0.8(74) = 258°F$
$\therefore \quad T = 2258°F$
Check: By Figure 8.7a, $\Delta T = 142°C = 256°F$.

8.2 $T = 1768°C = 2041 \text{ K}$ for solid platinum of $\epsilon = 0.3$

$$\Delta T \approx \frac{\lambda T^2}{c_2} \ln \frac{1}{\epsilon'}$$

$$\approx \frac{0.655 \times 10^{-6} \, m \times 2041^2 \, \text{K}^2}{0.014388 \, m \cdot \text{K}} \ln \frac{1}{0.3} = 228.32°C$$

$\therefore \quad T_B \sim 1768 - 228 = 1540°C = 1813 \text{ K.}$
Check: At $T_B = 1813 \text{ K}$

$$\Delta T \approx \frac{0.655 \times 10^{-6} \times 1813^2}{0.014388} \ln \frac{1}{0.3} = 180°C.$$

Conclude $\Delta T \sim 200°C$.

8.3 $T_B = 1000°C$ for solid steel of $\epsilon = 0.35$
By table 8.4, $\Delta T = 71 + 0.5(24) = 83°C$
$\therefore \quad T = 1083°C$
Check: By Figure 8.7a $\Delta T = 82°C$.

8.4 $T = 1200°C = 1473$ K

     $T_B = 1100°C = 1373$ K

$$\frac{1}{T} - \frac{1}{T_B} = \frac{\lambda}{c_2} \ln \epsilon$$

$$\ln \epsilon = \frac{\left(\dfrac{1}{1473} - \dfrac{1}{1373}\right)}{\left(\dfrac{0.655 \times 10^{-6}}{0.014388}\right)} = -1.08614$$

$\therefore$ $\epsilon = e^{-1.08614} = 0.3375.$

Check: At $\epsilon = 0.3375$ $T_B = 1100°C$

By Figure 8.7a, $\Delta T = 100°C$.

Also: By Table 8.4:

$$\Delta T = 83 + \frac{0.0625}{0.1000}(29) = 101°C.$$

8.5 $J_{b,\lambda}$, Planck $= J_{b,\lambda}$, Wien

$$e^{c_2/\lambda T_p} - 1 = e^{c_2/\lambda T_w}$$

where $c_2 = 0.014388$ $m \cdot$K, $\lambda = 0.655 \times 10^{-6}$ m

Let

$$A_{\text{planck}} = \exp\left[\frac{0.014388 \times 10^6}{0.655 \times T_p}\right] - 1$$

$\therefore$

$$T_{\text{wien}} = \frac{0.014388 \times 10^6}{0.655 \times \ln A_p}$$

A_p	$\ln A_p$	T_{wien}	error
58876.0182	10.983189	2000.0031	0.0031 K
1512.4354	7.321476	3000.271	0.271 K
241.6458694	5.4874733	4003	3 K

9.1   From Table 7.4:

$$T = 19.750953E - 0.185426E^2 + 0.0083683683958E^3 - 1.3280568 \times 10^{-4}E^4, °C$$

at $E = 1.521$, $T = 29.642°C = 85.35°F$, $\Delta T = T_{\text{actual}} - T_{\text{calculated}} = -0.35°F$,

at $E = 3.270$, $T = 145.18°F$, $\Delta T = -0.18°F$,

at $E = 4.743$, $T = 194.60°F$, $\Delta T = -0.60°F$,

at $E = 6.292$, $T = 245.86°F$, $\Delta T = -1.86°F$,

at $E = 7.865$, $T = 297.38°F$, $\Delta T = -2.38°F.$

9.2   By Figure 9.14 with $y = \Delta E$ and $x = E_c$

$$\Sigma \, \Delta E = 6a + b\Sigma E_c$$

$$\Sigma E_c \, \Delta E = a\Sigma E_c + b\Sigma E_c^2$$

In terms of numerical values available

$$-0.165 = 6a + 23.691b$$

$$-1.034033 = 23.691a + 1.36949879b$$

11.2 $T_{\text{stag}} = T + ST_v$

$$T_t = T + T_v$$
$$\overline{T_{\text{stag}} - T_t = (S-1)T_v}$$

$$\Delta T = (S-1)\left(\frac{V^2}{2Jg_c c_p}\right)$$

From Figure 11.3, $S = 1.74$

$$\Delta T = 0.74\left(\frac{1.2^2 \times 10^6}{2 \times 778 \times 32.174 \times 0.75}\right) = 28.38°\text{R}.$$

11.3 $T_{\text{adi}} = T + rT_v$

$r_{\text{lam}} = Pr^{1/2} = 0.833$

$$T_{\text{adi}} = (200 + 460) + 0.833\left(\frac{25 \times 10^4}{2 \times 778 \times 32.174 \times 0.25}\right)$$

$\therefore$ $T_{\text{adi}} = 660 + 16.64 = 676.64°\text{R}$

11.4 $T_{\text{adi}} = T + RT_v$

$R = \bar{P}S + (1 - \bar{P})r_{\text{turb}}$

$r_{\text{turb}} = Pr^{1/3} = 0.879$

$\bar{P} = -0.7$ according to text, $S_{\text{air}} \approx 1$

$R = -0.7 \times 1 + (1 + 0.7)(0.879) = -0.7 + 1.4943 = 0.79$

$$T_{\text{adi}} = (500 + 460) + 0.79\left(\frac{64 \times 10^4}{2 \times 778 \times 32.174 \times 0.25}\right)$$

$\therefore$ $T_{\text{adi}} = 960 + 40.4 = 1000.4°\text{R}$

11.5 $T_t - T_p$ at $K = -10$, $V = 100$, $c_p = 0.25$

$T_t = T + T_v$

$$T_p = T + KT_v$$
$$\overline{T_t - T_p = (1-K)T_v}$$

$$\Delta T = (1 + 10)\left(\frac{V^2}{2Jg_c c_p}\right)$$

$$\Delta T = 11\left(\frac{10^4}{2 \times 778 \times 32.174 \times 0.25}\right) = 8.79°\text{R}$$

12.1 Figure 12.7, $h_c' = 35$

Figure 12.10, $h_c/h_c' = 0.59$

$\therefore$ $h_c = 0.59 \times 35 = 20.6$ BTU/hr ft^2°F.

12.2 Figure 12.9, $h_c' = 85$

Figure 12.10, $h_c/h_c' = 0.624$

$\therefore$ $h_c = 0.624 \times 85 = 53$ BTU/hr ft^2°F.

12.3 Figure 12.9, $h_c' = 85$

Figure 12.12, $h_c/h_c' = 0.97$

$\therefore$ $h_c = 0.97 \times 85 = 82.45$ BTU/hr ft^2°F.

12.4 Figure 12.8, $h'_c = 35$

Figure 12.10, $h_c/h'_c = 0.62$

∴ $h_c = 0.62 \times 35 = 21.7$

$$x' = \left(\frac{h_c D}{12}\right)\left(\frac{L^2}{D^2 - d^2}\right) = \left(\frac{21.7}{15 \times 48}\right)(16 \times 16) = 7.7156$$

$x' < 20$ ∴ get x

$$x = \frac{(h_c + h_r)D}{k}\left(\frac{L^2}{D^2 - d}\right) = \left(\frac{21.7 + 4.4275}{15 \times 48}\right)(256) = 9.236$$

where h_r/ϵ' from Fig. 12.13 = 4.5, hence $h_r = 0.95 \times 4.5 = 4.275$
Figure 12.6 at $x = 9.236$ yields $y = 0.0029$

$$T_{adi} = \left(\frac{21.7 + 4.275}{21.7}\right)\left(\frac{960 - 0.0029 \times 760}{1 - 0.0029}\right) - \left(\frac{4.275}{21.7}\right)760$$

$T_{adi} = 1000.092°R$

$T_{adi} = 540.1°F$ versus 500°F for T_{ind}

∴ installation not satisfactory.

12.5 Figure 12.8, $h'_c = 35$

Figure 12.11, $h_c/h'_c = 0.78$

∴ $h_c = 0.78 \times 35 = 27.3$

$$x' = \frac{27.3 \times 256}{15 \times 48} = 9.7067 < 20 \quad ∴ \quad \text{get} \quad x$$

$$x = \frac{(27.3 + 2.779)256}{15 \times 48} = 10.695$$

where $h_r/\epsilon' = 4.5$ and $\epsilon' = \epsilon_{well}(1 - \epsilon_{fluid})$

$\epsilon' = 0.95(1 - 0.35) = 0.6175$

and $h_r = 0.6175 \times 4.5 = 2.779$

Figure 12.6 at $x = 10.695$ yields $y = 0.0019$

$$T_{adi} = \left(\frac{27.3 + 2.779}{27.3}\right)\left(\frac{960 - 0.0019 \times 760}{1 - 0.0019}\right) - \left(\frac{2.779}{27.3}\right)760$$

$T_{adi} = 980.783°R = 520.8°F$ versus 500°F

for $T_{indicated}$ ∴ installation not satisfactory

13.1 $T = \Delta T\left[1 - \left(\frac{r_1}{r_1 - r_2}\right)e^{-r_2 t} + \left(\frac{r_2}{r_1 - r_2}\right)e^{-r_1 t}\right]$

$$T = 300\left[1 - \left(\frac{0.2}{0.2 - 0.02}\right)e^{-0.02 \times 120} + \left(\frac{0.02}{0.2 - 0.02}\right)e^{-0.2 \times 120}\right]$$

$T = 300[1 - 0.1002 + 4.2 \times 10^{-12}] = 269.76°F$

13.2 $T = \Delta T[1 - e^{-t/\tau_2}]$ where $\tau_2 = \frac{1}{r_2} = 50$ sec.

∴ $T = 300[1 - 0.090718] = 272.78°F$

13.3

$$L = \tau_1 + \tau_2 = \frac{1}{r_1} + \frac{1}{r_2} = 5 + 50 = 55 \text{ sec.}$$

$$T_e - T = \epsilon_T = RL$$

$$\epsilon_T = 0.5 \times 55 = 27.5°F$$

13.4

$$L = \tau_2 = 50 \text{ sec.}$$

$$\epsilon_T = 0.5 \times 50 = 25°F$$

13.5

$$\tau_{\text{steam}} = \tau_{H_2O}\left(\frac{h_{H_2O}}{h_{\text{stream}}}\right)$$

$$t_{95\%} = 3\tau_{\text{steam}} = 3 \times 10\left(\frac{100}{1500}\right) = 2 \text{ sec.}$$

15.1

$$P_{\text{DW}} = P_{\text{ind}}\left[1 - \frac{w_{\text{air}}}{w_{\text{WTS}}} - 2.637 \times 10^{-3}\cos 2\theta - 5 \times 10^{-5}\right], \text{ psig}$$

$$P_{\text{DW}} = 500\left[1 - \frac{0.075}{495} - 2.637 \times 10^{-3}\cos 50° - 5 \times 10^{-5}\right]$$

$$P_{\text{DW}} = 500(1 - 0.0001515 - 0.001695 - 0.00005)$$

$$\therefore \quad P_{\text{DW}} = 500 \times 0.9981 = 499.052 \text{ psig}$$

15.2

$$P_1 = P_2'$$

$$P + w_{H_2O}\left(A + \frac{\Delta h}{2}\right) = P_{\text{baro}} + w_{\text{Hg}}\Delta h$$

$$P = P_{\text{baro}} + w_{\text{Hg}}\Delta h - w_{H_2O}A - w_{H_2O}\frac{\Delta h}{2}$$

$$\therefore \quad P = P_{\text{baro}} + w_{H_2O}\left[\left(\frac{w_{\text{Hg}}}{w_{H_2O}} - \frac{1}{2}\right)\Delta h - A\right]$$

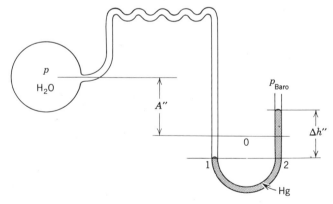

15.3   From Table 15.3 $C_t = -0.139''\text{Hg}$

$$h_{to} = h_{tI} + C_t = 30 - 0.139 = 29.861''\text{Hg}$$

$$P_{\text{baro}} = (w_{\text{Hg}})_{g,t_0} \times h_{t0}$$

where

$$(w_{\text{Hg}})_{g,t_0} = w_{s,t_0}(1+C_g) = 0.491154 \times \frac{g_l}{g_s}$$

$$\therefore \quad P_{\text{baro}} = 0.491154 \times \frac{32}{32.174} \times 29.861$$

$$P_{\text{baro}} = 14.5871 \text{ psia.}$$

15.4

$$C_z = P_{\text{site}} - P_{\text{baro}} = P_{\text{baro}}[\exp(z_{\text{baro}} - z_{\text{site}})/(RT)_{\text{air}} - 1]$$

$$P_{\text{site}} = P_{\text{baro}}\, e^{-(z_{\text{site}} - z_{\text{baro}})/(RT)_{\text{air}}}$$

$$P_{\text{site}} = 14.5871 \times e^{-100/(53.35 \times 539.6)}$$

$$\therefore \quad P_{\text{site}} = 14.5365 \text{ psia.}$$

15.5

$$h_1 = \frac{ah^2}{\mathbf{V}_1} = \frac{\pi d^2 h^2}{4\mathbf{V}_1} = \frac{\pi}{4} \times \frac{1}{64} \times \frac{3.2^2}{350}$$

$$\therefore \quad h_1 = 3.5904 \times 10^{-4}{''}\text{Hga @ } 76°\text{F}$$

a. $P_1 = (w_{\text{Hg}})_{s,t}h_1 = 0.488981 \times 3.5904 \times 10^{-4}$

$$\therefore \quad P_1 = 1.75564 \times 10^{-4} \text{ psia}$$

b. $h_1 = 3.5904 \times 10^{-4}{''}\text{Hg} \times \dfrac{25.4 \text{ mm}}{\text{inch}} \times \dfrac{1\mu}{10^{-3} \text{ mm}}$

$$\therefore \quad h_1 = 9.12\, \mu$$

16.1 $P_1 = P_2$

$$P + w_{\text{H}_2\text{O}}A = P_{\text{baro}} + w_{\text{Hg}}\Delta h$$

$$P = P_{\text{baro}} + w_{\text{H}_2\text{O}}\left[\frac{w_{\text{Hg}}}{w_{\text{H}_2\text{O}}}\Delta h - A\right]$$

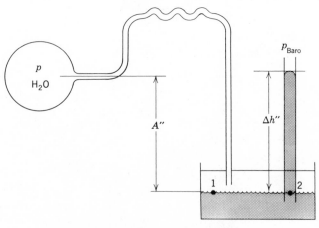

16.2

$$h = 2''\text{Hg} \times \frac{25.4 \text{ mm}}{1 \text{ inch}} = 50.8 \text{ mm Hg}$$

$$h = 50.8 \text{ mm Hg} \times \frac{1\mu}{10^{-3} \text{ mm Hg}} = 50800\, \mu$$

17.1

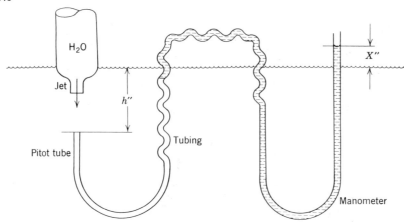

$$P_{t3} - P_3 = \frac{w_{H_2O} V^2}{2g} = w_{H_2O} X$$

$$\therefore \ X = \frac{V^2}{2g}$$

17.2

Method A

$$\frac{\Delta P}{P_v} = f\left(R_D, \frac{d}{D}\right)$$

$$R_D = \frac{VD}{\nu} = 15 \times \frac{6}{12} \times \frac{10^5}{0.74}$$

$$= 10.135 \times 10^5$$

$$\frac{d}{D} = \frac{1}{8 \times 6} = 0.02$$

Fig. 17.6 $\quad \dfrac{\Delta P}{P_v} = 0.82\%$

Method B

$$\frac{\Delta P}{\tau_0} = f(R_d^*)$$

$$R_d^* = \left(\frac{f}{8}\right)^{1/2}\left(\frac{d}{D}\right) R_D$$

By (17.18) $\quad f = 0.01165$

$$R_d^* = \left(\frac{0.01165}{8}\right)^{1/2} \times 0.02 \times 10.135 \times 10^5$$

$$R_d^* = 805.75$$

Fig. 17.5 $\dfrac{\Delta P}{\tau_0} = 2.9$

$$\frac{\Delta P}{P_v} = \frac{f\Delta P}{4\tau_0} = \frac{0.01165}{4} \times 2.9$$

$$\therefore \quad \frac{\Delta P}{P_v} = 0.84\%$$

17.3 $$P_v = \frac{\rho V^2}{2g_c} = \frac{0.035877 \times 15^2 \times 12}{2 \times 32.174} = 1.5054 \text{ psi}$$

$$\Delta P = 0.0084 \times 1.5054 = 0.01264 \text{ psi}$$

$$\Delta h = \frac{\Delta P}{w_{H_2O}} = \frac{0.01264}{0.036045} = 0.3507''H_2O$$

17.4 $$\frac{S}{C} = \frac{2d}{\pi D} = \frac{2 \times \frac{1}{4}}{\pi \times 4} = 0.03979$$

at $M = 0.2, \qquad C_D = 1.15$

By (17.31) $$\frac{\delta P}{P_v} = \left(\frac{-2}{1-0.2^2}\right)\left(\frac{1.15}{2}\right)0.03979 = 4.77\%$$

By Fig. 17.17, $$\frac{SP}{P_v} \sim 5\%$$

17.5 $\dfrac{\delta}{D} = 0.62K - 0.5876K^3$

where $$K = \frac{D(\Delta V/\Delta y)}{2V} = \frac{1 \times 100}{2 \times 300} = 0.1667$$

$$\therefore \quad \frac{\delta}{D} = 0.62 \times 0.1667 - 0.5876(0.1667)^3 = 0.100634$$

$$\theta_s = \sin^{-1}(0.9004 \times 0.1667) - 1.9357(0.1667)^3$$

$$\therefore \quad \theta_s = \sin^{-1} 0.14112973 = 8.11°$$

17.6 $$\Delta\theta_T = 2\sin\left(\frac{\theta_{\text{taps}}}{2}\right)\frac{KP_{\theta 0}}{\left|\dfrac{\Delta P_\theta}{\Delta\theta}\right|_0}$$

$$\Delta\theta_T = 2\sin\left(\frac{90}{2}\right) \times \frac{0.3 \times (-0.2)}{0.04} = -2.12°$$

18.1 $L_e = 0.5 + 12 \times 10\left(\dfrac{0.03}{0.12}\right)^4 = 0.96875$ in.

$$K = \frac{128 \times 4 \times 10^{-7} \times 0.96875 \times 10}{\pi \times 81 \times 10^{-8} \times 15 \times 144} = 0.09024 \text{ sec}$$

$$\tau_{95\%} = 3K = 0.2707 \text{ sec}$$

18.2 $p_i = 15$ psia, $p_f = 1.5 \times 15 = 22.5$ psia

$p_t = 22.5 - (7.5)0.01 = 22.425$ psia

$$B = \frac{128(4 \times 10^{-7})(0.96875)10}{\pi \times 81 \times 10^{-8}} = 194.9157$$

$a_t = 0.01$ for 99% recovery

$$b_t = \frac{22.5 + 15}{22.5 + 22.425} = 0.83472$$

$$t_{NL} = \left(\frac{B}{P_f}\right) \ln \left(\frac{1}{a_t b_t}\right)$$

$$= \left(\frac{194.9157}{22.5 \times 144}\right) \ln \left(\frac{1}{0.01 \times 0.83472}\right)$$

$$= 0.06016 \times 4.78583$$

$t_{NL} \atop 99\%} = 0.288$ sec.

18.3 $T = \dfrac{L}{a} = \dfrac{10}{1160} = 0.008621$ sec

$$\tau = \frac{Tv}{r^2} = \frac{0.008621 \times 0.18 \times 10^{-3} \times 144}{36 \times 10^{-4}} = 0.06207$$

$$\frac{P_t}{P_i} = 1 - \mathrm{erf}(x) = 0.95, \qquad \mathrm{erf}(x) = 0.05$$

From Table 18.1, $X \sim 0.04434 = \dfrac{1}{2}\left(\dfrac{8\gamma\tau_0^2}{\tau - \tau_0}\right)^{1/2}$

$$\tau = \tau_0 + \frac{8\gamma\tau_0^2}{(2X)^2} = 0.06207 + \frac{8 \times 1.4 \times 0.06207^2}{(2 \times 0.04434)^2}$$

$$\tau = 5.54901$$

$$\therefore \quad t_{95\%} = \frac{\tau r^2}{v} = \frac{5.54901 \times 36 \times 10^{-4}}{0.18 \times 10^{-3} \times 144} = 0.7707 \text{ sec}$$

18.4 $\dfrac{P_t}{P_i} = \mathrm{erfc}\left[\dfrac{1}{2}\dfrac{\tau_0}{\sqrt{\tau - \tau_0}}\right] = 0.95, \qquad \mathrm{erf}(x) = 0.05$

$$X \sim 0.04434 = \frac{1}{2}\frac{\tau_0}{\sqrt{\tau - \tau_0}}; \qquad T = \frac{L}{a} = \frac{100}{5000} = 0.02 \text{ sec}$$

$$\tau_0 = \frac{Tv}{r^2} = \frac{0.02 \times 10^{-5} \times 144}{3.90625 \times 10^{-3}} = 0.0073728$$

$$\tau = 0.0073728 + \left(\frac{0.0073728}{0.08868}\right)^2 = 0.014285$$

$$\therefore \quad t_{95\%} = \frac{\tau r^2}{v} = \frac{0.014285 \times 3.90625 \times 10^{-3}}{10^{-5} \times 144} = 0.03875 \text{ sec}$$

20.1 $\quad q = \dfrac{Q}{L} = \dfrac{4}{3}$ ft/sec

$$V = \dfrac{q}{D} = \dfrac{4}{3}\left(\dfrac{12}{2}\right) = 8 \text{ ft/sec}$$

$$E = D + \dfrac{V^2}{2g} = \dfrac{2}{12} + \dfrac{64}{2 \times 32.174} = 1.16126 \text{ ft}$$

$$D_c = \left(\dfrac{q^2}{g}\right)^{1/3} = \left(\dfrac{16}{9 \times 32.174}\right)^{1/3} = 0.380883 \text{ ft} = 4.57 \text{ in.}$$

$$V_c = (gD_c)^{1/2} = (32.174 \times 0.380883)^{1/2} = 3.5 \text{ ft/sec}$$

$$F_r = \dfrac{V}{G} = \dfrac{8}{(gD)^{1/2}} = \dfrac{8}{\left(32.174 \times \dfrac{2}{12}\right)^{1/2}} = 3.455$$

Check: $D = 2'' < D_c \quad \therefore \quad$ Rapid Flow

$$V = 8 \text{ ft/sec} > V_c \qquad \therefore \text{Rapid}$$

$$F_r = 3.455 > 1 \qquad \therefore \text{Rapid}$$

20.2 $\quad D_4 = \dfrac{D_2}{2}\left[-1 + \left(\dfrac{16E_2}{D_2} - 15\right)^{1/2}\right]$

$$D_4 = \dfrac{2}{2}\left[-1 + \left(\dfrac{16 \times 1.16126}{2/12} - 15\right)^{1/2}\right] = 8.822 \text{ in.}$$

20.3 $\quad \underset{\text{jump}}{h_{\text{loss}}} = E_2 - E_4 = \dfrac{(D_4 - D_2)^3}{4D_2D_4}$

$$\underset{\text{jump}}{h_{\text{loss}}} = \dfrac{(8.822 - 2)^3}{4 \times 2 \times 8.822 \times 12} = 0.374884 \text{ ft}$$

20.4 $\quad C_F = 0.604 + 0.02\left(\dfrac{1}{4}\right) + 0.045\left(\dfrac{1}{4}\right)^2 = 0.6118$

$$q'_F = \dfrac{1}{2}\left[\dfrac{2 \times 32.174 \times 4}{(1 + \frac{1}{4})12}\right]^{1/2} = 0.3452 \text{ ft}^3/\text{sec}$$

$$Q_F = (Lq'_F)C_F = 2 \times 0.3452 \times 0.6118 = 0.4224 \text{ ft}^3/\text{sec}$$
$$\therefore \quad \dot{w} = wQ = 62.4 \times 0.4224 = 26.358 \text{ lbf/sec}$$

20.5 $\quad C_{SE} = 0.49 + 1.376\left(\dfrac{1}{20}\right) - 1.43\left(\dfrac{1}{20}\right)^2 = 0.5624$

$$q'_S = \dfrac{1}{12}\left[\dfrac{2 \times 32.174(20 - 12)}{12}\right]^{1/2} = 0.5458$$

$$Q_S = (Lq'_S)C_{SE} = 2 \times 0.5458 \times 0.5624 = 0.6139 \text{ ft}^3/\text{sec}$$

$$\therefore \quad \dot{w} = wQ = 62.4 \times 0.6139 = 38.308 \text{ lbf/sec}$$

21.1 $\quad \dot{m}' = A_2\left[\dfrac{2g_c\rho(p_1 - p_2)}{144(1 - \beta^4)}\right]^{1/2}$

$$\therefore \quad \dot{m}' = \dfrac{\pi \times 4}{4}\left[\dfrac{2 \times 32.174 \times 60 \times 5}{144 \times 0.9375}\right]^{1/2} = 37.5674 \text{ lbm/sec}$$

21.2 $\dot{m}' = A_2 \left[\left(\dfrac{2\gamma}{\gamma-1} \right) \dfrac{g_c p_1 \rho_1 (r^{2/\gamma} - r^{(\gamma+1)/\gamma})}{144(1 - r^{2/\gamma}\beta^4)} \right]^{1/2}$

$r^{2/\gamma} = 0.9^{1.42857} = 0.860265$
$r^{(\gamma+1)/\gamma} = 0.9^{1.71428} = 0.834754$
$r^{2/\gamma} - r^{(\gamma+1)/\gamma} = 0.02551$
$(1 - r^{2/\gamma}\beta^4)^2 = 0.8953577$

$\dot{m}' = \pi \left[\dfrac{7 \times 32.174 \times 20 \times 0.076 \times 0.025511}{144 \times 0.946233} \right]^{1/2}$

$\therefore \quad \dot{m}' = 0.795 \ \text{lbm/sec}$

21.3 $\dot{m}' = \dfrac{A_2 p_1}{\sqrt{\dfrac{RT_t}{g_c}}} \left[\left(\dfrac{2\gamma}{\gamma-1} \right) \dfrac{(r^{2/\gamma} - r^{(\gamma+1)/\gamma})(1 - r^{(\gamma+1)/\gamma}\beta^4)}{(1 - r^{2/\gamma}\beta^4)^2} \right]^{1/2}$

and $\dot{m}' = A_2 \left[\left(\dfrac{2\gamma}{\gamma-1} \right) \dfrac{g_c p_1 \rho_1 (r^{2/\gamma} - r^{(\gamma+1)/\gamma})}{144(1 - r^{2/\gamma}\beta^4)} \right]^{1/2}$

Canceling: $T_t = \dfrac{144 p_1}{\rho_1 R} \left(\dfrac{1 - r^{(\gamma+1)/\gamma}\beta^4}{1 - r^{2/\gamma}\beta^4} \right)$

$T_t = \dfrac{144 \times 20}{0.076 \times 53.35} \left(\dfrac{0.9478279}{0.946233} \right) = 711.497°\text{R}$

21.4 $\dot{m}' = \dfrac{A_2 p_t}{\sqrt{T_t}} \Gamma_2 K_\Gamma$

From Table 21.1 $\Gamma_2 = f\left(\dfrac{p_2}{p_t} \right) = 0.63056$

$K_\Gamma = 0.40865 \ \dfrac{\text{lbm °R}^{1/2}}{\text{lbf sec}}$

$\therefore \quad \dot{m}' = \dfrac{4\pi \times 50 \times 0.63056 \times 0.40865}{23.2379} = 6.967 \ \text{lbm/sec}$

21.5 $r^{*(1-\gamma)/\gamma} + \dfrac{\gamma-1}{2} \beta^4 r^{*2/\gamma} = \dfrac{\gamma+1}{2}$

For $\gamma = 1.4$ $\beta = 0.4$;

$r_*^{-0.2857143} + 0.2 \times 0.0256 r_*^{1.42857} = 1.2$

Assume $r_{*1} = 0.528$

$1.200183 + 0.2 \times 0.0256 \times 0.401572 = ?1.2$

$1.202239 \neq 1.2, \quad \epsilon_1 = 0.002239$

Assume $r_{*2} = 0.5$

$1.219014 + 0.2 \times 0.0256 \times 0.371499 = ?1.2$

$1.220916 \neq 1.2, \quad \epsilon_2 = 0.020916$

By straight line interpolation (Newton-Raphson Scheme)

$$r_{*3} = r_{*2} - \epsilon_2 \left(\frac{r_{*1} - r_{*2}}{\epsilon_1 - \epsilon_2} \right)$$

$$= 0.5 - 0.020916 \left(\frac{0.528 - 0.5}{0.002239 - 0.020916} \right)$$

$r_{*3} = 0.5314$,   Try this,

$$1.197984 + 0.2 \times 0.0256 \times 0.405271 = ?1.2$$

$$1.20006 \sim 1.2$$

$\therefore \quad r_* \cong 0.5314$

22.1 $\text{Re}_1 = \dfrac{\rho V D}{\mu} = 62.4 \times 10 \times \dfrac{6}{12} \times \dfrac{10^5}{60.6653} = 5.14297 \times 10^5$

$$\text{Re}_2 = \frac{48(\rho A V)}{\pi D \mu} = \frac{48 \times 62.4 \times \dfrac{9\pi}{144} \times 10 \times 10^5}{\pi \times 6 \times 60.6653} = 5.14297 \times 10^5$$

22.2 $C_D = \left[\dfrac{1 - \beta^4}{A - \beta^4 + B + C - 0.4505\beta^{3.8} R_d^{-0.2}} \right]^{1/2}$

where $\beta = 0.5$, $R_d = 10^6$, $A = 1$, $B = 0$, $C = 0.296 R_d^{-0.2}$

$$C_D = \left[\frac{1 - 0.0625}{1 - 0.0625 + 0.296(10^6)^{-0.2} - 0.4505(0.5)^{3.8}(10^6)^{-0.2}} \right]^{1/2}$$

$\therefore \quad \underset{\text{turb}}{C_D} = \left[\dfrac{0.9375}{0.9541356} \right]^{1/2} = 0.99124$

22.3 $C_D = 0.19436 + 0.152884(\ln R_d) - 0.0097785(\ln R_d)^2 + 0.00020903(\ln R_d)^3$

where $R_d = 10^6$, $\ln R_d = 13.81551$, $(\ln R_d)^2 = 190.8683$,

$(\ln R_d)^3 = 2636.94335$

$\therefore \quad C_D = 0.991325$, whereas Table 22.4 yields $C_D = 0.99132$

Difference between Pb2 and Pb3;

$$\frac{\Delta}{\text{level}} = \frac{(0.99132 - 0.99124)}{0.99124} \times 100 = 0.008\%$$

22.4 $\dot{m}' = \dfrac{\pi}{4 \times 0.968246} \left[\dfrac{2 \times 32.174 \times 62.4 \times 10}{144} \right]^{1/2} = 13.545 \text{ lbm/sec}$

at $R_d = 10^5$, $C_{D1} = 0.6076$ from Table 22.5

$\dot{m}_1 = 0.6076 \times 13.545 = 8.2299 \text{ lbm/sec}$

$$R_{d_1} = \frac{48 \times 8.2299}{\pi \times 1 \times 0.60665} \times 10^3 = 2.07 \times 10^5,$$

$C_{D2} = 0.6066$ from Table 22.5

$\dot{m}_2 = 0.6066 \times 13.545 = 8.2164 \text{ lbm/sec}$

$$R_{d2} = \frac{48 \times 8.2164}{\pi \times 1 \times 0.60665} \times 10^3 = 2.07 \times 10^5$$

$\therefore \quad \dot{m} = 8.216 \text{ lbm/sec}$

23.1 $\dot{m} = C_{Di} Y \dot{m}'_i$

$\therefore \quad Y = \dfrac{\dot{m}}{C_{Di} \dot{m}'_i} = \dfrac{3.5}{0.98 \times 4} = 0.89286$

23.2 $Y' = \left[\left(\dfrac{\gamma}{\gamma - 1} \right) \dfrac{r^{2/\gamma} (1 - r^{(\gamma-1)/\gamma})(1 - \beta^4)}{(1 - r)(1 - r^{2/\gamma} \beta^4)} \right]^{1/2}$

where $r^{2/\gamma} = 0.9^{1.42857} = 0.860265$

$1 - r^{(\gamma-1)/\gamma} = 1 - 0.9^{0.2857143} = 0.0296544$

$1 - r = 0.1$

$\therefore \quad Y' = \left[\dfrac{3.5 \times 0.860265 \times 0.0296544}{0.1} \right]^{1/2} = 0.9449193$

23.3 $Y_{2-D} = 1 - \dfrac{(0.41 + 0.35 \beta^4)(1 - r)}{\gamma}$

where $\beta = 0, \qquad 1 - r = 0.1, \qquad \gamma = 1.4$

$\therefore \quad Y_{2-D} = 1 - \dfrac{0.41 \times 0.1}{1.4} = 0.97071$

NAME INDEX

509

SUBJECT INDEX

513